RISK GOVERNANCE OF OFFSHORE OIL AND GAS OPERATIONS

This book evaluates and compares risk regulation and safety management for offshore oil and gas operations in the United States, United Kingdom, Norway, and Australia. It provides an interdisciplinary approach with legal, technological, regulatory, behavioral, and sociological perspectives on these efforts to prevent major accidents and improve safety performance offshore. Presented in three parts, the book begins with discussion of the concept of risk governance, the role and modes of safety regulation, and the behavioral and other factors involved in developing an effective regulatory regime for industrial safety. It then discusses the four regimes for offshore safety, the industry's role, cultural and other contextual influences, and use of safety performance indicators. The final section focuses on the Norwegian regime, which features self-regulation and worker rights, and its capacity to respond to new offshore technologies, emerging risks, near-miss incidents, and other challenges. Discussion throughout the book provides insights about differing types of rules, inspection methods, enforcement, and other issues relevant to the quest for robust regulation and the development of a safety culture for preventing major accidents offshore. This book will be informative for those in government, industry, academia, and elsewhere in society who are interested in industrial safety in general and offshore safety in particular.

Preben Hempel Lindøe is Professor of Societal Safety at the University of Stavanger, Norway. He has an MS and a PhD on the implementation of "enforced self-regulation." He has worked within applied research for twenty-five years, including action-research methodology, occupational health and safety, risk regulation, and safety management. His publications include various books (in Norwegian) as well as articles, papers, and chapters in professional and academic journals, books, and other media.

Michael Baram is Professor Emeritus at Boston University Law School where he directed the Center for Law and Technology. He was also a partner in the law firm Bracken and Baram for twenty-five years, and previously was Associate Professor of Engineering and Assistant Dean at MIT. His research, publications, legal work, and advisory activities have dealt with risk regulation, environmental law, product liability law, occupational safety, and risk management in several technological sectors, including the chemicals, biotech, nuclear, and oil and gas industries. He is the author or editor of eight books including *Safety Management* (with Andrew Hale) and has published numerous articles.

Ortwin Renn is Professor and Chair of Environmental Sociology and Technology Assessment at Stuttgart University, Germany. He directs the Stuttgart Research Center for Interdisciplinary Risk and Innovation Studies at Stuttgart University (ZIRIUS) and the nonprofit company DIALOGIK, a research institute for the investigation of communication and participation processes in environmental policy making. Renn is primarily interested in risk governance, political participation, and technology assessment. He has published more than 30 books and 250 articles, most prominently the monograph *Risk Governance* (2008).

Risk Governance of Offshore Oil and Gas Operations

Edited by

PREBEN HEMPEL LINDØE
University of Stavanger

MICHAEL BARAM
Boston University Law School

ORTWIN RENN
University of Stuttgart

CAMBRIDGE
UNIVERSITY PRESS

CAMBRIDGE
UNIVERSITY PRESS

32 Avenue of the Americas, New York NY 10013-2473, USA

Cambridge University Press is part of the University of Cambridge.

It furthers the University's mission by disseminating knowledge in the pursuit of education, learning and research at the highest international levels of excellence.

www.cambridge.org
Information on this title: www.cambridge.org/9781107025547

First published 2014

A catalogue record for this publication is available from the British Library

Library of Congress Cataloguing in Publication data
Risk governance of offshore oil and gas operations / [edited by] Preben Hempel Lindøe, Michael Baram, Ortwin Renn.
pages cm
Includes bibliographical references and index.
ISBN 978-1-107-02554-7 (hardback)
1. Offshore oil industry – Risk management. 2. Offshore gas industry – Risk management. 3. Offshore oil industry – Safety measures. 4. Offshore gas industry – Safety measures. 5. Offshore oil industry – Government policy. 6. Offshore gas industry – Government policy. I. Lindøe, Preben. II. Baram, Michael S. III. Renn, Ortwin.
HD9560.5.R48 2014
622′.338190681–dc23 2013015854

ISBN 978-1-107-02554-7 Hardback

Contents

Contributors

Paul Bang has been working as a senior advisor in the Norwegian Petroleum Directorate (NPD) and Petroleum Safety Authority Norway (PSA) (from 2004) since 1987. He works particularly with regulatory issues giving advice for institutional arrangements, regulation making, and enforcement of regulation in petroleum activities. He holds a BSc in Organizational and Political Science from Universities in Bergen and Stavanger. He is also engaged in the Norwegian government's work in third-world countries through the Norwegian aid program Oil for Development (OfD).

Michael Baram is Professor Emeritus, Boston University Law School, where he directed the Center for Law and Technology. He was also a partner in the law firm Bracken and Baram for twenty-five years and was Associate Professor of Engineering and Assistant Dean at Massachusetts Institute of Technology. His research, publications, legal work, and advisory activities have dealt with risk regulation, environmental law, product liability law, occupational safety, and risk management issues in several technological sectors, including the chemicals, biotech, nuclear, and oil and gas industries.

Helene Cecilie Blakstad is Senior Safety and Quality Advisor in the Norwegian National Rail Administration. From 2006 until 2012, she worked as a Senior Scientist at SINTEF Technology and Society, Department of Safety Research. Earlier, she worked within occupational health services and with the Norwegian Labour Inspectorate. She has a PhD on adapting hierarchical and risk-based approaches to safety rule modifications in the Norwegian railway system.

Ole Andreas Engen is Professor at University of Stavanger within the field of Risk Management and Societal Safety. He holds an MPhil in Economics and a PhD in Sociology. He is also employed part time as a senior research scientist at the International Research Institute of Stavanger (IRIS).

Ulla Forseth is Associate Professor in the Department of Sociology and Political Science, Norwegian University of Science and Technology (NTNU), Trondheim, Norway. She holds a Dr.polit. in Sociology. Until recently she worked as a senior scientist in the SINTEF Department of Industrial Management. Her research interests include work and organization, power and resistance, change processes in society, and occupational health and safety.

Andrew Hale has a background in occupational psychology. He is Emeritus Professor of Safety Science from the Delft University of Technology in the Netherlands, where he worked for twenty-five years. Since 2008 he has been chairman of the consultancy HASTAM (Health & Safety Technology & Management) in the United Kingdom. One of his areas of study has been the management of safety rules and regulations at both company and national levels across a range of industries including transport, oil and gas, and other major hazard areas.

Jan Hayes has twenty-five years' experience in safety and risk management. Her current activities cover academia, consulting, and regulation. She holds a Senior Research Fellow appointment at the Australian National University where she is Program Leader for the social science research activities of the Energy Pipelines Co-operative Research Centre. Dr. Hayes is a member of the Advisory Board of the Australian National Offshore Petroleum Safety and Environmental Management Authority.

Knut Kaasen is Professor of Law at the Scandinavian Institute of Maritime Law, University of Oslo, and is presently Head of the Institute's Department of Petroleum and Energy Law. He has served as acting Justice of the Supreme Court of Norway and Dean and Associate Dean of the Faculty of Law. Prior to becoming professor in 1989, he served as an assistant district judge and a counsel in the legal department of Norsk Hydro a.s. The subject of his doctoral dissertation (1984) was safety regulation of offshore petroleum activities. Over the last twenty-five years he has been closely involved in developing agreed forms of contract in the petroleum industry.

Jacob Kringen is senior advisor at the Norwegian Directorate of Civil Protection and Emergency Planning and an associate professor at the University of Stavanger. He has an MSc in Anthropology and a PhD in Science and Technology studies from the University of Oslo. His main fields of interest have been Norwegian government administration, risk regulation and governance, and public management.

Preben Hempel Lindøe is Professor at University of Stavanger in the field of Risk Management and Societal Safety. He has a PhD on the implementation of "enforced self-regulation" in Norway. For a period of twenty-five years he has worked within applied research, including action research methodology, occupational health and safety, risk regulation, and safety management.

Kathryn Mearns spent twenty years working in academia, before taking up a new role as a Human Factors Specialist Inspector with the Offshore Division of the UK Health and Safety Executive in 2012. She has published extensively in the area of human factors in health and safety in a range of industries including offshore oil and gas, air traffic management, and health care. She continues to be involved in human factors research through a part-time professorship at the University of Bergen.

John Paterson is Professor of Law and Co-Director of the Centre for Energy Law at the University of Aberdeen, United Kingdom. His research has focused on the regulation of risk, not least in the offshore hydrocarbon industry, and he has been involved in a number of multidisciplinary research projects dealing with risk governance. He is the co-editor of *Oil and Gas Law: Current Issues and Emerging Trends* and is active in teaching, PhD supervision, training, and consultancy in this field.

Ortwin Renn serves as full professor and Chair of Environmental Sociology and Technology Assessment at Stuttgart University, Germany. He directs the Stuttgart Research Center for Interdisciplinary Risk and Innovation Studies at Stuttgart University (ZIRUS) and the non-profit company DIALOGIK, a research institute for the investigation of communication and participation processes in environmental policy making. Renn also serves as Adjunct Professor for "Integrated Risk Analysis" at Stavanger University, Norway; as Affiliate Honorary Professor at the Technical University of Munich; and as Affiliate Professor for Risk Governance at Beijing Normal University. Renn is primarily interested in risk governance, political participation, and technology assessment. He has published more than 30 books and 250 articles, most prominently the monograph *Risk Governance* (2008).

Ragnar Rosness is senior scientist in the Department of Safety Research at SINTEF in Trondheim, Norway. He holds an MSc in psychology and a PhD in industrial engineering. He conducts research mainly in the petroleum industry and the transportation sector. His research interests focus on the links between organizing, decision making, power, and resilience.

Helge Ryggvik (PhD) is an economic historian, working as a researcher at the Centre for Technology, Innovation and Culture at the University of Oslo. He has been working on business history, workplace relations, and safety themes for years. He is the author of several books on oil-related issues, including a history of Norwegian offshore contractors, and he published a history of the Norwegian Railways. Most recently, he has completed a report comparing Norwegian and U.S. offshore regulations.

Olaf Thuestad previously worked in the Norwegian Petroleum Directorate (NPD) and then as director of Regulatory Development at the Norwegian Petroleum Safety Authority (PSA) from 2004, when it was established, until 2010. He is now senior strategic advisor at PSA. He holds a BSc in Civil Engineering and has had a central role in the formulation of strategies for, and development of, HSE regulations for the petroleum sector from the early 1980s.

Emre Üşenmez is a lecturer in oil and gas law and an associate at the Aberdeen University Centre for Energy Law (AUCEL). He is the co-editor of *Oil and Gas Law: Current Issues and Emerging Trends* and the director of Lex Petrolea Limited, a business consultancy firm in oil and gas. Previously he worked in the natural resources sector in Azerbaijan. He holds an LLM in Oil and Gas Law from the University of Aberdeen, and he is currently working toward an MSc in Petroleum, Energy Economics, and Finance and a joint PhD in Oil and Gas Law and Petroleum Economics.

Preface

Modern society is increasingly dependent on technological strength and its careful application by complex enterprises to ensure public well-being and the achievement of other important goals. A prime example is the use of new technologies to meet global energy needs that are essential for the fulfillment of these goals.

The exploration and production of offshore oil and gas resources have become important means of meeting global energy needs. More than 90 percent of the oil and gas produced by EU member states and Norway is now derived from offshore production, mainly in the North Sea and the Norwegian Sea. Offshore operations also account for a substantial part of the oil and gas produced by the United States and Australia and are of growing importance to Canada, Brazil, and many other countries.

These offshore activities involve sophisticated analytic methods, heavy engineering, large-scale investment, and complex projects, and they must be managed appropriately to ensure that benefits are gained without incurring major accidents and other unacceptable harms to the public, the workers involved, and the human and natural environments. This requires partnership between public regulators and industry, the involvement of labor and other stakeholders, a supporting role for researchers, mutual trust that best practices will be used and continuously improved, and much more.

However, major accidents such as the Macondo blowout and oil spill in the Gulf of Mexico in 2010 demonstrate that simultaneously ensuring productivity and safety is a major challenge, particularly in deepwater regions and other difficult locales. To meet this challenge, leading countries have developed regulatory regimes that differ in several respects, particularly with regard to supervising and fostering self-regulation by industry, and all are engaged in a continuing quest for increasingly robust regulation.

From economic, environmental, and safety perspectives, the robustness of the regulatory regimes that govern the safety of offshore activities is of critical importance to all. In this regard, the research community plays a key role by contributing knowledge from many disciplines, which can improve the performance of the regimes and the offshore industry.

This book is the result of a common effort by a select group of researchers over several years. Their research and its integration has been enabled by a major research project, "Robust Regulation in the Petroleum Sector," which has been funded by the Research Council of Norway. The aim of the project has been to provide knowledge that can be used to develop and support robust regulation and inclusive risk governance for the complex and dynamic field of offshore oil and gas operations by addressing four objectives:

- conceptualization and better understanding of the resilience of the Norwegian risk regulation regime, which many other countries look to as a model;
- comparison of the Norwegian regime with other leading regimes to further improvements in an international context;
- examination of the interface between a regulatory regime and industrial safety management systems; and
- dissemination of research-based knowledge to promote learning about robust risk regulation and the challenges involved in its implementation.

The Macondo disaster in 2010 increased the significance of the project because it put the quest for robust regulation on the public agenda in all countries with offshore oil and gas resources. This led the core group of Norwegian researchers to extend their project by inviting researchers from the United Kingdom, United States, Germany, and Australia to join them and produce results of international relevance.

Finally, we believe this book demonstrates the value of learning from various disciplines and from safety research done in other technological and industrial domains, and we believe it provides knowledge in return and hope it contributes to risk reduction and improved safety performance in offshore operations and these other domains.

Preparing this book has been an inspiring and learning process. We are thankful for the cooperation of the contributing authors and of Cambridge University Press, as well as for the financial support of the Research Council of Norway.

Introduction: In Search of Robustness

Michael Baram, Preben Hempel Lindøe, and Ortwin Renn

I.1. OFFSHORE SAFETY

Offshore oil and gas resources are an important part of the global energy system and an asset of high value to many developed and developing nations. Technological advances have enabled the exploration and production of these resources initially in shallow coastal areas and now far offshore in the much deeper waters of the outer continental shelf regions that extend from Europe, North and South America, Africa, and Asia. As a result, vast reservoirs of oil and gas in previously inaccessible seabed regions are being discovered, extracted, processed, and distributed to meet the ever-increasing demand for fuel and energy.

Exploration and production are complex and costly operations that require the coordinated performance of drilling-rig owners and operators with many contractors and service firms. Their success provides significant benefits to the countries that have jurisdiction over such activities, and to the participating companies and investors. However, many aspects of these operations are intrinsically hazardous and thereby pose risks to health, safety, and the environment, as well as to other societal and commercial interests. In addition, there is synergistic risk potential because of extreme weather conditions and other natural hazards that can interfere with operations and emergency response at many drilling sites. Thus, ensuring that operations are safely conducted is a major concern of the governments and industrial organizations involved, and of many others who as stakeholders enjoy the benefits but fear the risks.

Of most concern are major accidents. The history of offshore operations is marred by the sporadic occurrence of blowouts, explosions, and fires at drilling rigs and other incidents that caused multiple injuries and deaths among the workforce, destroyed company assets and other property, and caused major spills that contaminated vast offshore and coastal areas, killed wildlife, and disrupted fishing, transport, recreation, and other activities.

Among the most notorious accidents are the capsizing of the Alexander Kielland flotel (1980), which caused the deaths of 123 of the workers it housed, and the fire that destroyed the Piper Alpha rig (1988) and killed 167 workers, in the Norwegian and British sectors of the North Sea; the Montara blowout and oil spill (2009) that continued unabated for 74 days in the Timor Sea off the Australian coast; and the blowout and fire on the Deepwater Horizon drilling rig (2010) in the U.S. sector of the Gulf of Mexico, which killed 11 workers, caused the worst of all recorded spills, which could not be stopped for 86 days, and contaminated a vast region of the Gulf and its shoreline.

Other accidents on a smaller scale occur more frequently and also take their toll, such as the Frade well leak and oil spill (2012) in the south Atlantic off the Brazilian coast, and the gas well blowout in the Scottish Elgin Field sector of the North Sea (2012), which created a hazardous zone and necessitated suspension of neighboring activities. In addition, numerous near-miss incidents also occur and command attention because they may be precursors to an accident or indicate a systemic or industrywide safety problem that may lead to multiple accidents.

These incidents, safety performance data, and particularly major accidents, prompt the countries that authorize offshore operations to undertake reforms in the ways by which they govern the risks of such operations. Risk governance may encompass several forms of institutional control over a hazardous industry (e.g., civil liability law, criminal prosecution, insurance requirements, etc.), but a regulatory regime is the primary means of ensuring offshore safety because of its capacity to fully investigate and address specific problems, and thereby improve performance and *prevent* accidents by mandating or promoting improvements in the design, equipment, and maintenance of operations and the safety management systems and practices of the firms involved.

When improvements in governance are sought, a critical assessment of the existing regulatory regime and industrial performance is undertaken. The assessment may then lead to incremental change in the existing regime and its implementation in the form of a specific recommendation or rule calling, for example, for more stringent testing of well control equipment by offshore operators or for improvements in inspection, maintenance, and training. But when triggered by a major accident, the reform is more likely to involve reconsideration of the premises, design, and implementation of the existing regime and the enactment of a new regulatory regime. In either case, what takes place is a *quest for robust regulation*, for a regime that will be more effective at drawing on industrial expertise and improving the safety performance of offshore operators.

I.2. THE QUEST FOR ROBUST REGULATION

The main purpose of this book is to illuminate the issues involved in conceptualizing, designing, and implementing a robust regulatory regime for preventing major accidents and improving the safety performance of industrial operators during offshore exploration and production of oil and gas resources. Because quests for robust regulation are also undertaken in other technological sectors that experience or have the potential to cause major accidents, and which encounter similar issues, the book is also intended to contribute to the global body of knowledge about improving safety regulation and industrial safety management of highly hazardous technologies.

The Deepwater Horizon disaster and other major accidents offshore have intensified the quest in producing countries, just as major accidents incurred by other types of industrial activities have done in their host countries. Among such events are the lethal explosions at petroleum facilities in Texas City and Toulouse, the accidental release of toxic materials at chemical facilities in Bhopal and Seveso, the radioactive contaminations caused by cooling system failures at nuclear reactors in Chernobyl, Three Mile Island, and Fukushima, and accidents arising from malfunctions in the operation of pipelines, waste disposal facilities, high-speed rail, passenger aircraft, and the *Challenger* and *Columbia* spacecraft.

The quest for robust regulation that follows such an event usually involves two levels of analysis: investigation to determine *how* the accident occurred, which produces specific findings about contributing factors such as defects in design, equipment, operation, human performance, safety management, and contingency plans; and more far-reaching and speculative analysis to discern the institutional, cultural, economic, behavioral, and other contextual factors which may explain *why* it occurred. This leads to consideration of the lively global discourse about competing concepts of regulation, inspection, and enforcement, issues raised by increasing reliance on industrial self-governance, and accommodation of emerging norms for information disclosure and stakeholder participation.

In order to take a coherent approach to the many issues raised by the quest, this book addresses four main themes. The first is risk governance and the process by which it leads to the design of a robust regulatory regime. This theme is discussed at a generic level applicable to hazardous industrial activities and then in more detail as it applies to offshore operations. The second theme is the critical importance of issues that arise at the interface between government and industry, the "sphere" wherein the regime encounters the private domain of offshore industry and its self-regulatory practices. The third theme is the value of comparing different regimes that have highly developed approaches to offshore safety in order to discern their respective strengths and weaknesses. Finally, the fourth theme is learning from experience, which involves consideration of governmental and industrial processes for gaining and applying knowledge from major accidents, near-misses, and safety performance data.

1.3. THIS BOOK

Part 1 of the book illustrates the need for generic and systemic approaches to technical safety. Given the dimensions of potential accidents, the strategy of trial and error is not acceptable from a safety perspective. It needs to be replaced by anticipatory risk analysis and simulation. Projecting trial and error in the virtual space places, however, many demands on risk modeling and accident prevention and necessitates reliable concepts for robust risk management and regulation. One of these concepts is the risk governance framework of the International Risk Governance Council (IRGC), which has been used as a reference model in this book. This framework is the main topic of Chapter 1. Based on this theoretical insight, Chapter 2 investigates the various patterns of safety regimes and highlights the importance of co-regulatory models. Those models underlie the need for inclusive governance, providing room for stakeholders to be an integral part of the regulatory process. Chapter 3 widens the perspective from governance to the role of safety and risk culture in regulatory practice. The chapter emphasizes the relevance of norms, values and beliefs that govern institutional and individual responses to risk and crisis management. The last chapter in Part I deals with the current literature on public and private regulatory regimes and demonstrates the need for public and private actors to cooperate.

Part II describes and evaluates four notable regulatory regimes governing oil and gas operations offshore. The chapters also consider the cultural, legal, and other contextual factors that have influenced the design and implementation of the regimes, and their reliance on industrial expertise and standards. Of these, the U.S. regime, which has remained essentially unchanged over decades, requires and enforces operator compliance with detailed government-developed prescriptive rules and numerous industrial standards. The U.K. regime, inspired by the Cullen report on the Piper Alpha disaster, requires company analyses of hazards and risks and preparation of a "safety case" for each operation, in order to determine and ensure company awareness and preparedness for addressing the risks. The Australian regime, which is patterned after the UK regime, reflects an engineering approach to safety and is called upon to incorporate knowledge from the social and behavioral sciences. The fourth and most innovative is the Norwegian regime, which calls for each company to develop an internal safety management system, self-determine how it will fulfill essential safety functions, and self-audit its performance, all pursuant to governmental oversight. Part II concludes with a chapter on the role and value of safety performance indicators for a regime, current usage, and the difficulties in garnering comparable data for comparative analyses of different regimes.

Part III presents six case studies based on empirical research of the Norwegian regime and its paradigm shift from prescriptive regulation towards a system of government supervised self-regulation with risk assessment. The robustness and resilience of the regime are tested by looking at its capacity to keep pace with new technologies and emerging risks, encourage industrial safety culture, maintain the involvement and vested rights of labor, respond to near miss incidents, and ensure performance of inspection and self-audit functions. The tripartite model for cooperation and the responsibilities and accountability of the players with shifting patterns of adversarial and cooperative modes of industrial relations and social partnerships are characterized as "boxing and dancing." The chapters discuss how purpose and principles-based regulation may evaporate, dissolve, or become irrelevant, unless industrial realities, management practices, and regulatory expectations are considered. Technological transformations and cost efficiency programs have created premises for the present regime with some of its current vulnerabilities. Finally, inspections, audits and enforcement measures are compared with the hard law approach of the U.S. offshore regime.

The concluding chapter provides a synthesis and summation of the major themes of the book, lessons learned from the case studies and analyses, and a succinct comparison of the major features of the Norwegian, UK and U.S. regimes. It also presents a clarification of the term "Robust Regulation" and considers future application of our insights by the regimes.

Throughout the book, the authors suggest functional features and attributes of a robust regulatory regime that will enable the regime to meet the diverse safety challenges that arise in offshore exploration and production, and to do so in a manner consistent with enlightened principles of democratic risk governance. Among the attributes discussed in detail from several disciplinary perspectives are the capacity of the regime to foster learning from operating experience and external sources of expertise, address the uncertainties and risks of technological change and the conduct of hazardous activities in new offshore locales, ensure effective collaboration between government regulators and industry, and between management and labor, promote improvements in industrial self-regulation and cooperation for the continuous reduction of residual risks, evaluate safety performance and ensure industrial accountability, apply knowledge from the behavioral and social sciences, and meet the transparency and stakeholder participation norms of democratic regulatory processes.

PART I

Regulatory Frameworks and Concepts

INTRODUCTION TO PART I

The emphasis on robust regulation in a comparative review of different regulatory regimes requires a general framework for the analysis of commonalities, differences, and peculiarities of each regime. Andrew Hale defines robustness in Chapter 16 as a regime "that has survived for a considerable period with its principles intact, but with adaptation in its detail to changing situations and priorities" (Chapter 16). Based on this definition, the three introductory chapters develop a common perspective on risk governance and robustness. Ortwin Renn in Chapter 1 sets the stage for the entire book: he explores the applicability of the risk governance framework of the International Risk Governance Council (IRGC) to the analysis of technical performance and safety and explains its use as a reference model for the oil and gas exploration industry. The stages of risk governance – pre-estimation, interdisciplinary risk estimation, risk characterization, risk evaluation, risk management, monitoring and control, and, embracing all these phases, communication/participation – constitute benchmarks for characterizing different regulatory regimes but also for judging "good governance."

Chapter 2 picks up this analytic as well as normative aspect. Michael Baram and Preben Hempel Lindøe investigate the various patterns of safety regimes and highlight the importance of co-regulatory models. Those models underlie the need for inclusive governance, providing room for stakeholders to be an integral part of the regulatory process. The chapter also explores how these different regulatory principles and practices manifest themselves in different countries, most notably Norway, the United Kingdom, and the United States.

Chapter 3, by Kathryn Mearns, widens the perspective from governance to the role of safety and risk culture in regulatory practice. Her emphasis is on the norms, values, and beliefs that govern institutional and individual responses to risk and crisis management. In accordance with the two preceding chapters, she advocates a structured dialogue between regulators and a wider public, including experts, stakeholders, and affected communities. This request for inclusiveness and adaptive management needs to be efficiently organized, however.

This is the topic of the fourth and final chapter of Part I: Emre Usenmez reviews the current literature on public and private regulatory regimes and demonstrates the need for public and private actors to cooperate. Managing such complex interactions requires, however, new decision-making and communication tools. The chapter discusses the effectiveness and efficiency of several such tools such as nominal group technique and Delphi surveys. None of these tools are perfect but they are capable of facilitating group input and multi-actor involvement.

The four chapters illustrate the need for generic and systemic approaches to technical safety. Given the dimensions of potential accidents, the strategy of trial and error needs to be replaced by anticipatory risk analysis and simulation. Projecting trial and error in the virtual space, however, places many demands on risk modeling and accident prevention and necessitates an excellent monitoring system that provides fast and valid feedback to the modeler and risk managers as to what seems to work well and what not (before it is too late). Parts II and III provide ample illustrations, empirical data, and case studies on the success and failure of safety regimes before Andrew Hale in Chapter 16 draws general conclusions about the basic lessons learned from the conceptual and empirical investigations.

1

A Generic Model for Risk Governance

Concept and Application to Technological Installations

Ortwin Renn

1.1. INTRODUCTION

Risk governance refers to a complex of coordinating, steering, and regulatory processes conducted for collective decision making involving uncertainty (Rosa et al. 2013). Risk sets this collection of processes in motion whenever the risk impacts multiples of people, collectivities, or institutions. Governance comprises both the institutional structure (formal and informal) and the policy process that guide and restrain collective activities of individuals, groups, and societies. Its aim is to avoid, regulate, reduce, or control risk problems.

The general process of making and implementing collective decisions – governance – is as old as the human species itself. It encompasses the traditions and institutions that are the vehicle and outcome of these decisions and has a long past.[1] Until very recently, the broad charge embedded in the idea of governance devolved into a much narrower idea, one referring to the administrative functions of government bodies and formal organizations.

Recent events have changed all that. Entirely new forms of coordination and regulation have emerged in response to rapidly changing societal conditions, such as globalization. Boundaries between the public and private spheres, between formal governmental bodies and informal political actors, especially nongovernmental organizations (NGOs), and between markets and business interests and the regulatory needs of society are all blurred. At the same time, as a result of the growing recognition of the increased scale of collective problems, the domains of sovereignty shifted upward to supranational bodies. Owing to these and other changes, the idea of governance has been re-elevated to its original – broad – scope (Rosa et al. 2013).

[1] The etymology of the term dates back to the Ancient Greek times (Halachmi 2005; Kjaer 2004). Plato used the term "kuberman" as a reference to leadership, which assimilated in Latin to "gubernare." This notion is evaluated along various trajectories. In addition to its meaning in English, it is part of, among others, the French, Spanish, and Portuguese vocabularies.

A number of key events are responsible for that elevation. Among them is a general rejection of the word "government" in favor of "governance" in postmodern thought on political and economic institutions. Others include the adoption of governance in the official parlance of the European Union. Still more specific actions include the prominent place (including its own title) the term holds in the prestigious independent organization in Geneva, Switzerland: the International Risk Governance Council (IRGC 2005; 2007).

This governance framework established by the IRGC provides guidance for the development of comprehensive assessment and management strategies to cope with risk. The framework integrates scientific, economic, social, and cultural aspects and includes the engagement of stakeholders. The concept of risk governance comprises a broad picture of risk: not only does it include what has been termed risk management or risk analysis; it also looks at how risk-related decision making unfolds when a range of actors are involved, requiring coordination and possibly reconciliation between a profusion of roles, perspectives, goals, and activities (Renn 2008, p. 366).

The shift from government to governance signals a crucial change in the process of how collectively binding decisions are being made: from traditional state-centric approaches, with hierarchically organized governmental agencies as the dominant locus of power, to multilevel governance systems, where the political authority for handling risk problems is distributed among separately constituted public bodies (cf. Lidskog et al. 2011). These bodies are characterized by overlapping jurisdictions that do not match the traditional hierarchical order of state-centric systems (cf. Skelcher 2005). They consist of multi-actor alliances that include traditional actors such as the executive, legislative, and judicial branches of government, but also socially relevant actors from civil society. Prominent among those actors are industry, science, and nongovernmental organizations (NGOs). The result of the governance shift is an increasingly multilayered and diversified sociopolitical landscape. It is a landscape populated by a multitude of actors whose perceptions and evaluations draw on a diversity of knowledge and evidence claims, value commitments, and political interests (Rosa et al. 2013). Their goal, of course, is to influence processes of risk analysis, decision making, and risk management. This is also evident in industrial safety. For a long time, safety was an exclusive topic for corporate management and public regulators (Rassmussen 1997). Not so any more!

In recent times, this dual relationship has emerged into a multi-player arena involving corporate safety managers, representatives of unions and labor, various governmental agencies, and NGOs. How these players interact and how their exchange is organized have been subsumed under the term "regulatory regime" (see Chapter 2 and Chapter 14 in this volume). Andrew Hale defines it as a regime "that has survived for a considerable period with its principles intact, but with adaptation in its detail to changing situations and priorities" (Chapter 14). Robustness relates to an optimal combination of stability and change and responds to the need of inclusiveness, that is, the openness to new developments and stakeholder input.

The shift toward more inclusive governance has manifested itself, for example, in the tripartite risk regulation regime in Norway (see Chapter 12 and Chapter 13 in this volume) and in the reforms of the regulatory system in the United Kingdom (see Chapter 6 in this volume). In particular the Norwegian regulatory framework of RNNP is organized through a collaborative network embracing research institutions, industry, employers and unions, and the government. The actor-network analysis conducted by Rosness and Forseth (Chapter 12) was able to demonstrate that the capacity and willingness to enroll new actors had been a prerequisite for the revitalization of the tripartite collaboration in the Norwegian regulation.

For making the IRGC model akin to industrial safety issues and dynamic-adaptive management practices, Klinke and Renn (2012) proposed some alterations to the original IRGC risk governance model, because it appeared too rigid and standardized for application to complex technological risks. They developed a comprehensive risk governance model with additional adaptive and integrative capacity. The modified framework suggested by Klinke and Renn (2012) consists of the following interrelated activities: pre-estimation, interdisciplinary risk estimation, risk characterization, risk evaluation, and risk management, monitoring, and control. This requires the ability and capacity of risk governance institutions to use resources effectively (see Figure 1.1). Appropriate resources include:

- institutional and financial means as well as social capital (e.g., strong institutional mechanisms and configurations, transparent decision making, allocation of decision-making authority, formal and informal networks that promote collective risk handling, and education);
- technical resources (e.g., databases, computer software and hardware, research facilities, etc.); and
- human resources (e.g., skills, knowledge, expertise, epistemic communities, etc.).

Hence the adequate involvement of experts, stakeholders, and the public in the risk governance process is a crucial dimension to produce and convey adaptive and integrative capacity in risk governance institutions (Pelling et al. 2008; Stirling 2008).

This work features the risk governance process as designed by the IRGC and modified by Klinke and Renn. It introduces each stage (pre-estimation, interdisciplinary risk estimation, risk characterization, risk evaluation, risk management, and communication/participation) and points to the application of each stage for technical risks handling and regulations, in particular oil installations. This chapter concludes with some basic lessons for risk governance in the field of technological risks.

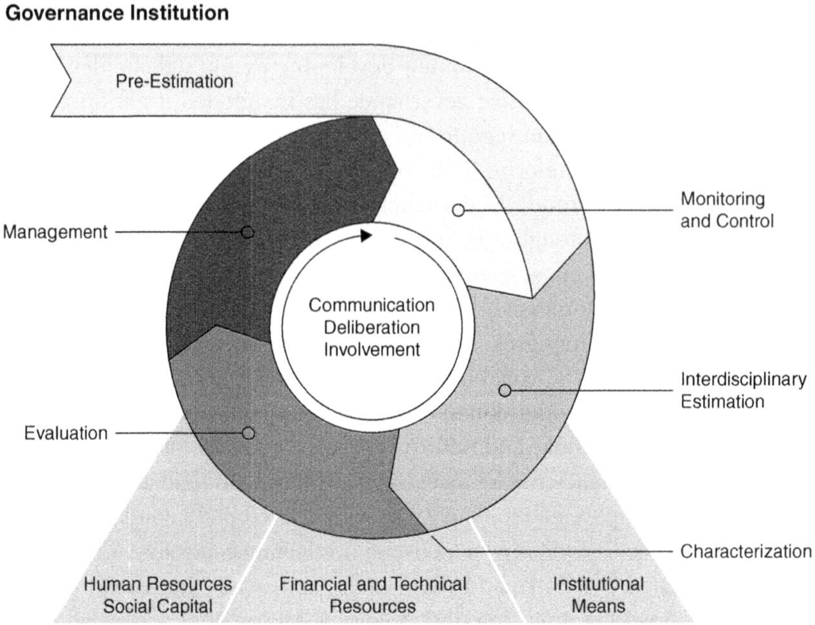

FIGURE 1.1 Adaptive and integrative risk governance model.
Note: The adaptive and integrative risk governance model is based on a modification and refinement of the IRGC framework (IRGC 2005). It has been published in Klinke and Renn (2012).

1.2. PRE-ESTIMATION

Risk, while a real phenomenon, is only understood via mental constructions resulting from how people perceive uncertain phenomena (Rosa et al. 2013). Those perceptions, interpretations, and responses are shaped by social, political, economic, and cultural contexts (IRGC 2005, p. 3; OECD 2003). At the same time, those mental constructions are informed by experience and knowledge about events and developments in the past that were connected with real consequences (Renn 2008, p. 2f.). That the understanding of risk is a social construct with real consequences is contingent on the presumption that human agency can prevent harm. The understanding of risk as a construct has major implications on how risk is considered. While risks can have an ontological status, understanding them is always a matter of selection and interpretation. What counts as a risk to someone may be destiny explained by religion to a second party or even an opportunity to a third party. Although societies over time have gained experience and collective knowledge of the potential impacts of events and activities, one neither can anticipate all potential scenarios nor be worried about all of the many potential consequences of a proposed activity or an expected event. At the same time, it is impossible to include all possible options for intervention. Therefore, societies always have been and will always be *selective* in what they choose to be worth considering and what they choose to ignore (IRGC 2005).

Pre-estimation, therefore, involves *screening* to choose from a large array of actions and problems that are risk candidates. Here it is important to explore what political and societal actors (e.g., governments, companies, epistemic communities, and NGOs) as well as citizens identify as risks. Equally important is to discover what types of problems they identify and how they conceptualize them in terms of risk. This step is referred to as *framing*, how political and societal actors rely on schemes of selection and interpretation to understand and respond to those phenomena that are relevant risk topics (Nelson et al. 1997; Kahneman and Tversky 2000; Reese 2007). According to Robert Entman (1993, p. 52 [emphasis in original]), "to frame is to *select some aspects of a perceived reality and make them more salient in a communication text, in such a way as to promote a particular problem definition, casual interpretation, moral evaluation, and/or treatment recommendation* for the item described." Perceptions and interpretations of risk depend on the frames of reference.

Framing implies that pre-estimation requires a multi-actor and multi-objective governance protocol. Governmental authorities (national, supranational, and international agencies), risk producers, opportunity takers (e.g., industry), those affected by risks and benefits (e.g., consumer organizations, local communities, and environmental groups on behalf of the environment), and interested parties (e.g., the media or experts) are all engaged. They will often debate about the appropriate frame to conceptualize the problem. What counts as risk may vary greatly among these actor groups.

Within the domain of technological risks, the pre-estimation phase includes the choice of an appropriate frame and the establishment of institutions and procedures to deal with emerging threats or events. Aven and Renn (2012) claim that many decisions to site hazardous facilities and to install adequate safety devices depend on the underlying frames of the actors involved. For the deployment of offshore oil facilities in the Barents Sea off the coast of Norway, the two authors identified three main frames:

1. *Political parties to the left and partly in the center in conjunction with environmental NGOs*: they have a focus on the environmental and social values at stake and they find the risk and uncertainties to be unacceptable (the point is made that we cannot rule out the possibility that a disaster will happen).
2. *Parties in the center of the political spectrum*: they would like to have more information before making a decision. They believe in principle that one can

balance benefits and risks, but are unsure whether the balance will result in benefits outweighing the risks or vice versa.

3. *Parties to the right in conjunction with industry*: they believe in the legitimacy of balancing pros and cons in a systematic way and they are convinced that such balancing would result in a judgment that the economic benefits outweigh the environmental and social risks.

A central issue in this debate on oil platform safety has been the legitimacy of trading off one set of values (environmental) against another set of values (economic); the second issue refers to the relative weights given to each of the values once the need for a balanced judgment is accepted. A third issue is the treatment of remaining uncertainties. Depending on which frame one uses, the risks look different and the balance between benefits and risks tends to vary a great deal. This was also highlighted by a study conducted by Rosness and Forseth (Chapter 12 in this volume). They report that in the public discourse about safety, "the actors used a variety of rhetoric devices to promote their positions during the controversy. We identified different discourses that embedded contrasting epistemological and ethical premises" (Chapter 12). Each of these premises constituted different frames that in the beginning of the discourse were incompatible with each other. Over time, however, these different frames could be reconciled and the actors were able to develop at least a working joint strategy.

It is therefore essential in any risk governance cycle to start with the frames of those who are involved in the issue. If these frames are not taken seriously, the risk governance effort may create major controversies and conflicts right from the beginning. Once frames are ignored, it is very difficult to develop a risk assessment and management strategy that is supported and accepted by all the major stakeholders.

1.3. INTERDISCIPLINARY RISK ESTIMATION

The interdisciplinary risk estimation comprises two activities: (1) *risk assessment* – producing the best estimate of the physical harm that a risk source may induce (including all scenarios that could compromise the safety of the facility under review); and (2) *concern assessment* – identifying and analyzing the issues that individuals or society as a whole link to a certain risk. For this purpose the repertoire of the social sciences, such as survey methods, focus groups, econometric analysis, macroeconomic modeling, or structured hearings with stakeholders may be used (Renn et al. 2011).

Why are two types of assessments needed in risk governance? For political and societal actors to arrive at reasonable decisions about risk and safety in the public interest, it is not enough to consider only the results of risk assessments, scientific or otherwise. In order to understand the concerns of affected people and various stakeholders, information about their risk perceptions, their behavior in crisis situation and their worries and concerns about the direct consequences if a crisis situation evolves, is essential and should be taken into account by risk managers.

Interdisciplinary risk estimation consists of a systematic assessment not only of the risks to human health and the environment but also of related concerns as well as social and economic implications (cf. IRGC 2005; Renn and Walker 2008). The interdisciplinary risk estimation process should be informed by scientific analyses; yet, in contrast to traditional risk regulation models, the scientific process includes not only the natural sciences but also the social sciences, including economics.

In 2000, the German Advisory Council on Global Environmental Change (WBGU 2000) suggested a set of criteria to characterize risks that go beyond the classic components probability and extent of damage. The Council identified and validated eight measurable risk criteria through a rigorous process of interactive surveying. Experts from both the natural sciences and the social sciences were asked to characterize risks based on the dimensions they would use for substantiating a judgment on tolerance to risk. Their input was subjected, through discussion sessions, to a comparative analysis. To identify the eight definitive criteria, the WBGU distilled the experts' observations down to those that appeared most influential in the characterization of different types of risk. In addition, alongside the expert surveys, the WBGU performed a meta-analysis of the major insights gleaned from existing studies of risk perception and evaluated the risk management approaches adopted by countries including the United Kingdom, the United States, the Netherlands, and Switzerland. The WBGU's long exercise of deliberation and investigation pinpointed the following eight physical criteria for the evaluation of risks:

1. *Extent of damage*, or the adverse effects arising from an accident or the occurrence of a damaging event – measured in natural units such as deaths, injuries, production losses, and so forth.
2. *Probability of the event occurring and probability of the extent of damage induced by the event* (relative frequency of a discrete or continuous loss function).
3. *Incertitude*, an overall indicator of the degree of remaining uncertainties inherent in a given risk estimate.
4. *Ubiquity*, which defines the geographic spread of potential damages and considers the potential for damage to span generations.
5. *Persistency*, which defines the duration of potential damages, also considering potential impact across the generations.
6. *Reversibility*, the possibility of restoring the situation, after the event, to the conditions that existed before the damage occurred (for example, restoration techniques including reforestation and the cleaning of water).

7. *Delay effect*, which characterizes the possible extended latency between the initial event and the actual impact of the damage it caused. The latency itself may be of a physical, chemical, or biological nature.
8. *Potential for mobilization*, understood as violations of individual, social, or cultural interests and values that generate social conflicts and psychological reactions among individuals or groups of people who feel that the consequences of the risk have been inflicted on them personally. Feelings of violation may also result from perceived inequities in the distribution of costs and benefits.

Subsequently, the UK Treasury Department (2004) recommended a risk classification that includes hazard characteristics, the traditional risk assessment variables such as probability and extent of harm, indicators on public perception, and the assessment of social concerns. In addition to the aforementioned eight criteria, the Department made an extra effort to define criteria for measuring concern. The list of concerns includes:

- perception of familiarity and experience with the hazard;
- understanding the nature of the hazard and its potential impacts;
- repercussions of the risk's effects on (intergenerational, intragenerational, social) equity;
- perception of fear and dread in relation to a risk's effect;
- perception of personal or institutional control over the management of a risk; and
- degree of trust in risk management organizations.

Helene Cecilie Blakstad in her contribution (Chapter 9) to this volume lists more specific indicators for risk evaluation related to accident prevention and crisis management. She compares the British and Norwegian safety indicators and comes to the conclusion that a multiple indicator set is preferable over a single index solution. This applies to both Norway and the United Kingdom. The United States, in contrast, has been less attentive to indicators of accident prevention and more focused on risk characterization.

When applied to technological risks, in particular oil and gas exploration, the phase of interdisciplinary risk estimation includes four major steps:

- First, there is a need to develop scenarios that include plausible sequences of accidents or other pathways of harm (pollution, waste production).
- Second, these scenarios need to be augmented with assumptions about human behavior one can expect in such situations, including crisis identification, crisis management, domino effects, perception-driven responses, and human errors. It is important that these behavioral components are integrated into the technical analysis because the interaction of both the technical and the human

sphere creates the combined threats to human health and the environment (IAEA 1995).

- Third, each scenario needs to be assessed according to its probability of occurrence within the uncertainty ranges in which these estimates are embedded.
- Fourth, these scenarios need to be tested for stakeholder and public concerns with respect to their consequences and its implications. There may be equity violations involved or special symbolic meanings affected.

These four steps of generating knowledge and insights provide the data and information base for the next step: risk evaluation.

1.4. RISK EVALUATION

A heavily disputed task in the risk governance process concerns the procedure of how to evaluate the societal acceptability or tolerability of a risk. In classical approaches, risks are ranked and prioritized based on a combination of probability (how likely is it that the risk will be realized) and impact (what are the consequences if the risk does occur) (Klinke and Renn 2002; Renn 2008, pp. 149ff). However, as described earlier, in situations of high uncertainty risks cannot be treated only in terms of likelihood (probability) and (quantifiable) impacts. This standard two-dimensional model ignores many important features of risk. Values and issues such as reversibility, persistence, ubiquity, equity, catastrophic potential, controllability, and voluntariness should be integrated in risk evaluation. Furthermore, risk-related decision making is neither about physical risks alone nor usually about a single risk. Evaluation requires risk-benefit evaluations and risk-risk trade-offs (Graham and Wiener 1995). If the benefits and potential substitutes were not included in the assessment phase, this needs to be done in the evaluation phase. Judgments about risk tolerance require a comparative review of alternatives and benefit-risk ratios. So, by definition, risk evaluation is multidimensional. In order to evaluate risks, the first step is to characterize the risks on all the dimensions that matter to the affected populations. Once the risks are characterized in a multidimensional profile, their acceptability can be assessed.

Furthermore, there are competing, legitimate viewpoints about evaluations over whether there are or could be adverse effects and, if so, whether these risks are tolerable or even acceptable. Drawing the lines between "acceptable," "tolerable," and "intolerable" risks is one of the most controversial and challenging tasks in the risk governance process. The UK Health and Safety Executive developed a procedure for chemical risks based on risk-risk comparisons (Löfstedt 1997). Some Swiss cantons such as Basel County experimented with round tables comprising industry, administrators, county officials, environmentalists, and neighborhood groups (Risko 2000). As a means

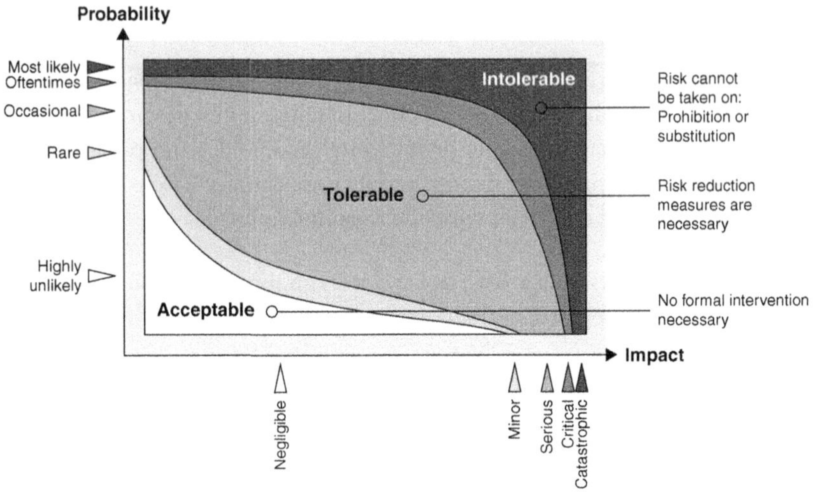

FIGURE 1.2 Areas of acceptable, tolerable and intolerable risks (adopted from Renn 2008, p. 150).

for reaching consensus, two demarcation lines were drawn between the area of tolerable and acceptable risk and between acceptable and intolerable risks (Figure 1.2). Irrespective of the selected means to support this task, the judgment on acceptability or tolerability is contingent on making use of a variety of different knowledge sources; in other words, it requires taking the interdisciplinary risk estimation seriously.

Looming above all risks is the question of what is safe enough, implying a normative or moral judgment about acceptability of risk and the tolerable burden that risk producers can impose on others. The results of the risk and concern assessment can provide hints as to what kind of mental images are present and which moral judgments guide people's perceptions and choices. Yet the decision on what is safe enough and which risk is tolerable or acceptable is a political decision that needs public legitimization.

As Kringen (Chapter 11 in this volume) points out, legitimacy issues can tentatively be clustered around two sets of problems: (1) the legitimacy of the regulatory goals in terms of how much risk the regime is prepared to tolerate, and (2) the legitimacy of the regulatory means and instruments designed for achieving these goals. The first issue is related to risk evaluation, the second one to risk management, which is described in the next section. Kringen argues that the issue of tolerance is often embedded and hidden in legal decision making, although a clear value judgment (how safe is safe enough?) is at stake. Obscuring this value judgment often leads to effects of delegitimization, because the affected stakeholders have no information about the trade-offs that the risk managers have applied to balance risks and benefits or risks and opportunity costs.

Of particular importance is the perception of just or unjust distribution of risks and benefits. How these moral judgments are made and justified depends to a large degree on social position, cultural values, and worldviews (Fiorino 1989). They also depend on shared ontological (a belief in the state of the world) and ethical (a belief of how the world should be) convictions. This collection of forces influences thinking and evaluation strategies. The selection of strategies for risk management is, therefore, understandable only within the context of broader worldviews (Rayner 1987; see also Chapter 3 in this volume). Hence, society can never derive acceptability or tolerability from the assessment evidence alone. Facts do not speak for themselves. Furthermore, the evidence is essential not only to reflect the degrees of belief about the state of the world regarding a particular risk, but also to know whether a value might be violated or not (or to what degree).

In sum, risk evaluation involves the deliberative effort to characterize risks in terms of acceptability and tolerability. Such contexts often imply that neither the risks nor the benefits can be clearly identified. Multiple dimensions and multiple values are at work and have to be considered. Finally, risk evaluations may shift over time. Notwithstanding large uncertainties, decisions need to be made. It may well be possible at a certain point in time to agree whether risks are acceptable, tolerable, or intolerable. When the tolerability or acceptability of risks is heavily contested, that too is highly relevant input to the decision-making process.

With respect to technological risks, the judgment of acceptability or tolerability is usually related to three issues: *occupational safety*; *routine emissions of waste into air soil or water*; and *accidents with sudden emission of energy and/or material*. For all three aspects of technical risks, there are normally regulatory standards that need to be adhered to. For sudden events such as accidents, often deterministic (safety provisions) and probabilistic (safety goals) standards are in effect; for controlling emissions, maximum tolerance levels for certain time intervals (daily, yearly) are specified (Aven and Renn 2010; Chapter 7 in this volume). In addition, regulatory agencies can issue flexible standards based on ALARA (as low as reasonably achievable) or BACT (best available control technology). Regardless of what methods for standard setting are applied, it always involves political judgments about the tolerability of statistical losses (in terms of lives, health, environmental damage, and money). These judgments need to be legitimized in an open and democratic process. This is further discussed in the section on risk management.

1.5. RISK MANAGEMENT

Risk management starts with a review of the output generated in the previous phases of interdisciplinary risk estimation, characterization, and risk evaluation. If the risk is acceptable, no further management is needed. Tolerable risks are those where the benefits are judged to be worth the risk and further risk reduction measures are not necessary but can be accomplished by voluntary measures. If risks are classified as tolerable but not as acceptable, risk management needs to design and implement actions that render these risks either acceptable or sustain that tolerability in the longer run by introducing risk reduction strategies, mitigation strategies, or strategies aimed at increasing societal resilience at the appropriate level (Renn 2008, pp. 173ff.). If the risk is considered intolerable, notwithstanding the benefits, risk management should be focused on banning or phasing out the activity creating the risk. If that is not possible, management should be devoted to mitigating or fighting the risk in other ways or to increasing societal resilience. If the risk is contested, risk management can be aimed at finding ways to create consensus. If that is impossible or highly unlikely, the goal would be to design actions that increase tolerability among the parties most concerned or to stimulate an alternative course of action (van Asselt and Renn 2011).

Risk management is based on different regimes, that is, the set of rules and standards that govern the handling of the risk in a specific regulatory context (see Baram and Lindøe, Chapter 2 in this volume, and Bang and Thuestad, Chapter 10 in this volume). The main goal of these regimes is to ensure robust regulatory results. Robust regulation is highly dependent on how risk managers and regulators deal with three major challenges: complexity, uncertainty, and ambiguity (Renn 2008, pp. 187ff; Renn et al. 2011). Complexity refers to the difficulty of identifying and quantifying causal links between a multitude of potential candidates and specific adverse effects. Uncertainty denotes the inability to provide accurate and precise quantitative assessments between a causing agent and an effect. Finally, ambiguity denotes either the variability of (legitimate) interpretations based on identical observations or data assessments or the variability of normative implications for risk evaluation (judgment on tolerability or acceptability of a given risk). Based on the distinction between complexity, uncertainty, and ambiguity, one can distinguish four risk management routes (Figure 1.3).

The easiest task is to deal with simple, linear risks. Simple does not mean small or negligible. It refers to the assessment and evaluation process. If the risk is well known to the actors and there is hardly any uncertainty and ambiguity, the acceptable risk management strategy is to assess the risk-benefit ratio and develop measures that move the risk into the acceptable area of the risk evaluation diagram. The risk management actions are straightforward and include technical standards, safety manuals, and routinized monitoring.

In a case where scientific complexity is high and uncertainty and ambiguity are low, the challenge is to invite experts to deliberate with risk managers to understand complexity. Understanding the risks of oil platforms may be a good example of this. Although the technology is highly complex and many interacting devices lead

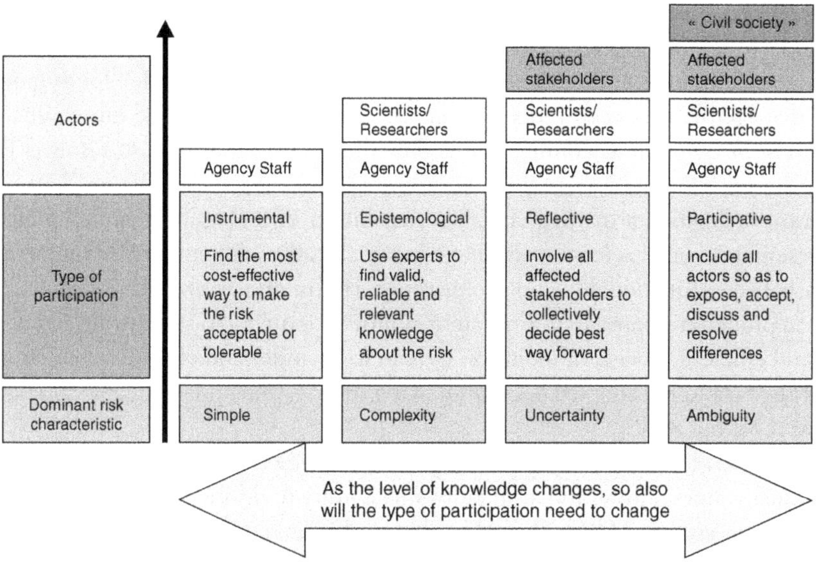

FIGURE 1.3 Relationship between stakeholder participation and risk categories in risk governance (adopted from Renn 2008, p. 280).

to multiple accident scenarios, most possible pathways to a major accident can be modeled well in advance. The major challenge is to determine the limit to which one is willing to invest in resilience.

The third route concerns risk problems that are characterized by high uncertainty but low ambiguity. Expanded knowledge acquisition may help reduce uncertainty. If, however, uncertainty cannot be reduced (or only reduced in the long run) by additional knowledge, a "precaution-based risk management" is required. Precaution-based risk management explores a variety of options: containment, diversification, monitoring, and substitution. The focal point here is to find an adequate and fair balance between over-cautiousness and insufficient caution. This argues for a reflective process involving stakeholders to ponder concerns, economic budgeting, and social evaluations.

For risk problems that are highly ambiguous (regardless of whether they are low or high in uncertainty and complexity), the fourth route recommends "discourse-based management." Discourse management requires a participatory process involving stakeholders, especially the affected public. The aim of such a process is to produce a collective understanding among all stakeholders and the affected public about how to interpret the situation and how to design procedures for collectively justifying binding decisions on acceptability and tolerability that are considered legitimate. In such situations, the task of risk managers is to create a condition where those who believe that the risk is worth taking and those who believe otherwise are willing to respect each others' views and to construct and create strategies acceptable to the various stakeholders and interests. But deliberation is not a guarantee for a smooth risk management process. Lidskog et al. argue that complexity and ambiguity are grounds for continuous conflict that is difficult, if not impossible, to resolve. The reduction of complexity simultaneously implies reducing the number of actors as relevant or legitimate participants. The resolution of ambiguity requires a broad representation of all actors involved in the case. It is thus difficult to find the perfect path between functionality and inclusiveness (Renn and Schweizer 2009). In any case, a prudent response to this inherent conflict is to invest in structuring an effective and efficient process of inclusion (whom to include) and closure (what counts as evidence and the adopted decision-making rules) (Aven and Renn 2010, pp. 181ff; Renn 2008, pp. 284ff).

In addition to structuring risk management strategies according to the extent of complexity, uncertainty, and ambiguity, one can try to distinguish different regulatory styles or cultures (Löfstedt and Vogel 2001). Kringen (Chapter 11 in this volume) differentiates between self-regulation, enforced self-regulation, meta-regulation, purpose-based regulation, principles-based regulation, performance-based regulation, systems-based regulation, and management-based regulation. Given that some of these concepts overlap, one can draw a line between self-regulation and governmental regulation on one hand, and between experience-based and principles-based management on the other hand. All existing regulatory regimes are probably mixtures of the two dimensions (Fioritto and Simoncini 2011). However, it seems that Norway is aiming toward a more pragmatic, experience-based yet government-controlled regime, whereas the United States tend to be more principles-based and reliant on self-regulation.

Risk management for technological systems requires *technological*, *organizational*, and *behavioral* measures for reducing risks that are not regarded as acceptable in the first place (Hood et al. 2002):

- Technological measures relate to the inclusion of active and passive safety devices, inclusion of filters and purifiers, and waste-handling technology.
- Organizational measures include emergency and contingency plans, guidelines for daily operations and safety checks, monitoring requirements, and provisions for assuring accountability and competence.
- Behavioral measures extend to all educational and training efforts to improve personal performance, increase sensibility for safety issues, and strengthen the feeling of responsibility and accountability among the staff (safety culture).

The historic record about technological accidents and failures shows that both the integration of and the compatibility between technological, organizational, and behavioral measures was often the main reason for the events that lead to accidents and disasters (Cohen 1996; Hutter 2010). Michael Baram (Chapter 7 in this volume) points to the problem of integration among these three factors:

> In supporting its proposed SEMS rule, MMS also discussed data on 1,443 incidents that occurred over the same years, which involved 41 fatalities, 302 injuries, 10 losses of well control, 11 collisions, 476 fires, 356 "pollution events", and 224 crane and hoist mishaps. It concluded that the majority of these incidents were related to operational and maintenance procedures or human error, and that operating procedures and mechanical integrity accounted for the greatest number of spills. It also found there had been no discernible trend of industrial improvement over the seven year study period despite its inspection program's issuance of corrective orders for some 150 findings of operator non-compliance per year. (Chapter 7)

Ole Andreas Engen (Chapter 14 in this volume) reports on another example of the lack of compatibility between these three factors:

> The direct cause of the Gullfaks C and Snorre A incident was a leakage in a well casing. Although, understood as an organizational matter, failure to comply with company procedures and guidelines for drilling operations; inadequate use of specialized competence of relevance for the drilling method being utilized; and failure to communicate and transfer prior knowledge and lessons were among the most prominent explanations for why the incident occurred. In addition, it is apparent that management on all levels failed to ensure proper safety, risk and quality while planning the well. (Chapter 14).

There have been major debates, in particular with respect to low-probability, high-consequence risks, to develop regulatory regimes that would be able to control for a robust integration of technological, organizational, and behavioral safety factors (Graham and Rhomberg 1996). One issue is the creation of organizational safety culture that provides the mental links between technology, organization, and behavior (see Mearns, Chapter 3 in this volume). Another major issue is leadership: Knut Kaasen (Chapter 5 in this volume) concludes that in order to avoid "fragmentation of responsibility … it might be easier to place the responsibility on a person who is entrusted with a particular safety function than to rely on the assumption that everyone who is somehow involved in any activity can be roused to implement safety thinking every hour of the day" (Chapter 5). Multiple actors and multiple involvements should not be mistakened for diffusion of accountability or lack of clear command structure. Complexity requires the input of many disciplines and many sources of expertise, but the command structure needs to be functional and unambiguous, particularly in crisis situations. Lack of leadership and inability to create a learning environment are, according to Paterson (Chapter 6 in this volume), two of the main problems found by a health, safety and environment (HSE) review of safety performance in the United Kingdom. "In an effort to account for these problems, the regulator highlighted three underlying issues relating to learning, the engineering function and leadership. In relation to learning, there was seen to be a problem both of inadequate auditing and monitoring and of a lack of processes to allow learning to be embedded within the organizations concerned" (Chapter 6).

Of specific interest is the issue of private-public partnerships as a means of sustaining a high safety record over time (Weintraub 1997). In this volume, Emre Üşenmez discusses the pros and cons of private self-regulation. His conclusions emphasize the complementarity of public and private risk management. A skeptical view on privatizing regulatory functions has been put forth by Kathryn Mearns (Chapter 3 in this volume). She believes that the competitive nature of the oil industry may still prevent a learning culture from being developed, and the fear of blame and litigation is still prevalent within the industry. According to her view, a strong regulatory oversight is necessary to develop and sustain an effective and robust risk management record. However, she acknowledges that different national cultures have developed specific incentives and barriers for the emergence of safety cultures (or safety climates) that cannot be generalized beyond the culture in which they evolved. She comments: "While it is likely that no 'national' culture possesses all the optimum components necessary for safety, it is possible that certain combinations of national dimensions (based on the arguments above) have the potential to create cultural norms that will determine the propensity to engage in particular types of safety relevant behaviour at work" (Chapter 3). According to her analysis, a proper arrangement of checks and balances between the private and public sectors is a major step toward better safety performance irrespective of native cultural traditions and constraints.

Within the phase of risk management, monitoring and control play a major role in sustaining high safety levels. Risks often occur as a consequence of relaxed inspections, incomplete or insufficient monitoring, and overconfidence of the staff. Often accidents are caused or aggravated by mere lack of oversight. The book lists several examples of this problem. Baram (Chapter 7 in this volume) cites several fallacies in the long-term monitoring and control process. One issue is the normalization of deviance. "This concept provides that many small behavioral and technical deviations commonly occur over time within a regulatory program or safety management system because of convenience, cost reduction, or other reasons, without being seen as safety-significant. Eventually, the deviations become norms which weaken the program or system to the point where it is incapable of preventing accidents under certain conditions" (Chapter 7). This and other problems of maintaining high safety levels are paramount to most of the case studies covered in this volume. For example, Hayes (Chapter 8 in this volume) complains that there is a lack of leadership, professionalism, competence, experience, and judgment in both the private public sectors, which prevents the development of a learning environment. Baram and Lindøe (Chapter 2 in this volume) point out that public scrutiny and open information exchange are key to improving safety records and preventing laxness in safety control.

It is contested in the literature whether monitoring and control should be performed by the operator of technical facilities (with random inspections by the public authorities) or closely observed and enforced by public authorities. In recent times the idea of public-private partnership (joint responsibility) has also been raised and recommended as a potential solution to this question (Sikor 2008). It appears that Norway favors close inspection cycles with clear mandate for public oversight and trust among the main actors, the United Kingdom favors more self-regulation but with strong public oversight, and the United States favors models of self-regulation with occasional public inspections (see Paterson, Chapter 6 in this volume). However, all three regulator regimes are experiencing rapid changes. Norway has adopted the perspective of government-supervised self-regulation (Engen, Chapter 13 in this volume), whereas the United States has adopted more regulatory responsibilities after the Deep Water Horizon accident (Ryggvik, Chapter 15 in this volume). There is no statistical evidence as to which of the two regimes works best, which may also depend on the political culture (Mearns, Chapter 3 in this volume.). However, several chapters in this book provide ample evidence for mismanagement by private actors, which seems to enforce the idea of at least more public oversight and a "checks and balances" approach to sustaining an appropriate safety record. Yet there is also room for improvement as Ryggvik states in Chapter 15: "These are questions which can be approached with some objective measures like high wages, good educational programs, institutional formations that underpin independence and a focus on ethical rules" (Chapter 15).

1.6. RISK COMMUNICATION AND PARTICIPATION

Effective communication among all relevant interests is one of the key challenges in risk governance. It is not a distinct stage (in contrast to how it is often treated in the risk literature), but it is central to the entire governance process (IRGC 2005). Positively framed, communication is at the core of any successful risk governance activity. Negatively framed, a lack of communication destroys effective risk governance. Early on, risk communication was predicated on the view that disagreements between experts and citizens over risks stemmed from the lack of accurate knowledge by citizens. The solution was sought in the education and persuasion of the deficient public (Fischhoff 1995). Implied in this solution was the belief that an educated public would perceive and evaluate risks the same way as experts. However, this deficit model has been subject to considerable criticism. For one thing, increased knowledge often elevated citizens' concerns about risk, creating an even greater divergence between them and experts. For another, as Pidgeon et al. (2005, p. 467) phrased it, "One of the most consistent messages to have arisen from social science research into risk over the past 30 years is that risk communication … needs to accommodate far more than a simple one-way transfer of information … the mere provision of 'expert' information is unlikely to address public and stakeholder concerns or resolve any underlying societal issues." Third, research on risk controversies has demonstrated that in general the public does not always misunderstand science. Furthermore, experts and governments may also misunderstand public perceptions (Horlick-Jones 1998; Irwin and Wynne 1996).

The important point to emphasize is that risk communication and trust are delicately interconnected processes, and there is a large volume of literature demonstrating the connection between trust in the institutions managing risks and citizen perceptions of the seriousness of risks (Earle and Cvetkovich 1996; Löfstedt 2005; Whitfield et al. 2009). Communication breakdowns can easily damage trust. At the same time, communication strategies that misjudge the context of communication, in terms of the level of and reasons for distrust, may boomerang, resulting in an increased distrust (Löfstedt 2005).

Communication strategies proliferate. Communication refers to meaningful interactions in which knowledge, experiences, interpretations, concerns, and perspectives are exchanged (Löfstedt 2003). In the context of risk governance, exchanges among policy makers, experts, stakeholders, and affected publics are of special interest. The aim of communication is to provide a better basis for responsible risk management. Its aim is also to enhance trust and social support (Poortinga and Pidgeon 2003). Depending on the nature of the risks and the context of governing choices, communication will serve various purposes. It might serve for sharing information about the risks and possible ways of handling them. It might support building and sustaining trust among various actors where particular arrangements or risk management measures become acceptable. It might result in actually engaging people in risk-related decisions, through which they gain ownership of the process.

However, communication in the context of risk governance is not simple. It is not just a matter of having accurate assessments of risks. It is not just a matter of bringing people together. It is not just a matter of effective communication. It requires all these features and more. Also required is a set of procedures for facilitating the discourse among various actors from different backgrounds so they can interact meaningfully in the face of uncertain outcomes (Rosa et al. 2013).

Proper communication features multiple actors. The U.S. National Research Council report (Stern and Fineberg 1996) is an important milestone in the recognition of the need for risk decision making as an inclusive multi-actor process. It also was a germinal precursor to the idea of risk governance with its emphasis on the coordination of risk knowledge and expertise with citizen and other stakeholder priorities (cf. Jasanoff 2004; Stirling 2007).

One key challenge to risk governance is the question of inclusion: Which stakeholders and publics should be included in governance deliberations? The inclusion challenge has deep implications. Contrary to the conventional paradigm where risk topics are usually identified by experts, with the analytic-deliberative process underpinning risk governance, public values and social concerns are key agents for identifying and prioritizing risk topics. Inclusion means more than simply including relevant actors. That is the outmoded practice of "public hearings" where relevant actors are accorded a fairly passive role. Inclusion means that actors play a key role in framing (or pre-assessing) the risk (IRGC 2005; Renn and Schweizer 2009; see also Roca et al. 2008). Inclusion should be open to input from civil society and adaptive at the same time (Stirling 2004; 2007). Crucial issues in this respect are (see also Renn and Schweizer 2009): Who is included? What is included? What is the scope and mandate of the process?

Inclusion can take many different forms: round tables, open forums, negotiated rule-making exercises, mediation, or mixed advisory committees including scientists and stakeholders (Renn 2008, pp. 332ff ; Rowe and Frewer 2000; Stoll-Kleemann and Welp 2006). Because of a lack of agreement on method, social learning promoted by structured and moderated deliberations is required to find out what level and type of inclusion is appropriate in the particular context and for the type of risk involved. What methods are available? They all have contrasting strengths and weaknesses (Pidgeon et al. 2005).

A focus on inclusion is defended on several grounds (Roca et al. 2008). First, one can argue that in view of uncertainty, there is a need to explore various sources of information and to identify various perspectives. It is important to know what the various actors label as risk problems and which most concern them. Here inclusion is interpreted to be a means to an end: a procedure for integrating all relevant knowledge and for including all relevant concerns. Second, from a democratic perspective, actors affected by the risks or the ways in which the risks are governed have a legitimate right to participate in deciding about those risks. Here inclusion is interpreted as not just a means, but also an end in itself. At the same time, inclusion is a means to agree on principles and rules that should be respected in the processes and structures of collective decision making. Third, the more actors are involved in the weighing of the heterogeneous pros and cons of risks, the more socially robust the outcome. When uncertainty is prevalent, there is no simple decision rule. In that view, inclusion also is a way to organize checks and balances between various interest and value groups in a plural society. Inclusion thus is intended to support the co-production of risk knowledge, the coordination of risk evaluation, and the mutual design of risk management.

Social learning is also required here. And it is not simply a matter of degree where more inclusion equals better risk governance. The degree and type of inclusion may vary depending on the phase of governance and the risk context. In each phase and context, careful thought is needed on the kind and degree of inclusion that is needed. Hence, differentiation is not an exception but rather the rule. The benefits of mutual cooperation and social learning provide tangible improvements in safety records, as Ragnar Rosness and Ulla Forseth (Chapter 12 in this volume) are able to prove. "By establishing arenas such as Safety Forum and at the same time promoting the formation of common perception of the state of safety in the petroleum sector, the authorities promoted a process where converging sense-making and the development of organisational structures for collaboration mutually reinforced each other" (Chapter 12).

The task of inclusion is to organize productive and meaningful communication among a range of actors, who have divergent interests but complementary roles. The cumulative empirical analyses suggest that providing a platform for the inclusion of a variety of stakeholders – to deliberate over their concerns and exchange arguments – can help to de-escalate conflicts and legitimize the final decision. Nevertheless, no matter how careful the establishment of the platform and the decision rule about inclusion, there will always be some disappointed actors in society (Beierle and Cayford 2002).

With respect to technological risks, communication and participation are key to improved risk and crisis management as well as public acceptance of facilities and operations (Chakraborty 2011). Jacob Kringen (Chapter 11 in this volume) points out that the style of communication and inclusion defines much of what is known as regulatory style in the literature. This may vary between educational, accommodative, conciliatory, persuasive, insistent, and legalistic enforcement practices. Yet each of these styles demands a specific approach to internal and external communication and stakeholder involvement.

Renn and colleagues, for example, interviewed major stakeholders for the Barents Sea oil exploration. As a result of major efforts to communicate with and involve major stakeholders, most respondents were quite convinced that the risk management efforts of the companies involved were sufficient. They did worry, however, about the sharing of benefits with the affected communities (Renn et al. 2013):

> None of the interviewed persons demanded a new risk reduction plan or a reappraisal of the risk in the aftermath of the accident at the Deepwater Horizon drilling rig.... They were convinced that the knowledge basis is sufficient to reach a positive judgment and that petroleum operations in these areas will not pose different challenges from those that the Norwegian public is familiar with further south. In contrast to the lack of ambiguity in the perception of risks, there was a

> lot of conflict and controversy about benefits. Therefore all local interview partners expected the companies to provide additional benefits to the communities in exchange for taking any potential risk from the oil development.

Other examples, as outlined in this book, demonstrate that a lack of public involvement increases the feeling of distrust, which impedes crisis management efforts in the cases of disasters or serious accidents. The less people are aware of and involved with potential crisis situations, the more one can expect that the extent of damage is more serious than necessary.

1.7. CONCLUSIONS

This chapter has described the genesis and analytical scope of risk governance. It argued for a broader, paradigmatic turn from government to governance. In the context of risk, the idea of governance is used in both a descriptive and normative sense: as a description of how decisions are made and as a normative model for improving structures and processes of risk policy making. Risk governance draws the attention to the fact that many risks, particularly pertaining to large technological systems, are not simple; they cannot all be calculated as a function of probability and extent of damage. Many risks embed complex trade-offs of costs and benefits. Risk governance underscores the need to ensure that societal choices and decisions adequately address these complicating features. However, conventional risk characterization typically treats, assesses, and manages such risks as if they were simple. This practice has led to many failures to deal adequately with risks. This has been echoed by many case studies that are described in this volume. Hayes, in Chapter 8 in this volume, concludes: "Attempts that have been made to date to incorporate these issues into consideration of risk have not been very successful, primarily because of the three issues described above –difficulty in linking organisational factors into the current risk management tools, lack of a universally agreed framework for organisational issues and a narrow understanding of the concept of risk leading to a misconception about what can and cannot reasonably be addressed in a safety case" (Chapter 8). It is our conviction that the risk governance framework advocated in this chapter provides an adequate and practically proven concept for dealing with complex, uncertain, and ambiguous risk problems and could serve as a model for future regulatory reforms aimed at improving the robustness of the risk management performance.

In a pluralistic society where the pressure to legitimize political action is always high, the process of developing and locating potentially dangerous technologies such as oil drilling installations often encounter widespread skepticism and deep distrust. More than in other policy arenas, decisions on risks must be made plausible to a wider audience (i.e., based on intuitively understandable reasoning) and depend on trust in the major actors involved, the respective industry, and the regulatory agencies. Hence, risk governance can only be successful if there is an intense, communication-oriented dialogue with the major actors and the interested public. The larger the number of individuals and groups that are impacted by a technology development or deployment, the more likely it is that conflicts will arise. These conflicts deal with issues of risk acceptability or tolerable risk levels as well as notable equity issues such as a just distribution of risks among the affected population and, even more important, the distribution of benefits. If equity issues are ignored or not given due attention, people tend to amplify their experience of risk and lower the thresholds of tolerability as an expression of their discontent with the process rather than the resulting risk. Hence the timely and mutual participation of social actors in managing risks and crisis is technically appropriate, as they may bring important local knowledge to the decision-making process and democratic imperative, as the distribution of risks and benefits demands a legitimate key for designing a fair risk/benefit-sharing initiative. Effective participation helps technology providers, users, and political decision makers secure greater legitimacy in siting processes, thus contributing to the democratic culture of a country.

Risk governance is not simply a timely buzzword; it is a disciplined argument for a paradigm shift. Paradigms and reforms do not just shift in the abstract; they also shift in practices. Such fundamental transitions are not easy. Yet, the chapters in this volume, by combining the insights of many case studies with an argument for governance, are meant to stimulate and facilitate that shift.

REFERENCES

Aven, T. and Renn, O. (2010) *Risk Management and Governance*. Springer: Heidelberg and New York.

(2012) "On the risk management and risk governance for petroleum preparations in the Barents Sea Area", *Risk Analysis*, 32 (9): 1561–1575.

Beierle, T. C. and Cayford, J. (2002) *Democracy in Practice: Public Participation in Environmental Decisions*. Resources for the Future: Washington, DC.

Chakraborty, S. (2011) "The challenge of emergency risk communication: Lessons learned in trust and risk communication from the volcanic ash crisis", in: A. Alemanno (ed.): *Governing Disasters. The Challenges of Emergency Risk Regulation*. Edward Elgar: Cheltenham, pp. 80–100.

Cohen, A. V. (1996) "Quantitative risk assessment and decisions about risk: An essential input into the decision process", in: C. Hood and D. K. C. Jones (eds.): *Accident and Design: Contemporary Debates in Risk Management*. UCL Press: London, pp. 87–98.

Derby, S. L. and Keeney, R. L. (1981) "Risk analysis: Understanding 'How safe is safe enough'", *Risk Analysis*, 1 (3): 217–224.

Earle, T. C. and Cvetkovich, G. (1996) *Social Trust Toward a Cosmopolitan Society*. Praeger: Westport, CT.

Entman, Robert. M. (1993) "Framing: Toward clarification of a fractured paradigm," *Journal of Communication*, 43 (4): 51–58.

Fiorino, D. J. (1989) "Technical and democratic values in risk analysis", *Risk Analysis*, 9 (3): 293–299.

Fioritto, A. and Simoncini, M. (2011) "If and when. Towards standard-based regulation in the reduction of catastrophic risks", in A. Alemanno (ed.): *Governing Disasters. The Challenges of Emergency Risk Regulation*. Edward Elgar: Cheltenham, pp. 115–136.

Fischhoff, B. (1995) "Risk perception and communication unplugged: Twenty years of process," *Risk Analysis*, 15 (2): 137–145.

Goldstein, J. and Keohane, R.O. (1993) "Ideas and foreign policy. An analytical framework', in J. Goldstein and R.O. Keohane (eds.): *Ideas and Foreign Policy. Beliefs, Institutions, and Political Change*. Cornell University Press: Ithaca, pp. 3–30.

Graham, J.D. and Rhomberg, L. (1996) "How risks are identified and assessed", in H. Kunreuther and P. Slovic (eds.): *Challenges in Risk Assessment and Risk Management*. Annals of the American Academy of Political and Social Science, Sage: Thousand Oaks, pp. 15–24.

Graham, J.D. and Wiener, J.B. (1995) "Confronting risk tradeoffs", in J.D. Graham and J.B. Wiener (eds.): *Risk vs Risk: Tradeoffs in Protecting Health and the Environment*. Harvard University Press: Cambridge, MA., pp. 1–41.

Halachmi, A. (2005) "Governance and risk management: Challenges and public productivity", *International Journal of Public Sector Management*, 18 (4): 300–317.

Hutter, B.M. (2010) *Anticipating Risks and Organizing Risk Regulation*. Cambridge University Press: Cambridge.

Hood, C., Rothstein, H. and Baldwin, R. (2002) *The Government of Risk: Understanding Risk Regulation Regimes*. Oxford University Press: Oxford.

Horlick-Jones, T. (1998) "Meaning and contextualization in risk assessment", *Reliability Engineering and Systems Safety*, 59: 79–89.

International Atomic Energy Agency (IAEA) (1995) *Guidelines for Integrated Risk Assessment and Management in Large Industrial Areas*. Technical Document, IAEA-TECDOC PGVI-CIJV, IAEA: Vienna.

International Risk Governance Council (IRGC) (2005) *Risk Governance: Towards an Integrative Approach*. White Paper No. 1, O. Renn with an Annex by P. Graham. IRGC: Geneva.

International Risk Governance Council (IRGC) (2007) *An Introduction to the IRGC Risk Governance Framework*. Policy Brief. IRGC: Geneva

Irwin, A. and Wynne, B. (eds) (1996) *Misunderstanding Science? The Public Reconstruction of Science and Technology*. Cambridge University Press: Cambridge.

Irwin, I. (2008) 'STS perspectives on scientific governance', in E. Hackett, O. Amsterdamska, M. Lynch and J. Wajcman, J. (eds.): *The Handbook of Science and Technology Studies*. MIT Press: Cambridge, MA.

Jasanoff, S. (2004) "Ordering knowledge, ordering society', in S. Jasanoff (ed.): *States of Knowledge: The Co-Production of Science and Social Order*. Routledge: London.

Kahneman, D. and Tversky, A. (eds.) (2000) *Choices, Values, and Frames*. Cambridge University Press: Cambridge.

Kjaer, A.M. (2004) *Governance- Key Concepts*. Polity Press: Malden.

Klinke, A. and Renn, O. (2002) "A new approach to risk evaluation and management: Risk-based, precaution-based and discourse-based management", *Risk Analysis*, 22 (6): 1071–1094.

(2012) "Adaptive and integrative governance on risk and uncertainty", *Journal of Risk Research*, 15 (3): 273–292.

Lidskog, R. (2008) "Scientised citizens and democratised science. Re-assessing the expert-lay divide", *Journal of Risk Research*, 11 (1): 69–86.

Lidskog, R., Uggla, Y. and Soneryd, L. (2011) "Making transboundary risks governable: Reducing complexity, constructing spatial identity and ascribing capabilities", *AMBIO*, 40 (2): 111–120.

Löfstedt, R.E. (1997) *Risk Evaluation in the United Kingdom: Legal Requirements, Conceptual Foundations, and Practical Experiences with Special Emphasis on Energy Systems*. Working Paper No. 92. Center of Technology Assessment: Stuttgart.

(2003) "Risk communication: Pitfalls and promises", *European Review*, 11 (3): 417–435.

(2005) *Risk Management in Post-Trust Societies*. Palgrave Macmillan: Hampshire and New York.

Löfstedt, R.E. and Vogel, D. (2001) "The changing character of regulation: A comparison of Europe and the United States". *Risk Analysis*, 21 (3): 398–405.

Nelson, T.E., Oxleay, Z.M. and Clawson, R.A. (1997) "Toward a psychology of framing effects", *Political Behavior*, 19 (3): 221–246.

Organisation for Economic Co-Operation and Development (2003) *Emerging Systemic Risks. Final Report to the OECD Futures Project*. OECD: Paris.

Pelling, M., High, C., Dearing, J. and Smith, D. (2008) "Shadow spaces for social learning: A relational understanding of adaptive capacity to climate change within organisations", *Environment and Planning*, 40: 867–884.

Pidgeon, N.F., Poortinga, W., Rowe, G., Jones T.-H., Walls, J. and O'Riordan, T. (2005) "Using surveys in public participation processes for risk decision making: The case of the 2003 British GM Nation public debate", *Risk Analysis*, 25 (2): 467–479.

Poortinga, W. and Pidgeon, N.F. (2003) "Exploring the dimensionality of trust in risk regulation", *Risk Analysis*, 23: 961–972.

Rassmussen, I. (1997) "Risk management in a dynamic society: A modeling problem", *Safety Science*, 27: 183–213.

Rayner, S. and R. Cantor (1987) "How fair is safe enough? The cultural approach to societal technology choice", *Risk Analysis*, 7: 3–10.

Renn, O. (2008) *Risk Governance. Coping with Uncertainty in a Complex World*. Earthscan: London.

Renn, O., Grieger, K., Øien, K. and Andersen, H.B. (2013) "How major stakeholders perceive the decision making process in the Goliat oil field development in the Barents Sea", Risk Research, DOI:10.1080/13669877.2012.761266.

Renn, O. and Klinke, A. (2013) "Space matters! Impacts for risk governance", in: D. Müller-Mahn (ed.): *The Spatial Dimension of Risk*. Earthscan and Routledge: London, pp. 5–27.

Renn, O., Klinke, A. and van Asselt, M. (2011) "Coping with complexity, uncertainty and ambiguity in risk governance: A synthesis", *AMBIO*, 40 (2): 231–246.

Renn, O. and Schweizer, P. (2009) "Inclusive risk governance: Concepts and application to environmental policy making", *Environmental Policy and Governance*, 19: 174–185.

Renn, O. and Walker, K. (2008) "Lessons learned: A re-assessment of the IRGC framework on risk governance", in O. Renn and K. Walker (eds.): *The IRGC Risk Governance Framework: Concepts and Practice*. Springer: Heidelberg and New York, pp. 331–367.

Reese, S.R. (2007) "The framing project: A bridging model for media research revisited", *Journal of Communication*, 57, 148–154.
RISKO (June 2000) "Mitteilungen für die Kommission für Risikobewertung des Kantons Basel-Stadt: Seit 10 Jahren beurteilt die RISKO die Tragbarkeit von Risiken", *Bulletin*, 3: 2–3.
Roca, E., Gamboa, G. and Tàbara, J.D. (2008) "Assessing the multidimensionality of coastal erosion risks: Public participation and multicriteria analysis in a Mediterranean coastal system", *Risk Analysis*, 28 (2): 399–412.
Rosa, E., Renn, O. and Mccright, A. (2013) *The Risk Society Revisited. Social Theory and Governance*. Temple University Press: Philadelphia.
Rowe, G. and Frewer, L.J. (2000) "Public participation methods: A framework for evaluation", *Science, Technology and Human Values*, 225 (1): 3–29.
Sikor, T. (2008) "New public and property rights", in T. Sikor (ed.): *Public and Private in Natural Resource Governance*. Earthscan: London, pp. 215–228.
Skelcher, C. (2005) "Jurisdictional integrity, polycentrism, and the design of democratic governance", *Governance*, 18 (1): 89–110.
Stern, P.C. and Fineberg, H.V. (1996) *Understanding Risk: Informing Decisions in a Democratic Society*. U.S. National Research Council. The National Academy Press: Washington, DC.
Stirling A. (2003) "Risk, uncertainty and precaution: Some instrumental implications from the social sciences", in F. Berkhout, M. Leach and I. Scoones (eds.): *Negotiating Change*. Edward Elgar: Cheltenham, pp. 33–76.
(2004) "Opening up or closing down: Analysis, participation and power in the social appraisal of technology", in F. Berkhout, M. Leach and B. Wynne (eds.): *Science, Citizenship and Globalisation*. Zed: London, pp. 218–231.
(2007) "Risk assessment in science: Towards a more constructive policy debate", *EMBO Reports*, 8: 309–315.
(2008) "Pluralism in the Social Appraisal of Technology", *Science Technology Human Values*, 33 (4): 262–294.
Stöhr, C. and Chabay, I. (2010) "Science and participation in governance of the Baltic Sea fisheries", *Environmental Policy and Governance*, 20: 350–363.
Stoll-Kleemann, S. and Welp, M. (eds.) (2006) *Stakeholder Dialogues in Natural Resources Management: Theory and Practice*. Springer: Heidelberg and Berlin.
UK Treasury Department (2004) *Managing Risks to the Public: Appraisal Guidance*. Draft for Consultation, HM Treasury Press: London, October, http://www.hm-treasury.gov.uk, accessed November 11, 2012.
U.S. National Research Council of the National Academies (2008) *Public Participation in Environmental Assessment and Decision Making*. The National Academies Press: Washington, DC.
van Asselt, M.B.A. and Renn, O. (2011): "Risk governance", *Risk Research*, 1 (4): 431–449.
WBGU, German Advisory Council on Global Change (2000) *World in Transition: Strategies for Managing Global Environmental Risks*. Springer: Heidelberg and New York.
Weintraub, J. (1997) "The theory and politics of the public/private distinction", in J. Weintraub and K. Kunar (eds.): *Public and Private in Thought and Practice: Perspectives on a Grand Dichotomy*. University of Chicago Press: Chicago, pp. 1–42.
Whitfield, S.C., Rosa, E.A., Dan, A. and Dietz, T. (2009) "Nuclear power: Value orientation and risk perception", *Risk Analysis*, 29: 425–437.

2

Modes of Risk Regulation for Prevention of Major Industrial Accidents

Michael Baram and Preben Hempel Lindøe

2.1. INTRODUCTION

This chapter discusses the evolution of new modes of risk regulation for preventing major industrial accidents, their co-regulatory and self-regulatory features and implementation issues, the global discourse about their legitimacy, its mutation into a discourse on whether the legitimacy question can be resolved by ensuring accountability, and then considers the legal standards and norms relevant to achieving accountability.[1]

Many social controls such as regulation, self-regulation, liability law, corporate governance principles, and market forces influence industrial performance of hazardous activities. As discussed, each has potential for reducing risks posed by such activities to health, safety, and the environment. But various contextual factors continuously shape the social controls, creating a dissonance that limits and confuses their efficacy. As a result, societal efforts to reduce the risks have been focused on improving risk regulation and stimulating improved self-regulation by companies and their industrial associations as the most promising means of preventing industrial accidents and other harmful events.

These efforts have led many countries to reject further use of traditional governmental regulatory regimes which rely on prescriptive standards and hard law to enforce compliance, and to enact new co-regulatory regimes. In the co-regulatory regimes, the government regulator supervises and assists self-regulatory activities by companies in a particular industrial sector to ensure they fulfill safety management functions defined by performance-based rules, encourages company reliance on industrial voluntary standards, and fosters development of improved standards

[1] For an all-encompassing discussion of regulation, see Baldwin, B. Cave, C., and Lodge, L. (2012) *Understanding Regulation. Theory, Strategy, and Practice*. Oxford University Press: Oxford; and Baldwin, B. Cave, C., and Lodge, L. (2010) *The Oxford Handbook of Regulation*. Oxford University Press: Oxford.

by industry associations. These regimes are described and implementation issues examined with reference to the co-regulatory models of the Seveso Directive and the Norwegian regime governing the safety of offshore drilling for oil and gas. Such regimes constitute the operational rules and standards of a special risk governance approach. As Ortwin Renn (Chapter 1 in this volume) pointed out, the quest for robust regimes lies at the core of different approaches to organize risk governance structures and procedures.

Discussion then focuses on concerns about the legitimacy of these approaches to sustain or reform regimes which involve privatizing government's moral obligation to ensure public and workplace safety and environmental protection. Examination of the extensive global discourse on these concerns indicates that the legitimacy issue has mutated into an accountability issue, namely whether government's supervisory role can be made sufficient to maintain its moral obligation for safety and bring about improved safety performance by the self-regulating companies, and do so in a manner consistent with other law and norms regarding worker rights, transparency and public participation, for example.

Because the regulator is responsible for ensuring accountability, and is itself politically accountable, how the co-regulation model performs in these contested terrains will ultimately determine its value and credibility As discussed, this has led in turn to a subsequent discourse, in which the accountability issue has mutated into a legality issue, namely whether the law establishing the co-and self-regulatory approaches provides sufficient legal standards for guiding the regulator in implementing the law and resolving conflicts with companies and stakeholders.

The chapter concludes with consideration of the view that reliance on legal standards and norms for accountability prevents a more responsive form of co-regulation in which evolving societal norms would add to the content of the governing law and enable the co-regulatory regime to keep pace with new attitudes, information, and other developments without causing unmanageable uncertainty.

2.2. SOCIAL CONTROL OF HAZARDOUS INDUSTRIAL ACTIVITIES

A constellation of social controls influence industrial management of hazardous activities and the risks they pose to workers, the public, property, and the environment.[2] Most prominent are government regulatory regimes that have been

[2] Baram, M. (1973) "Technology assessment and social control", *Science*, 180 (May 4): 4085; Rasmussen, J. and Svedung, I. (2000) *Proactive Risk Management in a Dynamic Society*. Swedish Rescue Services: Stockholm; Faure, M. (2012) "The complementary role of liability, regulation and insurance in safety management: Theory and practice", Presentation at Annual NeTWork Conference on Liability Law and Industrial Safety (October 11–13, 2012): Toulouse; and Baram, M. (1982) *Alternatives to Regulation*. Lexington Books: Lexington.

designed and authorized to prevent harm, usually by developing prescriptive rules and enforcing industry compliance, or more recently by providing performance-based rules and supervising and assisting industrial self-regulatory activities. These approaches, both of which involve reliance on industry expertise, are intended to ensure the safety of products, the workplace, and industrial activities (such as preventing major accidents during operation of chemical plants or drilling for offshore oil and gas resources).

Another important type of ex ante social control is voluntary self-regulation by individual companies, and by industrial associations that develop codes of conduct, voluntary standards, and recommended practices for companies in their business sectors. Both of these forms of self-regulation often provide the technical expertise and proven best practices needed to support governmental regulatory regimes.[3]

In addition, ex post social control in the form of various liability laws provide an independent basis for shaping industrial safety performance by posing the threat of liability when, for example, a company's behavior fails to meet a requisite standard of care and causes harm, or incurs harmful results which may lead to strict or no-fault liability under a prevailing legal doctrine. Although actual liability in either case necessarily requires the occurrence of harm and proof of causation, the threat of such liability may be sufficient to deter a company and others in its industrial sector from negligent or inferior performance of safety management functions, or cause them to reduce or eliminate hazardous aspects of their activities.[4]

Other social controls indirectly influence industrial safety performance, such as laws and norms for corporate governance that cause companies to inform shareholders and potential investors about corporate activities so they can make informed decisions about financial risks. If the activities are hazardous, these sources of financial support may need to be convinced that their financial risks are held to acceptable levels by evidence of effective safety management, which thereby makes it necessary for companies to develop and implement codes of conduct and safety management practices that adhere to industrial standards and comply with government regulations.[5]

Similarly, corporate governance principles also establish management accountability to these financial stakeholders, and causes companies to take the pragmatic step of securing insurance coverage for losses and liabilities which could arise from accidents and other mishaps. This induces companies to maintain their safety performance at a level sufficient to convince insurers to provide sufficient coverage at

3 See notes 31 and 32, infra, and EC Enterprise Web site at http://ec.europa.eu/enterprise/policies/european-standards/index_en.htm

4 Baram, M. (2007) "Liability and its influence on designing for product and process safety", *Safety Science*, 45 (1–2): 11.

5 OECD (2012) *Corporate Governance for Process Safety*. OECD: Paris; CERES.

affordable rates.[6] Thus, "corporate governance is not only a legal concept but is also embedded in organizational theory."[7] It creates a linkage between financial risk and risks to health, safety, property, and the environment and can be an important promoter of safety management.

In addition, companies are induced to effectively manage the safety of their products and activities by market forces and special interest groups. Companies with notable brand names or other public recognition are vulnerable to loss of consumers when their products are shown to be harmful, and are vulnerable to public hostility and opposition when siting a new activity or continuing its operation if the activity threatens or incurs a major accident, chronic illness, or other harms.[8] Because harmful outcomes can cause irreparable loss of public trust and company credibility, the economic consequences for the company may be ruinous. Further, progressive NGOs that demand the sustainability of industrial activities and other advances in corporate social responsibility, and organized labor which advocates workplace safety, play important roles in promoting the safety of industrial activities.

In examining these social controls, two common features are apparent: their dynamic quality and their organic relationship to context. Each is dynamic in the sense that it is responsive over time to new risk information and advances in the state of the art of safety engineering and management, to failures in the form of accidents and other harmful outcomes of operating experience, and to changes in values, perceptions, norms of business behavior, politics, technology, and many other societal, cultural, and economic factors. These factors and the changes they undergo cause the contextual shaping of social controls, which is apparent in several chapters in this book which review the evolution of regulatory regimes for offshore drilling of oil and gas resources, and the preceding chapter by Renn on risk governance.

Because the social controls and the contextual factors that shape them are not orchestrated or systematically managed at the national level, problems arise. One type of problem is the internal dissonance of a social control. An example is tort liability law that has the potential to deter negligent corporate behavior by threatening liability, but which also has the potential to deter a company from self-evaluation of its safety performance and impair its organizational learning process because its knowledge of near misses or other safety problems can be used, after a harm occurs, as evidence of its negligence or violation of regulation even if it earnestly tried to

6 Swiss Re (2011) *Operational Hazards in the Oil and Gas Industry*. Zurich.

7 De Groot, C. (2009) *Corporate Governance as a Limited Legal Concept*. Wolters Kluwer Law & Business: Amsterdam, p. 128.

8 See, for example, "Shell reverses decision to dump the Brent Spaar", at http://www.greenpeace.org/international/en/about/history/Victories-timeline/Brent-Spar/

remedy the problems. Thus, the internal dissonance of tort liability law negates its potential for promoting safety in several ways.[9]

Another type of problem created by the absence of orchestration at a national level is that the aggregation of social controls will be dissonant and incoherent and therefore difficult for a regulatory regime and industry to respond to and reconcile. This occurs for several reasons. One is that social controls are shaped by factors that arise from different contexts, as in the case of advances in safety engineering and management methods which usually evolve in a global context, and the societal factors which evolve in a national context. Thus a regime in responding to the technical factors must consider how to do so in a manner that is consistent with the compelling influence of the societal and political factors. An example from the Norwegian regime is the challenge of reconciling global recognition of behavior-based approaches to safety which are aimed at the work-force with societal norms and laws protecting the rights and interests of organized labor and preventing the transference of company responsibilities for safety to workers.[10]

The dissonant outcome problem may arise even when the social controls are shaped by factors arising in the same national context, such as the case when a company heeding principles of corporate governance that emphasize its responsibility to financial stakeholders is nevertheless able to secure insurance coverage for liability, losses, and the protection of top company officials despite its inferior safety performance and its violation of safety regulations. This frequently happens because the shaping role of corporate governance and the calculus used by insurers for coverage and pricing decisions are not adequately aligned with each other or with government regulation.[11] Another case occurs when a company is fully compliant with safety regulations, yet is found strictly liable for harms arising from its conduct of an ultra-hazardous activity under a liability law.[12]

As a result, the constellation of social controls provides an imperfect frame for advancing industrial safety. A regulatory regime is not empowered to cure this problem by orchestrating all social controls and the factors that shape them into

[9] Rosenthal, I. (1997) "Major event analysis in the US chemical industry", in A. Hale, B. Wilpert, and M. Freitag (eds): *After the Event: From Accident to Organizational Learning*. Pergamon Press: Oxford, p. 179; and Baram, M. (1997) "Shame, blame and liability: Why safety management suffers from organizational learning disabilities", in A. Hale, B. Wilpert, and M. Freitag (eds): *After the Event: From Accident to Organizational Learning*. Pergamon Press: Oxford, p. 168.

[10] See Engen (Chapter 13, Section 13.3) in this volume and Lindøe, P. H. and Engen, O. A. (2007) "Behaviour based safety and the nordic model", in T. Aven and J. E. Vinnem (eds.): *Risk, Reliability and Societal Safety*. Taylor & Francis: London, p. 1705.

[11] Baram, M. (1988) "Insurability of hazardous materials activities", *Statistical Science*, 3(3): 328.

[12] Baram, M. (2007) "Liability and its influence on designing for product and process safety", *Safety Science*, 45 (1–2): 11.

a coherent system for promoting safety because its statutory mandate does not provide it with authority to adjust corporate governance, liability law, private insurance, market forces, and other social controls. Thus, the creation of a "governance coordinating committee" for an emerging technology has been proposed to provide a forum for stakeholders to initiate dialogue and "harmonize existing requirements."[13] This fits well with the inclusive governance model discussed in Chapter 1.

However, vesting complete authority to fully orchestrate all social controls and their dynamic shaping factors in any single entity would not be consistent with principles and norms of democratic governance. Indeed, the argument can be made that preserving the dissonance or lack of coherence of the constellation of social controls preserves and respects higher democratic values, "that the larger principles of justice and policy that must underwrite our assessments of the administrative state and its reform will always be multiple, conceptually ambiguous, politically contested, and compromised in their implementation. To that extent, they will be incoherent. So much the better for democracy."[14]

Further, it can be argued that a vigorous liability law system, despite its problems, is actually convergent with, or constitutes a form of, safety regulation.[15] As in the United States, common law tort liability fills gaps and supplements regulation with qualitative standards of conduct that must be met in order for a company to avoid incurring compensatory and punitive damages and significant administrative penalties. In the aftermath of the BP blowout and spill in the Gulf of Mexico in 2010, it has been estimated that liability likely to be borne by BP and its partners in the Deepwater Horizon venture will exceed $30 billion, as discussed in Chapter 7.[16] It therefore seems reasonable to conclude that liability law has potential to promote a precautionary approach by industry to its hazardous undertakings and thereby supplements safety regulation.

In lieu of trying to orchestrate all social controls, many countries have sought to develop a coherent and efficient institutional system that will simultaneously foster technological advances and their application by industry while minimizing risks and other externalities, and that performs in a manner consistent with democratic principles, societal values, laws, norms, and other contextual factors. This quest has been

[13] Wallach, W. and Marchant, G. (forthcoming) "Proposal for a governance coordination committee", in G. Marchant, K. Abbott, and B. Allenby (eds.): *Innovative Governance Models*. Edgar Elgar: Cheltenham.

[14] Schuck, P. (2012) "Professor Rabin and the administrative state", *DePaul Law Review*, 61 (2012): 595.

[15] Halliday, S. (2012) *Liability as Regulation*. University of Strathclyde Working Paper No. 1, at http://www.ucd.ie/roads/roads_documents/Liability%20as%20Regulation%20WP%20final.pdf

[16] Further detail in Rustad, M. (2012) "Preventive law: Lessons from the BP oil spill to avoid punitive damages liability", Presentation at NeTWork Meeting on *Liability and Insurance and their Influence on Safety Management of Industrial Operations and Products* (October 11–13, 2012): Toulouse.

focused on the development of regulatory regimes, which involve the delegation of self-governing roles to industry.[17]

The following discussion examines this normative approach, the debate over effective modes of blending governmental regulation with industrial self-regulation for preventing major accidents, concerns about the legitimacy and accountability of co-regulation and self-regulation, the "soft law" approach to industrial safety, and the legal standards and norms that determine the legality and societal acceptability of decisions made by the regime and industry when implementing these approaches.

2.3. MODES OF REGULATION FOR PREVENTING MAJOR ACCIDENTS

The industrial revolution of the mid-nineteenth century was powered by steam boilers that exploded with "terrifying frequency" causing "50,000 dead and two million injured by accidents annually" in the United States. This led a professional society in 1915 to begin developing safety codes for the design, testing, operation, inspection, maintenance, and management of various types of boilers, many of which were then adopted as mandatory standards by local, state, and federal governments. Current boiler codes now fill twelve volumes.[18] Thus was born the prescriptive approach to industrial accident prevention, and the process by which private sector voluntary standards and recommended practices are created and then often adopted as mandatory prescriptive rules and standards by government, an approach that continues today in many countries for industrial sectors posing risks of major accidents, including the United States in its regulation of offshore drilling, as discussed in Chapter 7.

But for several decades, a vigorous debate throughout developed countries has been taking place regarding governmental use of inflexible command-and-control prescriptions for industrial safety and the suitability of alternative modes of regulation.[19] This debate, which is enveloped by the larger debate over deregulation and the privatization of government functions,[20] has been fueled by major accidents in several industrial sectors and the resulting demand for more effective means of managing risks to health, safety, and the environment. The proponents of better safety regulation have also been informed by research in the social and behavioral sciences

[17] See Frick, K. (2000) *Systematic Occupational Health and Safety Management: Perspectives on an International Development*. Pergamon: Amsterdam.

[18] http://en.wikipedia.org/wiki/National_Board_of_Boiler_and_Pressure_Vessel_Inspectors#A_Short_History

[19] See Kringen (in this volume) for an overview of the debate.

[20] Verkuil, P. (2007) *Outsourcing Sovereignty: Why Privatization of Government Functions Threatens Democracy and What We Can Do About It*. Cambridge University Press: Cambridge; and Kamarck, E. (2002) "The end of government", in J. Donohue and J. Nye (eds.): *Market Based Governance*, Brookings Institution: Washington, DC.

revealing aspects of individual and organizational behavior and management systems, as discussed by Mearns and Hayes in their chapters, that confound the engineering approach to safety that characterizes prescriptive regulation,[21] and by other evidence that prescriptive regimes have intrinsic limitations.

There are many limitations of the prescriptive approach to safety, of which some are discussed in more detail in Chapter 7 and several other chapters. Here we point at the following:

- inability of prescriptive rules to keep pace with rapidly emerging technologies and capture the continuous streaming of new risk information and lessons learned;
- governmental lack of technical expertise for prescribing how to design and operate complex industrial ventures in ways which ensure safety;
- the rigidity of "one size fits all" prescriptive rules that are unsuitable for companies with special operational circumstances and militate against their use of in-house expertise to develop more cost-effective or technically superior approaches to safety;
- the costs and resources involved in developing prescriptive rules and monitoring and inspecting for compliance.

Recognition of these limitations has led to the development of what many are convinced is a better pathway to safety, namely a collaboration between government and industry which features co-regulation for the fulfilment of performance-based rules. Supporting this development are several additional factors:

- changed values and attitudes that oppose authoritarian command-and-control policies and seek freedom from bureaucracies;
- fears of global competition and the transfer of industrial activity to less-stringently regulated venues abroad; and
- accidents that have undermined confidence in government regulators.

In addition, a co regulatory approach which authorizes company self-regulation is seen as a more responsive, bottom-up approach that when properly supervised by government, will capitalize on global networks of professionals and stakeholders who are continuously engaged in knowledge creation and the development of effective working routines, best practices, codes of conduct, and voluntary standards for improving safety performance.

As a result, the debate has been driven by the convergence of forces for improving industrial safety, deregulation, and privatization – a "perfect storm" that has

[21] Amir, O. and Lobel, O. (2012) *Liberalism and Lifestyle: Informing Regulatory Governance with Behavioral Research*. Research Paper 12–094, at http://ssrn.com/abstract=2145040

caused policy makers in Europe, and to a far lesser degree in the United States, to develop an alternative mode of safety regulation. This mode involves the use of performance-based rules to fulfil essential safety management functions, the delegation of self-regulatory responsibilities to companies for that purpose, the assignment of a supervisory and mentoring role to the governmental regulator, and a soft approach to ensuring compliance.[22]

Under this approach, companies are responsible for developing their own internal practices and procedures for performing the safety functions, and therefore need a more highly developed management system with internal controls for its deployment than what is needed for ensuring routine compliance with government-imposed prescriptive rules. Similarly, regulators overseeing company performance need more expertise to evaluate the sufficiency and legality of each company's approach than the simplistic checklist method that has commonly been used for prescriptive rules.[23]

The global debate has also considered the uncertainties for companies when faced with the need to comply with performance-based regulation. These uncertainties relate to the design, management, operation, procedures, training, maintenance, and many other aspects of carrying out complex projects such as offshore drilling, and determining "how safe is safe enough." To reduce these uncertainties, the chosen solution has been to recommend that companies conform to industry-wide standards and best practices to the extent these are sufficient for the safety of their activity, to use some form of cost-benefit analysis (CBA) to determine the "safe enough" issue, and to progressively reduce residual risk over time in accordance with the ALARP principle ("as low as reasonably practicable").[24]

This has become the preferred solution to the uncertainty issue in performance-based safety regulation as it has been for many years in prescriptive safety regulation, and made explicit in laws authorizing both types of regulatory modes for several industrial sectors (e.g., nuclear, chemical process, oil and gas, and food and drug industries). It is a central feature of the performance-based Norwegian and UK co-regulatory regimes, as well as the U.S. prescriptive regime, for offshore safety, which are discussed in the chapters on these regimes in Part II of this book. Nevertheless, it's CBA and ALARP features are considered as ethically flawed, utilitarian formulas for decision making on safety problems by progressive safety advocates.[25]

[22] Coglianese, C., Nash, J., and Olmstead, T. (2002) *Performance-Based Regulation: Prospects and Limitations in Health, Safety, and Environmental Protection*. Regulatory Policy Program Report No. RPP-03. Kennedy School of Government, Harvard University: Cambridge, MA.

[23] Aven, T. and Renn, O. (2010). *Risk Management and Governance*. Springer: Heidelberg and New York.

[24] See http://en.wikipedia.org/wiki/ALARP

[25] OECD (2006) *Cost-Benefit Analysis and the Environment*. OECD: Paris; Heinzerling, L. and Ackerman, F. (2002) *Pricing the Priceless*. Georgetown University: Washington, DC.

Further issues raised by performance-based co-regulation involve the supervisory or oversight and enforcement roles to be played by the governmental regulatory regime and whether these can be implemented in a manner consistent with administrative principles of fairness and transparency, meet applicable legal standards and norms, incur public trust, and lead to safety performance at levels that are superior or at least equivalent to what has been accomplished by the old regime it would replace. Thus, the debate has led to defining the anatomy and functions of an appropriate model for such a regime, and the outcome has been a co-regulation model, which involves a close working relationship between the governmental regime and the companies responsible for fulfilling its performance-based rules.[26]

Such models need further definition to ensure their appropriate implementation. Their official enactment should explicitly indicate standards to determine the boundaries of authority and the legality of decisions made by the regime and its industrial partners, issues that are subsequently discussed in this chapter. In addition, assigning any oversight and enforcement roles to the regime inevitably requires determining whether the main purpose of the government regime is ensuring compliance with rules and decisions, or improving safety performance – in other words, determining whether industrial failings should be punished by various means of enforcement and the imposition of sanctions or corrected by collaboration between government and industry.[27] As a result, the debate over modes of regulation for accident prevention has led to reconsideration of the purpose and basic regulatory functions of rule making, inspection, and enforcement.

Reinforcing the motivation to develop co-regulatory models is evidence that its self-regulation component can capitalize on various self-regulatory programs developed by companies independent of any governmental regime. A leading example is provided by the chemical process industry associations in sixty countries which have adopted the Responsible Care program. In this global program, the associations require their member companies to implement management practices for continuous improvement in employee safety and health, process safety, emergency planning, and security, and to audit and report on performance, secure independent certification of reports, and share best practices.[28]

[26] Sinclair, D. (1997) "Self-regulation versus command and control? Beyond false dichotomies", *Law & Policy*, 19 (4): 529; Short, J. L. and Toffel, M. W. (2010) "Making self-regulation more than merely symbolic: The critical role of the legal environment", *Administrative Science Quaterly*, 55: 361. Compare also: Gilad, S. (2010) "It runs in the family: Meta-regulation and its siblings", *Regulation & Governance*, 4: 485.

[27] Hopkins, A. (2011) "Risk-management and rule complience: Decision-making in hazardous industries", *Safety Science*, 49: 110.

[28] See http://www.cefic.org/Responsible-Care/Performance/ and http://responsiblecare.americanchemistry.com/Responsible-Care-Program-Elements/Management-System-and-Certification

Research on company performance under Responsible Care (RC) has shown a reduction of process facility accidents in member companies in the United States "by 5.75 percent per 100 plants in a given year or by 85.9%" overall. According to one study, RC "provides additional impetus for plants to improve their safety standards, beyond the incentives they face from government regulations and their potential liability from accidents," because its implementation involves "bringing safety shortcomings to the attention of management" and thereby improves senior management's "attention to safety and their ability to correct errors, the two key factors of improving plant safety." The authors conclude with the cautious recommendation that "self-regulation programs should be treated as complements, and not substitutes, to regulatory programs" because certain governmental regulations are essential baselines for gauging company performance.[29]

Self-regulation has also been advanced by many industrial, technical, and certification organizations, including the International Organization for Standardization (ISO), Det Norske Veritas, TUV Rheinland, and the American Petroleum Institute (API).[30] Some focus on certifying quality management systems and written procedures, but do not evaluate actual performance. Others go deeper and evaluate actual facilities, plans, and operations in conducting their certification processes. And others such as API, which is discussed further in Chapter 7 in this volume, develop voluntary technical standards and recommended practices based on the expertise and experience of their member companies. As the most influential private developer of voluntary standards and practices in the global oil and gas industry sector, API has enacted several hundred technical standards and practices, many of which have been officially adopted by government regulators and thereby made mandatory in many countries or officially recommended as potential means of fulfilling performance-based requirements.[31]

Given such evidence of industry readiness to self-regulate, policy-makers in Europe have enacted several co-regulatory models for industrial safety. The leading example is the multinational regime created by the EU Seveso Directive, enacted in 1982 and since expanded by several amendments following major accidents in Bhopal, Basel, Toulouse, and Enschede. The Seveso regime requires that companies in member states, whose storage and process activities involve several types

[29] Finger, S. R. and Gamper-Rabindran, S. (2012) *Does Industry Self-Regulation Reduce Accidents? Responsible Care in the Chemical Sector* (March 1), at http://ssrn.com/abstract=2014386 or http://dx.doi.org/10.2139/ssrn.2014386

[30] Zwesloot, G. (2000) "Development and debates on OHS system standardization and certification", in K. Frick, P. L. Jensen, M. Quinlan, and T. Wilthagen (eds): *Systematic Occupational Health and Safety Management: Perspectives on an International Development*. Pergamon: Oxford and Amsterdam.

[31] http://www.ihs.com/products/industry-standards/organizations/api/index.aspx

of dangerous chemicals at designated volumes, "take all measures necessary" to identify hazards, prevent major accidents, and limit their consequences.

The companies must document and report how they are fulfilling these functional requirements to the cognizant authorities, with information about the dangerous substances stored or used and the activities involved. They must also address significant changes as they arise as well as near misses, accidents, and new risk information, develop internal emergency plans, and inform the public to enable preparation of external emergency plans. The Directive requires that designated authorities review and evaluate the information provided, conduct inspections, require additional information or actions by the regulated companies as necessary to meet the Directive's goals, and under certain circumstances prevent continued operation of a company facility.[32] A similar approach to chemical accident prevention has been taken in the United States but with fewer functions for company performance and less oversight and intervention authority for the regime.[33]

Seveso is a co-regulatory model for chemical industry accident prevention in which the regime oversees and decides whether companies meet the goals set by the regime's performance-based requirements. It provides a detailed administrative framework intended to serve as a generic template for uniform procedures and equivalent risk outcomes by the sub-regimes of member states, which have widely differing governance systems. But it is far less explicit in defining goals and functions (e.g., what does "take all measures necessary" mean?) and the criteria needed by the sub-regimes of member states for evaluating company safety reports and determining the legality of their decisions. Thus, member states have had much discretion on these matters.[34]

A similar model has been developed in Norway and the United Kingdom where co-regulatory regimes for the offshore drilling industry use performance-based regulations, anticipate that companies will often rely on industrial standards and best practices, and provide for oversight and corrective measures by the governmental regime. However, their "regulatory gestalt" differs from Seveso in several respects.

[32] Seveso I, Council Directive 82/501/EEC on the major accident hazards of certain industrial activities (OJ No L 230 of 5 August 1982) was amended in 1987 and 1988. After several accidents, it was extended by Directive 2003/105/EC of the European Parliament and of the Council of 16 December 2003 amending Council Directive 96/82/EC. Seveso III: Further adaptation of the provisions on major accidents occurred on 4 July 2012 with publication of a replacement directive – 2012/18/EU.

[33] Emergency Planning and Community Right-to-Know Act of 1986, 42. U.S. Code ch. 116; and Risk Management Plan Rule, 40 CFR 68 (section 112(r)(7) of the Clean Air Act). Also see Baram, M., Dillon, P., and Ruffle, B. (1992) *Managing Chemical Risks*. Lewis Publishers: Boca Raton, FL.

[34] For a discussion of some differing national approaches, see Jongejan, B. (2009) *Industrial Safety*, at http://jongejanrmc.com/Externe%20veiligheid_eng.html; Trbojevic, V. M. (2005) *Risk Criteria in the EU*, at http://www.risk-support.co.uk/B26P2-Trbojevic-final.pdf; and Britton, T. J. (1999) *Examining Safety Reports and Evaluating Safety Management Systems*, at http://citeseerx.ist.psu.edu/viewdoc/summary?doi=10.1.1.195.4401

As shown by Kaasen and Paterson (Chapters 5 and 6, respectively), both regimes are more intimately involved in evaluating and improving company performance, implement a risk-based approach, and each has special features. As presented by Bang and Thuestad in Chapter 10, the Norwegian regime includes organized labor in virtually all aspects of its oversight and performance-improving initiatives, and uses performance indicators, which enable it to determine which companies lag and need intervention from the regime. In contrast, the U.S. regime for offshore drilling discussed by Baram in Chapter 7 remains primarily prescriptive, even though U.S. regulation for accident prevention in other industrial sectors has implemented performance-based rules.[35]

As one digs deeper into U.S. regulation, unofficial or ad hoc collaborations with industry can be found within several agency programs. According to the National Institute of Standards and Technology (NIST), "government agencies use externally developed standards in a wide variety of ways," including formal adoption as their own regulations or by officially referencing a voluntary standard, by permitting adherence to a voluntary standard or recommending it as an acceptable course of action for industry, or by otherwise deferring to the voluntary standard in lieu of taking other action.[36]

In addition, the Technology Transfer and Advancement Act of 1995 (TTAA) requires that all federal agencies use technical standards that are developed by voluntary consensus standards organizations, unless inconsistent with other law or impractical, in carrying out agency activities.[37] Although mainly intended to apply to standardization of the multitude of products and services procured by agencies, the statutory language is broad and has been construed by NIST and the Executive Office of the President as applicable to many other aspects of agency regulation.[38]

As discussed in Chapter 7, the U.S. regulatory regime for offshore drilling has routinely and heavily relied on voluntary standards and recommended practices developed by API and other industrial and professional organizations in enacting

[35] Baram, M. (2002) "Improving corporate management of risks to health, safety and environment", in B. Wilpert and B. Fahlbruch (eds.): *System Safety – Challenges and Pitfalls of Interventions*. Elsevier: Amsterdam, pp. 85.

[36] NTIS Web site at http://standards.gov/regulations.cfm

[37] 15 USC 3701. Section 12(d) on "Standards Conformity" of the act states that "all Federal agencies and departments shall use technical standards (defined as 'performance-based or design-specific technical specifications and related management systems practices') that are developed and adopted by voluntary consensus standards bodies, using such technical standards as a means to carry out policy objectives or activities determined by the agencies and departments." The act further states that "Federal agencies and departments shall consult with voluntary, private sector, consensus standards bodies, and shall … participate with such bodies in the development of technical standards." Also see http://standards.gov/nttaa.cfm for implementation information

[38] Circular A-119, Office of Management and Budget, Executive Office of the President (February 15, 2011) at http://standards.gov/a119.cfm#6

its prescriptive rules, and proclaimed its allegiance to the TTAA. Other agencies have followed a similar course. Thus it can be said that even prescriptive regulation in the United States has several co-regulatory features, but they differ from the European models.

To sum up, dissatisfaction with prescriptive regulation and its accompanying "hard law" features of inspection and enforcement to ensure compliance has led to widespread adoption of co-regulatory programs in Europe. These programs have common features: the regulator's role is to supervise and ensure that companies self-regulate in accordance with the performance-based rules and goals set forth in the legislation that created the programs, and when its evaluation of a company's self-regulatory efforts finds inadequacies, to first take a "soft law" approach by suggesting improvements and providing other assistance without lessening the company's self-regulatory obligations, and later take enforcement action to impose sanctions only if serious problems persist. The government regulator may also be authorized or given discretion to suggest that companies rely on industrial voluntary standards and practices when relevant and useful, foster the development of such voluntary measures by industrial and professional associations, and convene meetings with companies and stakeholders to address systemic risk and other problems. Other aspects of the regulator's role in the co-regulatory context are discussed subsequently.

2.4. LEGITIMACY, ACCOUNTABILITY, AND LEGALITY

Adoption of the co-regulation model and other models that authorize industry self-regulation has neither brought closure to the debate over modes of safety regulation nor completed the quest for robust regulation. Important issues remain, and societal regard for three of these issues will determine the quality of a co-regulation model and the scope and significance of its institutional and cultural implications:

- the legitimacy of delegating governmental responsibility for public and workplace safety to private organizations undertaking hazardous activities;
- the accountability of the governmental regime with regard to ensuring that companies self-regulate in a manner that fulfills their obligations; and
- the legality of decisions made by the regime in its implementation of the co-regulatory model.

With regard to legitimacy, there is no disputing that the power to control industrial activities is inherent in a nation's sovereignty and includes enactment of laws, regulations and adoption or acceptance of rules and practices developed by industrial, professional, and international organizations. In mature democracies, this power is exercised by either the chief executive, the parliament or other legislative body to create a regulatory regime by means of legally defined and transparent procedures, and is subject to review by the electorate. Such actions may also be subject to review and approval by a court if there is a national constitution or other law, which defines and limits the power of the executive and legislators.

Delegation of the power to regulate to a government agency rarely raises a viable legitimacy issue in a democracy because it is usually a transfer of authority within government that is taken on the basis of established procedures and political consensus. However, when regulatory authority is delegated to companies or other private entities in order to create a self-regulatory component of a co-regulation program, or an independently functioning program of self-regulation, and public safety is at stake, the legitimacy issue arises. This has occurred as new modes of regulation with self-regulatory features have been proposed and enacted, and stimulated an extensive discourse about their legitimacy.

Within this discourse are arguments why new regulatory models that involve self-regulation should be regarded as antithetical to public safety and trust in government. The main argument is that these models privatize and diminish the moral obligation of government to ensure public safety and that this obligation belongs only to government because it is responsible for permitting, facilitating, and in many cases subsidizing hazardous industrial activities. Further, outsourcing governmental responsibility for safety to companies and industry – a practice promoted by the Thatcher government and imported by others in Europe – increases risks because these private entities are historically shown to have seriously compromised public safety and caused many harms because of their economic interests and opportunism. It assigns the fox to guard the hens, a situation that is generally believed to have led to the collapse of the financial system and many industrial accidents. In addition, the "democracy deficit" of the unelected regulatory regime is increased when it includes private sector self-regulators who are not politically accountable and whose decision processes are not transparent or public-involving.[39]

The discourse also provides analyses and rationales, which negate or diminish the legitimacy issue.[40] The main rationale offered is defensive, namely that

[39] See Verkuil, P. (2007) *Outsourcing Sovereignty*. Cambridge University Press: Cambridge; Stafford, S. (2011) "Outsourcing enforcement: Principles to guide self-policing regimes", *Cardozo Law Review*, 32: 2293; Smith, D. and Tombs, S. (1995) "Beyond self-regulation: A critique of self-regulation as a control strategy for hazardous activities", *Management Studies*, 2 (5): 619; Molloy, T. (2010) "The social construction of regulation: Lessons learned from the war against command and control", *Buffalo Law Rev*, 58 (2): 267; Freeman, J. (2000) "The private role in public governance", *NYU Law Review*, 75: 543; Glynn, T. (2012) "Taking self-regulation seriously", *Berkeley Jnl. Employment & Labor Law*, 32 (2): 279.

[40] See Senden, L. (2005) "Soft law, self-regulation and co-regulation in European law", *Electronic Journal of Comparative Law*, 9.1: 23; Ayres, I. and Braithwaite, J. (1992) *Responsive Regulation: Transcending the Deregulation Debate*. Oxford University Press: Oxford; Dorf, M. and Sabel, C. (1998) "A constitution of democratic experimentalism", *Columbia Law Review*, 98 (2): 267; Cass, R. (1988) "Privatization, politics, law and theory", *Marquette Law Review*, 71: 449: Stewart, R. (2003) "Administrative law in the

government does not shed its moral obligation but retains it because co-regulation and self-regulation models are created by laws or decrees that authorize a governmental regulator to supervise the private entities involved, define the goals and standards of self-regulatory performance to be met, and provide the regulator with the means for evaluating, assisting and enforcing the requisite performance on a case-by-case basis by the regulator. Further, it is held that the models meet the need of a democratic and pluralistic society of competing interests and objectives for "networked governance," which enables the contextual shaping of policies, thereby replacing narrow economic and engineering optimization strategies with a satisficing strategy for societal problem solving – an issue that is discussed by Usemez in Chapter 4. In addition, it is claimed there is no clear legal basis for preventing such delegation of government regulatory authority to private entities. Other points made are that safeguards can be developed to ensure the integrity of the supervising regulator (e.g., to prevent self-dealing and bias), that research shows that private safety performance is stimulated and improved in self-regulatory contexts, and that safety performance can be further improved by, for example, holding top corporate officials personally responsible and liable for failings. Finally and perhaps most persuasively, there is the affirmative view that "privatization can be a means of publicization, through which private actors increasingly commit themselves to traditionally public goals."[41]

Taking these arguments into account, it is apparent that resolution of the legitimacy issue requires that laws creating the models indicate that they are intended to fulfill government responsibility for public safety and explicitly provide a framework for ensuring that self-regulators will be supervised and held accountable for the performance of their delegated functions. Thus, the legitimacy issue inevitably mutates into an accountability issue, that is, whether the government's moral obligation can be met by having a regulator oversee and ensure that the self-regulators fulfill performance-based rules and other responsibilities delegated to them.

The framework for ensuring accountability will have several essential features, which, for example, are present in the Seveso regime and in the U.S., UK, and Norwegian regimes for offshore safety discussed in this book:

- the goals to be met and the self-regulatory functions to be performed by the private entities;
- authorization of a government regulator to evaluate their performance;
- the criteria and standards for the regulator to apply when evaluating their performance;

twenty-first century", *NYU Law Review*, 78: 437; Glynn, T. note 35 supra; Vermeule, A. (2012) "Contra 'Nemo Iudex in Sua Causa'", *Yale Law Journal*, 122 (1): 384.

[41] Freeman, J. (2003) "Public values in an era of privatization: Extending public law norms through privatization", *Harvard Law Review*, 116: 1285.

- authority for the regulator to inspect and access other information needed for evaluation;
- authority for the regulator to take actions needed to have the private entities correct performance deficiencies, including assistance, enforcement, and imposing sanctions;
- documentation and reporting requirements for the regulator and private entities.

The regulator plays the central role in this accountability framework and must deal with many challenges in making it work as a solution to the legitimacy issue. First, there is the need for a high level of competency and technical expertise in evaluating company performance far beyond what is necessary for this purpose under prescriptive regimes. The performance-based rules used in co- and self-regulatory models to define the functions to be carried out by companies creates uncertainties that lead them to seek a safe harbor by showing in their response to the rules that they are adhering to the relevant voluntary standards and best practices of their industrial sector. But it is inevitable that each company's complete response will be unique because it will also use its own internal standards and practices for reasons of cost-effectiveness and address many special features of its activity for which there are no industry guidelines. Thus, each company's response differs, and the regulator needs considerable expertise, time, and resources for evaluating its sufficiency in each case.

Second, there is the challenge of establishing the fairness that is expected of the regulator in conducting these activities, either because it is an explicit legal requirement or a prevailing norm in its regulatory culture. This requires the regulator to consider whether fairness depends on its securing from each company a residual risk outcome that is quantitatively equivalent, or an approach and level of effort that is qualitatively equivalent (taking into consideration each company's size, resources, and other special characteristics), or to apply both concepts of fairness.

A third challenge is posed by the need to ensure that helping companies respond to performance-based rules does not confuse or negate company responsibility to fulfill functional requirements or undermine governmental ability to subsequently hold a company legally accountable – an issue discussed in chapters on the UK and Norwegian regimes for offshore safety. The regulator's assistance-providing function also creates ambiguity of purpose, that is, whether improving company performance is more important than ensuring compliance, and the extent to which a softer and less coercive path to fulfillment of performance-based rules is taken before resorting to enforcement action and sanctions in order to protect the credibility of the accountability framework.

Finally the regulator must address the need to coordinate with other regimes to ensure that the self-regulatory activities it supervises are consistent with their implementation of other laws, such as law protecting worker rights and natural resources. In addition, there is the democracy deficit created by self-regulation to be dealt with. Because self-regulating companies are unwilling to publicly disclose internal information and preclude public involvement in the conduct of their activities and operations, the regulator is obliged by democratic norms to counteract these deficiencies by making information available to the public regarding its evaluations of the companies and creating forums for public participation.

In meeting these challenges, the regulator encounters legislators, ministries responsible for enforcement of laws and prosecution of offenders, and other units of government that also have oversight functions but have different objectives and strategies. When bad things happen, these other overseers are prone to deflecting responsibility and blaming the regulator for making the difficult decisions they have avoided. Thus, the regulator operates "in a broader context of multi-level and polycentric regimes in which responsibilities are widely dispersed," a circumstance that ensures that its own accountability in a political sense "will always be complex and contested."[42]

Thus, the regulator's role involves complex evaluations and difficult decisions, many of which will be contested by self-regulators, industrial associations, and diverse stakeholders unless there is a high degree of trust between these parties. Conflicts erode the credibility of the regulator and the accountability framework, and revive the legitimacy issue. To avoid or mitigate such conflicts, the regulator must ensure that it adheres to the "legal standards" that are expressed, or implicit as legal norms in the law or decree, which authorizes its role as supervisor of self-regulatory activities.[43] The accountability issue therefore mutates into a legality issue, just as legitimacy had previously mutated into accountability.

The legal standards and legal norms must therefore be considered by the regulator when it sets goals and develops an implementation strategy, evaluates each company's self-regulation and makes a decision regarding its sufficiency, and undertakes soft law cooperation and subsequent enforcement. The legal standards and norms also help define the scope of the regulator's activities, and may also be relevant to

42 Black, J. (2012) *Calling Regulators to Account: Challenges, Capacities and Prospects*. LSE Law, Society and Economy Working Paper 15/2012, London School of Economics and Political Science, at http://ssrn.com/abstract=2160220

43 Braut, G. and Lindøe, P. (2010) "Risk regulation in the North Sea: A common law perspective on Norwegian legislation", *Safety Science Monitor*, 14 (1): no page numbers; and Lindøe, P., Baram, M., and Paterson, J. (forthcoming) "Robust offshore risk regulation. An assessment of US, UK and Norwegian approaches", in G. Marchant, K. Abbott, and B. Allenby (eds.): *Innovative Governance Models*. Edgar Elgar: Cheltenham. Also see L. Senden, note 35 supra.

defining other functions of the regulator such as inspection, reviewing company self-auditing and third-party certifications, determining the suitability of the industry voluntary standards and best practices being followed by companies and promoting their further reductions of residual risk, and addressing systemic, industrywide safety issues.

In addition, other laws that are generically applicable to all regulators also provide legal standards, which must be heeded with regard to, for example, the regulator's integrity (e.g., preventing bias and conflict of interest), its licensing and rule-making procedures, and its public participation and transparency initiatives. Kaasen in Chapter 5 discusses in more detail the use of binding and nonbinding legal norms and their relevance in the implementation of regulatory programs based on functional rules. Some of the challenges of using legal standards are discussed by Kringen and Ryggvik in Chapters 11 and 15, respectively.

Thus, when regulatory actions are contested by companies, industry, or stakeholders, the conformance of the actions to the relevant legal standards and norms will be a vital consideration in resolving issues, which, in most countries, is done informally and often not in a public forum. But in the United States, a formal public process is provided. Parties whose interests are impacted by the application of a rule or regulatory decision have the right to appeal to a federal court for determination of the legality of the regulatory action – that is, whether the action conforms to the legal standards set forth in the law that empowers the regulator. Appeals are often made and based on claims that the action exceeded the scope of the regulator's authority, was taken without proper procedure, or was arbitrary in that it lacked the factual support needed to fulfill a substantive requirement of the law. This highly defined approach to resolving disputes about regulation on the basis of the legal standards in a governing law is intended to prevent closed-door corporatist policy shaping. But it has fostered an "adversarial legalism," which involves regulators in costly litigation and exposes them to second-guessing by judges whose rulings often enjoin or invalidate their actions.[44]

In either type of system for resolving disputes over regulatory actions, the clarity of the legal standards and legal norms will be an important factor. Legal standards may be expressed in the governing law at various levels of specificity. At one extreme, highly detailed and prescriptive language that clearly binds the regulator may be expressed, such as language indicating that an action must be based on a cost-benefit analysis or a best available technology determination. At the other extreme, the legal standard may be expressed in vague or ambiguous language and resemble the qualitative standards of care or responsibility found in the common law, which are defined

[44] Kagan, R. (2002) *Adversarial Legalism: The American Way of Law*. Harvard University Press: Cambridge, MA.

and applied on a case by case basis. This latter type of legal standard may have several plausible interpretations that the regulator and the contesting parties will construe and argue to their advantage. And when no legal standard is discernible in the governing law and a legal norm is at issue, or a societal or moral norm is invoked, its characterization, relevance, and significance will be similarly contested.[45]

"Framework laws" authorizing performance-based rules and self-regulatory functions, as seen in the Norwegian offshore regime, are more likely to present vague standards and thereby turn attention to relevant norms. Whereas legal norms and the moral norms on which they may be based are slow to change over time, societal norms continuously evolve because they are responsive to advances in research and expertise, changes in economic and demographic conditions, new concepts of human rights and environmental sustainability, and many other factors.

Societal norms therefore give new content to the governing law and new meaning to its legal standards and thereby stipulate new behaviors and performance by regulators and industry, which keep pace with new knowledge and other developments. To the extent this occurs, regulation moves beyond the confines of legal logic and becomes increasingly responsive.[46] This development can be viewed as healthy for democracy but creates uncertainties for the regulator and industry and may lessen the potential of the governing law to provide for public safety unless carefully guided and the enforceability of the law is clarified and maintained. Thus, robust regulation remains a work in progress.

2.5. CONCLUSION

Most of the foregoing discussion has been about governmental responsibility for the safety of people, property, and natural resources and the issues and concerns raised by delegating a substantial share of this responsibility to industry under models of co-regulation. Although the models provide for governmental oversight and supervision of self-regulating companies, the supervisory role for the designated government regulator is extraordinarily complex. As discussed, it bears the full burden of ensuring the efficacy and credibility of co-regulation under circumstances, which inevitably involve disputes with and between industry, other units of government, labor, and stakeholders whose interests range from the personal to the global. Thus, co-regulation depends on trust in the regulator.

45 Note 39 supra.

46 Drobak, J. (ed.) (2006): *Norms and the Law*. Cambridge University Press: Cambridge; and Lindøe, P. and Engen, O. (2012) "Offshore safety regimes: A contested terrain", Presentation at the *Working on Safety* Conference (September 11–14, 2012): Sopot, Poland.

Companies are chartered by government on the basis that they will serve public interests as well as their own. But surprisingly absent from the discourse is any serious discussion of company responsibility and ethical principles for management in dealing with safety issues, despite the historical record of seriously harmful lapses in several industrial sectors. Further, the co-regulatory model with soft law features offers government assistance that diffuses company responsibility and ethical obligations and restrains government enforcement and sanctions until some indeterminate point when company behavior egregiously avoids fulfilling performance-based rules or incurs major harm and public outrage. And if there is no liability law with explicit provisions for its certain application to serve as a supplement to co-regulation, the social license for company activity created by co-regulation imposes few legal obligations. Thus, co-regulation depends on trust in the self-regulating companies.[47]

Self-regulation is authorized with the expectation that companies will develop internal rules and best practices, promote organizational learning and internal reviews of incidents and near misses, progressively reduce residual risks, and thereby create a safety culture. But in many industrial sectors, self-regulators simply adopt voluntary standards and practices developed by industrial, technical, and trade associations. Many of these third parties are motivated to protect the economic interests of their member companies, and their safety initiatives may be deferred or compromised. Thus, co-regulation depends on trust in the private organizations that enact voluntary standards and recommended practices for industrial self-regulation.[48]

The discourse also shows little concern for the democracy deficit incurred by the self-regulatory component of co-regulation. Self-regulation by industry is usually an opaque, closed process protective of management discretion and proprietary information. Companies preclude public involvement and access to company information, especially information that would illuminate the trade-offs made between safety and production or profit in the conduct of their activities. Self-regulation may thereby enable "business as usual" in determining "how safe is safe enough" unless remedial measures are provided by the government regulator. Much the same can be said for the industrial and professional associations that enact the voluntary standards and practices adopted by self-regulating companies. Co-regulatory models that do not emphasize the need for transparency of deliberative proceedings, disclosure of documentation, and informed public

[47] Baram, M. (2010) "Self-regulation and safety management", Presentation at the *Working on Safety* Conference (September 7–10, 2010): Roros, Norway, at http://www.wos2010.no/presentations.php (click on #87).

[48] Id.

involvement fail to meet norms of democratic governance. Co-regulation in this case therefore depends on trust in the regulator, the self-regulating companies, and private standards organizations.[49]

Thus, co-regulation with its soft law approach to advancing industrial safety rests on a fragile foundation of trust. The question that remains is whether more is needed to ensure that it will meet the inseparable needs of modern democracies for good governance and robust protection from industrial risks to public and workplace safety.[50]

49 Id.

50 For a discussion of soft law and "good governance" issues, see European Parliament, Directorate-General for Internal Policies, *Checks and Balances of Soft EU Rule-Making* (2012).

3

Values and Norms – A Basis for a Safety Culture

Kathryn Mearns

3.1. INTRODUCTION

For the past two decades there has been an increased focus on the role of norms, values, attitudes and beliefs in relation to safety in the UK oil and gas industry. Such concepts are believed to constitute components of 'safety culture', which became a key focus for the industry following the Piper Alpha disaster in 1988. The importance of bringing about a fundamental change in the values and norms of the industry was proposed in Lord Cullen's wide-ranging Public Inquiry (Cullen 1990). Values refer to what people believe is important in their lives – for example, is production more important than safety? Norms refer to the behaviour expected by people who are important to you – for example, do your workmates expect you to make safety a priority in relation to other organisational goals? The results of Lord Cullen's Public Inquiry indicated that safety was not the number-one priority in the UK oil and gas industry in 1988, a state of affairs that directly and indirectly led to failings on the night of the disaster. In the intervening years, regulators and operating and contracting companies on the UK Continental Shelf have made major changes to the legislation and the engineering infrastructure, thus improving safety performance; however, tackling values and norms has proved more challenging because they are more resistant to change and subject to a wide range of influences. Initiatives such as Step-Change in Safety[1] – an increased focus on implementing effective safety management systems and behavioural safety programs – have helped to improve the status of safety, but there are still some fundamental failings in understanding about the need and ways to address deeply rooted values and norms around what constitutes acceptable and unacceptable risks to safety in the higher echelons of oil and gas organisations and how this is communicated to the workforce. The ability of the oil and gas industry to learn lessons from accidents and disseminate these lessons

[1] See http://stepchangeinsafety.net/stepchange/

to the industry worldwide is another value and norm that needs to be developed. Although the Step-Change in Safety initiative was set up because 'safety was too important to be competitive about it anymore', the competitive nature of this industry may still prevent a learning culture from being developed, and the fear of blame and litigation is still prevalent within the industry.

The current chapter first presents some definitions of terms such as 'values', 'norms', 'attitudes' and 'beliefs'. It then focuses on how these factors relate to behavioural issues in the context and practice of safety management. Ortwin Renn has already discussed societal norms in relation to policy making by government and industry in Chapter 2 in this volume. In addition to the concept of societal norms, this chapter discusses national culture and how it can influence safety culture in the oil and gas industry. Finally, the chapter compares and contrasts the safety culture of the oil and gas industry with that of another industry, air traffic management (ATM), which is considered to be one the safest industries in the world. The possibility of learning lessons from this high-reliability industry are discussed.

3.2. VALUES, NORMS, ATTITUDES, AND BELIEFS

The history of safety culture is well established, having been discussed extensively in Public Inquiries into major accidents (e.g. Baker 2007; Cullen 1990; Sheen 1987) and also in the academic literature (Cox and Flin 1998; Hale 2000; Mearns and Flin 1999; Guldenmund 2000; Pidgeon 1998). Defining the constructs that contribute to safety culture has been more challenging, because safety culture is believed to encompass values, beliefs, attitudes, social norms, rules, practices, competencies and behaviours (Mearns and Flin 1999). Owing to problems with the theoretical conceptualisation of safety culture, many studies have focused instead on investigating the related concept of *safety climate*, which has been defined by Zohar (2003) as the perceived priority of acting safely based on shared assessments of what types of behaviour are most likely to be rewarded and supported by supervisors and managers. Zohar maintains that these shared perceptions are very valuable in ambiguous situations when there are competing operational demands because they orientate workers to the expected actions rather than the espoused policies.

Within the psychological literature, values are defined as cognitive representations about desirable end states and behaviour that transcend specific situations, guide selection of specific behaviour and are ordered by relative importance (see Schwartz and Bilsky 1987). Beliefs, on the other hand, are taken-for-granted psychological states where the individual holds a premise to be true – for example, religious beliefs. According to Eagly and Chaiken (1993, p. 1), attitudes constitute '*a psychological tendency that is expressed by evaluating a particular entity with some degree of favour or disfavour*'; furthermore, attitudes are believed to consist of a cognitive component, an emotional component and a behavioural component. Sunstein (1996, p. 914) defines social norms as '*social attitudes of approval and disapproval, specifying what ought to be done and what ought not to be done.*' Thus attitudes and social norms take slightly different perspectives on particular evaluations. Finally, according to the online version of the Oxford Dictionary, a rule is '*one of a set of explicit principles governing conduct or procedure within a particular area of activity*', a practice is the '*actual application of an idea, belief or method*' and a competency is the '*ability to do something effectively and successfully*' (emphasis added). This leads us to *behaviour* which is the ultimate enactment of our cognitive, social and emotional representations as constrained by the environment or situation in which we find ourselves. How one acts or conducts oneself, particularly in the company of others, could be seen as the final manifestation of all other cultural influences. Thus, culture can be considered a group rather than an individual phenomenon. Jacob Kringen discusses the behavioural aspects of safety and HSE culture in Chapter 9 in this volume.

3.3. THE ROLE OF VALUES AND NORMS IN SAFETY CULTURE

So what does this tell us about safety culture? Figure 3.1 has been further developed from the figure shown in Mearns and Flin (1999), based on an original model representing organisational culture by Kopelman, Guzzo and Brief (1990). The figure attempts to show how safety culture could be constituted and reflected throughout an organisation. At the very outset, the bedrock of any culture is based on the basic values, norms, beliefs, expectations and assumptions of the society in which people are embedded. In most advanced civilisations there is appreciation of the value of human life, strong expectations and norms about how people should behave and well-established beliefs about the moral and social fabric of the wider community. Within the workplace, wider environmental influences such as the current economic climate, government legislation and regulatory influences, for example the UK Health and Safety at Work Act, 1974, will have an impact on how people perform and behave in relation to health and safety matters. If the government and other external influences set a particular context, it then becomes the responsibility of the senior management to take rules and regulations and develop them into specific safety management practices, which will maintain the integrity of the organisation and protect people, plant and the wider environment from harm. These practices are integrated into the Safety Management System (SMS) and will encompass activities such as developing health and safety policies, organising for health and safety, communication strategies and health and safety auditing and monitoring procedures. The breadth and depth of such management activities will depend, to some extent, on the underlying values, beliefs and assumptions that wider society,

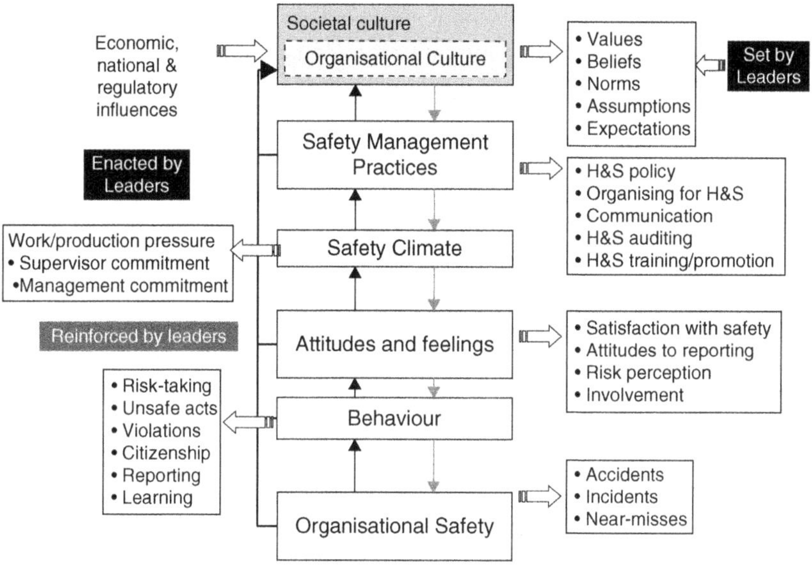

FIGURE 3.1 Proposed model for safety culture.

politicians and leaders of the organisation hold. This in turn sets the context in which people can form attitudes and perceptions and associated feelings regarding risk and safety, within the organisation. In Chapter 14 in this volume, Ole Andreas Engen discusses issues involving failure to implement the safety management system in two near-miss cases.

A central component of this model is safety climate. The concept of safety culture has been described as lacking theoretical clarity. However, it may simply be an all-encompassing construct that sets the context in which people are expected to behave in particular ways. The concept of a 'climate for safety' has been around for a long time (Zohar, 1980) and has been clearly conceptualised in subsequent studies. Mearns and Flin (1999) have defined safety climate as a 'snapshot' of workforce perceptions regarding the status of safety within the organisation. Zohar (2003) leaves us in little doubt when he clearly defines safety climate as the perceived priority of acting safely as verified by the daily actions of organisational leaders. 'Climate' is about shared perceptions of what happens at the workplace and what behaviours are most likely to be supported and rewarded. These perceptions can be shared at the group level with reference to the actions of supervisors or at the organisational level with reference to the actions of managers. Over the past three decades reviews of the literature have identified the management commitment and workforce involvement as key dimensions of safety climate with perceptions of management or supervisor commitment to safety emerging as the factor most commonly associated with positive safety behaviour and accidents and injuries across a range of industries (Flin et al. 2000; Guldenmund 2000; Zohar 2003). Recent meta-analyses involving large datasets collected from many industries in different countries have further supported a moderate relationship between safety climate and safety compliant behaviour, with a weaker relationship supported between safety climate and accidents and injuries (Christian et al. 2009; Clarke 2006). This weak relationship is perhaps not surprising given that a number of factors usually come together at a particular point in time to cause an accident. What is perhaps more interesting to note is that climate impacts on behaviour, which is usually one of the proximal influences on accidents and injuries, for example, someone routinely violating a procedure that has been put in place to prevent harm. Such behaviour would reflect a poor safety climate and by association a poor overall safety culture, except where such violations are necessary to avoid greater harm. Where there is a positive climate arising from the safety culture, reporting and learning behaviour and safety citizenship – for example, helping others at work – will be expected and rewarded. Such behaviour should help to reduce or even prevent accidents and injuries.

The safety culture model is developed in a linear fashion, with the initial route of influence moving from values, beliefs and norms through safety management practices to their influence on the behaviour of members of the organisation via perceptions of safety climate and the attitudes and feelings that the climate engenders. Although the model is described in this chapter through a top-down approach (functionalist), it is apparent that the influence can also move up through the system from front-line personnel to supervisors, middle management and senior management (interpretative, bottom-up approach) (see Glendon and Stanton 2000). Safety citizenship-type behaviour – for example, supporting others at work to behave safely – and reporting behaviours – for example, reporting dangerous situations or events – can have positive feedback back up through the system. However, whether or not personnel engage in positive or negative behaviour is highly dependent on what type of behaviour is being reinforced within the organisation and that tends to be under the influence of supervisors and managers, although coworkers and worker representatives will also have a role to play. It is also worth mentioning that this is not a static system; it is constantly shifting as the organisation adjusts to both external and internal influences. Finally, it is apparent from the way the model has been displayed that the values, norms and expectations as set by senior managers and then enacted through the system by successive layers of management and supervisors are critical to developing and sustaining a positive safety culture at all levels. 'Expectations' can be described as the *desired approach to work*, as imagined by management, and 'practices' describe *work as actually performed on the front line*. Often there is a mismatch between the two. If such problems are not highlighted and discussed and resolved up through the management chain, they can reside in the system ready to be realised when the appropriate circumstances arise.

3.4. SOCIETAL CULTURE: THE BEDROCK OF SAFETY CULTURE?

The oil and gas industry is global in nature, and has been around for a long time. Archaeological evidence indicates the existence of oil seeps on the banks of the Euphrates (now Iraq) from about 4000 BC. These seeps were known as the 'foundations of pitch' and the population used it as building material and a waterproofing agent. The earliest oil wells have been reported in China in about 350 AD. By the eighth century, oil was being developed in the Middle East and by the twelfth century it had been introduced to Europe via Islamic Spain. There is also evidence that natural gas has been exploited by mankind for a long time, with early references to *burning water* appearing in China and Japan in the seventh century. The first modern oil well was drilled in Baku in 1848 and in the nineteenth century the full potential of hydrocarbon reserves began to be exploited in North America, leading to the build-up of considerable wealth for the early pioneers.[2]

Utilising oil seeps would represent minimal risk in terms of safety, although the health risks of inhaling petroleum fumes (*petro* is Greek for rock and *oleum* is Latin for oil) may have been considerable for the developers of 'rock oil'. However, as drilling for oil became more prevalent in humankind's quest for what became dubbed as 'black gold', the likelihood of harm through major accidents, such as blowouts and fires, and through occupational injuries, such as falling from height or being crushed, increased considerably. Safety really seemed to start playing a major role in the minds of the oil and gas companies, when natural disasters such as hurricanes threatened new offshore installations in Louisiana and the Gulf of Mexico in the late 1930s and 1940s. This led to more detailed risk assessments by the engineers in order to establish how to build resilient structures in the face of powerful forces of nature. Although much of the history of oil field development is relayed in terms of protecting the commercial interests of prospectors and big business, the loss of life must have seriously preyed on the minds of many. Priest (2008) points out that while the dangers of working in the oil and gas industry, particularly offshore, are often described in terms of operating complex technologies in harsh and difficult environmental conditions, research in the UK sector indicates that many accidents arise from deviations from routine, normal procedures and practices, attributable to time pressure to complete projects and the ever present pressure to maintain and even increase production (Mearns et al. 1997; Wright 1986). The same problems have been apparent

[2] Dr George E. Totten presents a full timeline of the history of oil and gas at http://www.astm.org/COMMIT/D02/to1899_index.html

in the Gulf of Mexico in the United States and in the Norwegian Petroleum Sector (Hellesøy 1985; Rundmo 1992). According to Priest (2008), the U.S. industry and government began to take safety seriously after the loss of the *Sea Gem* jack-up drilling vessel on the United Kingdom Continental Shelf (UKCS) in December 1967. This tragedy led to the loss of thirteen lives and began a process of re-evaluation of safety that would eventually have ramifications throughout the U.S. industry. Within the United States itself the blowout of a Union Oil platform in the Santa Barbara Channel, just off California, in January 1969 had a national impact, with offshore operators suddenly being challenged in their safety policies, procedures and practices both by politicians and regulators. Thus major accidents both within and outside national oil provinces can mobilise society to demand more stringent measures to prevent harm to people, plant and the environment. This mobilisation seems to consist of an overhauling of regulatory requirements and addressing inadequate legislation, but, as noted earlier, many of the world's major accidents involving the oil and gas industry seem to be mired in a set of values where production was perceived to be 'king' and employees often felt under enormous pressure to get the job done in the minimum time for maximum profit ('time is money'). This seems to have been an overriding priority and an implicit (or even explicit) value in major accidents ranging from the Piper Alpha disaster in 1988 to Deepwater Horizon in 2010. Hollnagel (2009) describes this eloquently as the efficiency versus thoroughness trade-offs: the so-called ETTO principal. People routinely make a choice between being efficient (i.e. productive with less effort) and being thorough (i.e. careful, safe and reliable). It is often not possible to be both simultaneously, so, if productivity demands are high, thoroughness is sacrificed until productivity goals have been met. If safety demands are high, efficiency is reduced until safety goals have been met. The requirement for productivity can also have a political component, for example governments earn a great deal from oil and gas revenues and since we are all very dependent on oil for the quality of life, which we currently experience, falling production is bad news on all fronts. Most of the human race and human society is currently dependent on oil and gas for their well-being. We need production to be high to sustain our increasingly developed planet; witness the industrial rise of nations such as India and China. However, we also demand risks to be well-managed and levels of safety to be high to avoid fatalities, injuries and environmental pollution. When there is a major disaster such as Piper Alpha or Deepwater Horizon, the general public and politicians become mobilised requiring ever-more safety. Thus society could almost be accused of having double standards. We want more oil and gas production to live the lifestyles we enjoy at as cheap a price as possible, without realizing that there is this trade-off between production and safety, which needs to be carefully balanced. The question remains for society to answer – how much are we prepared to pay for our quality of life (through our addiction to oil) and what level of risk are we willing to tolerate?

3.5. THE ROLE OF NATIONAL CULTURE ON SAFETY CULTURE AND SAFETY PERFORMANCE

Clearly the oil and gas industry is multinational in nature, and, as in many other industries, there has recently been a recognition that a relationship could exist between national culture and safety (Helmreich and Merrit 1998; Havold 2005; Mearns and Yule 2009). Nevertheless, there are surprisingly few studies investigating the impact of national culture on safety attitudes, behaviour and safety performance. It seems reasonable to assume that people's attitudes to risk will vary according to deeply held values, beliefs and assumptions – that is, the foundations of national cultural differences. Hofstede (1991) defined national culture as '*the collective programming of the mind acquired by growing up in a particular country*' (p. 262). This collective programming is formed by values, which Hofstede (1984) defined as '*a broad tendency to prefer certain states of affairs over others*' (p. 281) (emphasis in the original). In other words, national culture represents a set of common mental programmes that are shared by a particular group of individuals. These common mental programmes are then reflected in acceptable behavioural norms to which the majority of people will subscribe.

Geert Hofstede (1984; 1991) developed one of the most influential psychological theories about national culture in the 1960s and 1970s when he conducted a cross-cultural study of a large number of employees at a multinational company (IBM) in many different locations throughout the world. The research data was analysed in such a way that it controlled for differences attributable to various policies and practices, thus distinguishing similarities and differences between the cultures at work across the different countries. Hofstede's research resulted in a framework of five dimensions for describing different national cultures: Power Distance, Individualism/Collectivism, Masculinity/Femininity, Uncertainty Avoidance and Long-Term Orientation, the last of which was added when the studies were extended to Far Eastern cultures (Hofstede and Bond 1988).

Power Distance refers to the extent to which members of the organisation *expect and accept* that power is distributed unequally. Low power distance is reflected in flat, decentralised organisations and *enacted* by a consultative style of decision making amongst the workforce; this style is representative of the Scandinavian countries such as Norway, Denmark and Sweden. In high power distance cultures, such as Mexico, India, Venezuela and Brazil, workers are more willing to accept directive management styles from their supervisors and managers and would not necessarily question such directives. Power distance has been found to correlate with income inequality.

Individualism refers to a cultural dimension where everyone is expected to look after themselves; the United States and, to a lesser extent, the United Kingdom would reflect this type of culture. On the other hand, *Collectivist* cultures have strong cohesive in-groups which protect the individuals who belong to them in return for the loyalty of those individuals to the other members of the group. Individualism has been found to correlate with national wealth.

Masculine cultures generally value material success and progress and value money over relationships. Japan has been shown to score highly on this dimension. In cultures low in masculinity (high in femininity), however, people and relationships are valued more than material success – for example, in the Scandinavian countries. Not surprisingly, there is a negative correlation between Masculinity and the percentage of women in senior management and in democratically elected government.

Uncertainty Avoidance is defined as the degree to which individuals feel threatened by uncertain or unknown situations, with high scorers showing a need for predictability and rules. France, for example, was shown to score highly on Uncertainty Avoidance, whereas Indonesia reflected a low score. This dimension has been shown to correlate with the requirement in certain countries to carry identity cards. It is important to note, however, that felt need does not necessarily go with a greater tendency to follow the rules.

Finally *Long-Term Orientation* concerns the time frame that people use to work towards achieving goals, with high scorers more likely to favour long-term over short-term gains. As noted earlier, long-term orientation is a particular characteristic of Far Eastern culture. Interestingly, this orientation is correlated with school results in international comparisons.[3]

Such deep-rooted enduring values are reflected by social and behavioural norms, which are very difficult to change and can exert their influence in subtle ways. Nevertheless, organisations can learn and progress from understanding cultural differences, particularly when operating globally. It is important to note that the scores shown by different countries are relative and not absolute in nature. Many forces exert their influence on national cultures, but these forces tend to be wide-ranging, affecting many different countries at the same time, thus cultures tend to shift together with their *relative* positions on the various value dimensions remaining the same. The specific links between these cultural dimensions and safety are discussed in the next section.

3.6. NATIONAL CULTURE AND SAFETY IN OTHER INDUSTRIES

Perhaps the most extensive study of the relationship between national culture and safety is a study of the international airline industry (Merrit and Helmreich 1996;

[3] See http://geert-hofstede.com/national-culture.html

Helmreich and Merrit 1998). In this study, more than 15,000 pilots from 23 countries responded to a questionnaire containing the Values Survey Module, which measures Hofstede's five cultural dimensions (Hofstede 1984), and the Cockpit Management Attitudes Questionnaire, which measures attitudes to safety (Helmreich 1984). The results showed national differences between the pilots, particularly regarding attitudes to command and control, attitudes to automation and attitude to rules and procedures. With reference to Hofstede's cultural dimensions, the analysis suggested that Power Distance and Individualism could influence pilots' attitudes to safety. For example, pilots from high Power Distance nationalities seemed to be more likely to follow orders and to adhere to standard operating procedures. Pilots from countries high on Individualism seemed to be more independent, more flexible and able to use more discretion when following company procedures.

More recently, two studies have addressed the issue of how national culture impacts on the effectiveness of safety training. Havold (2005) surveyed 287 personnel on 15 ships with a 40-item questionnaire designed to measure national cultural values, safety climate and management style. The sample also included sixty-two ships' officers. Power Distance and Individualism were negatively correlated with workers' and officers' attitudes to safety and both worker and management attitudes to safety accounted for 50 per cent of the variance in the proportion of ship nonconformities with certification (an international measure of ship seaworthiness). Burke et al. (2007) studied sixty-eight organisations from fourteen different nationalities and tested for the moderating effects of national culture and organisational safety climate on the transfer of safety training to the workplace. In this study Uncertainty Avoidance and organisational safety climate moderated the relationship between safety training and safety outcomes, showing that the positive relationship between training and safety outcome was weaker when Uncertainty Avoidance was high.

3.7. NATIONAL CULTURE AND SAFETY CULTURE IN THE OIL AND GAS INDUSTRY

The oil industry in the Western world has traditionally been dominated by a male-oriented, 'macho', 'can do' culture (Wright, 1994). The early offshore oil pioneers on the UKCS came from the United States and were 'rough', 'tough' guys who accepted risk-taking as part of the job and were provided with high wages to compensate for the inherent dangers and the inconvenience of living and working in an inhospitable environment far from home. The Piper Alpha disaster in 1988, in which 167 workers lost their lives after a gas explosion and oil fire, was a catalyst for culture change in the UK offshore oil and gas industry. The resulting Public Inquiry identified failures in maintenance, emergency procedures, training, the Permit-to-Work system and the overall commitment of managers to safety (Cullen 1990). Culture change began with engineering improvements, then the implementation of safety management systems and finally behavioural safety programmes. During the 1990s, the UK oil industry in particular focused its efforts on developing a positive safety culture throughout the whole industry.[4] The Norwegian sector developed its own version of this, called Samarbeid for Sikkerhet (working together for safety), at about the same time. It is interesting to reflect on the different language and meaning used to express a major initiative for safety in the two countries. In the United Kingdom, companies were seeking a Step-Change, defined by the online Oxford Dictionary as *a significant change in policy or attitude that results in an improvement or increase of some kind*. In Norway, the objective was to work together for safety. These differences in prevailing values and attitudes are discussed later with reference to national culture and safety.

As noted earlier, given the global nature of the oil and gas industry, understanding how national culture affects safety performance is becoming increasingly important. Mearns et al. (2004) compared how a sample of 1,138 Norwegian workers from 12 offshore installations and 622 UK offshore workers from 6 installations evaluated psychosocial and organisational factors regarding safety in the industry. The participants completed a questionnaire containing scales measuring risk perception, satisfaction with safety measures, perceptions of the job situation, perceptions of others' commitment to safety, perceptions of social support and attitudes to safety. There were significant differences between workers in the Norwegian and UK sectors on all these measures, but the between-installation variance was greater than the sector variance for all factors, with the exception of attitudes to safety. The attitudes to safety measure reduced to three subscales: Accident Causation, Fatalism and Production versus Safety Goals. Although the installations targeted in the study were conducting similar types of work – drilling, production activities – it is important to note that working arrangements between the installations in the two sectors were very different. For example, offshore workers in the Norwegian sector were (and continue to be) more strongly unionised and tended to work more favourable rotation patterns – for example, two weeks offshore, three weeks onshore. Two of these subscales – Fatalism' and Production versus Safety Goals – contributed almost equally to both sector and installation effects. Although responses to these subscales varied across installations, it was evident that workers on the Norwegian Continental Shelf tended to have more Fatalistic attitudes – an example statement is 'Accidents just happen, there is little one can do to avoid them' – whereas workers on the UK Continental Shelf tended to express the belief that production targets were more important than safety goals – for example, 'Rules and instructions relating to safety make it difficult to keep up with production targets'. There were no differences between UK and Norwegian workers

[4] See http://stepchangeinsafety.net/stepchange/

regarding attributions of the causes of accidents for the sector. Finally there were no significant differences in the self-reported accident rate between the two sectors. It would therefore appear that in this study, national culture accounted for very little variance in safety-related measures, and it seemed to be that characteristics of how safety was managed on the installations that were more important. The importance of good safety management – for example, policy, planning, organising and training – has a more proximal influence on safety performance than national culture, which is discussed in more detail later.

3.8. NATIONAL CULTURE, SAFETY CLIMATE, AND SAFETY PERFORMANCE

In a further study investigating the relationship between national culture and safety, Mearns and Yule (2009) examined the extent to which Hofstede's dimensions of national culture influence safety climate and safety-related behaviour in a multinational construction, maintenance and facilities management company servicing the oil and gas sector. A number of hypotheses were tested proposing relationships between Hofstede's cultural values, safety climate and the safety-related behaviour of employees. For example, it was proposed that high Power Distance could result in a one-way flow of communication from superiors to subordinates resulting in the knowledge and experience of front-line operators not being utilised to aid the development of working methods and procedures and a positive safety culture, that is, lack of workforce involvement. If Collectivism is too strong, individuals may refrain from offering a divergent point of view, which may be vital in critical safety decision-making situations. In contrast, it was hypothesised that Individualism would be related to more direct communication and speaking up about issues, an attribute that appears to be particularly important in developing a positive safety culture (Reason 1997). 'Femininity' is about valuing people and relationships, which could extend to concerns about others' health, safety and well-being. In contrast, Masculinity was expected to refer to a value where success, progress and monetary gain were key motivators for behaviour, not concern for other people, which could undermine interpersonal relations and again lead to lack of communication. While it is likely that no 'national' culture possesses all the optimum components necessary for safety, it is possible that certain combinations of national dimensions (based on the arguments presented earlier in the chapter) have the potential to create cultural norms that will determine the propensity to engage in particular types of safety relevant behaviour at work. A practical outcome of this is that all national cultures probably have the capacity to be 'safe', but it is important for the leaders of multinational (and national) organisations to understand what strengths need to be highlighted and what weaknesses need to be worked on to optimise safety performance amongst their personnel.

As noted earlier, the support and commitment of management may be an important influence when it comes to encouraging employees to behave safely, thus safety climate at different levels of the organisation (i.e. supervisors, middle management and senior management) was measured using a modified version of the Offshore Safety Questionnaire (OSQ) developed by Mearns et al. (1997). Two different types of safety behaviour were measured in this study. Self-reports of risk-taking and lack of safety compliance were measured because a number of studies have indicated that violations and unsafe acts are key antecedents for accidents and injuries. Safety-citizenship behaviours (e.g. putting pressure on managers to improve safety) were also measured.

The sample consisted of 845 respondents representing 6 different national groups: 303 from the Philippines, 216 from the United States, 87 from the United Kingdom, 83 Hispanic Americans, 83 Australians and 73 Malaysians. Respondents completed a questionnaire which included Hofstede's Value Survey Module, safety climate at different levels of the organisation (i.e. supervisor for group climate; management for organisational climate), safety noncompliance (e.g. 'I ignore safety rules to get the job done') and safety-citizenship behaviour (e.g. 'Employees put pressure on supervisors or management to improve safety').

National cultural profiles were found that were similar to those proposed by previous studies using Hofstede's measures. However, in this study, scores for some dimensions (e.g. Power Distance) were more homogenous than expected according to Hofstede's original data for these particular national groupings. This may reflect the fact that the data were collected from one organisation with a similarly flat management hierarchy structure regardless of the national context of its projects. Overall, perceptions of supervisor and management commitment to safety (safety climate) were generally positive, although there were significant differences between sites. For example, Filipino workers had significantly more positive perceptions of management than workers on other sites. Workers in the United Kingdom and the Philippines reported the fewest risk-taking and unsafe behaviours, and workers in Malaysia reported the highest levels. Filipino workers reported engaging in more safety-enhancing behaviours than workers on other sites. It is interesting to note that the Philippines have a very collectivist culture and this could be reflected in supporting and helping others to behave more safely.

Having established significant differences between the various national groups on cultural values, safety climate and self-reported risk-taking and safety-enhancing behaviours, Mearns and Yule (2009) considered whether aspects of national culture or site-specific safety climate were driving differences in behaviour. The first analysis focused on the factors associated with risk-taking behaviour and lack of safety compliance. Masculinity was the only cultural variable to significantly predict this type of behaviour, with respondents who scored higher on the Masculinity scale being more likely to engage in risk-taking/noncompliance at work. This relationship was small but significant, and Masculinity remained a significant contributor to risk-taking even when perceptions of safety climate were added to the analysis. Perceived safety climate on its own was a substantially larger predictor of risk-taking behaviour than masculinity, but in the opposite direction – that is, respondents who perceived managers to be committed to safety reported engaging in less risk-taking behaviours at work.

A second analysis established the statistical predictions of cultural values and safety climate on safety-enhancing behaviours at work. Cultural variables were tested first, and Power Distance emerged as the only significant cultural predictor. As Power Distance between workers and managers increased, workers were less likely to engage in behaviours that enhanced safety. Perceived safety climate was then added to the Power Distance analysis and emerged as a significant predictor, indicating that workers who perceived senior management to be more committed to safety were more likely to engage in proactive behaviours that enhanced the state of safety on site. Adding actual safety climate to this analysis removed the effect of Power Distance. To some extent, these results support the findings of Merrit and Helmreich (1996) and Helmreich and Merrit (1998), who found that Power Distance emerged as a potentially significant influence on behaviour and safety in their study of pilots.

Our attempt to establish the relationships between cultural values, perceived management commitment to safety and risk-taking behaviours suggests that these relationships are not uniform between cultures. With regard to the safety climate measures used, commitment of corporate (senior) managers was a more proximal predictor of self-reported safe and unsafe behaviours than aspects of national culture. The effects of senior management commitment to safety tended to override perceptions of supervisor commitment to safety. It was apparent that as perceptions about the positive commitment of senior managers to safety decreased, workers were more inclined to take risks, break rules and violate procedures. The opposite also held true as senior managers were perceived to be more committed to safety (for example, by genuinely being concerned for the well-being of employees and by acting promptly on safety concerns), propensity to take risks and violate procedures amongst workers appeared to decrease. Also, the lower the Power Distance between workers and management, the more behaviours directed at enhancing the status of safety tended to increase.

What does this tell us about culture and safety? The oil industry has been built around a 'macho culture', which means that 'macho'-type values, beliefs, attitudes and behaviours such as toughness and bravery have been selected and endorsed over time and have become more engrained as part of the culture. It is therefore very plausible that as workers become more 'macho', they will take more risks and break rules owing to the effects of bravado and a desire and willingness to get the job done despite technical or other obstacles. Perhaps more important, however, is the finding that organisational safety climate as represented through perceptions of senior management commitment to safety seemed to override the 'national' cultural influences. This provides some useful data on which to build regulatory requirements. For example, safety regulation should perhaps focus on detailing requirements for identifying and selecting senior managers who have the appropriate focus on safety in relation to other organisational goals, and training them how to enhance and demonstrate their commitment through their behaviour and statements.

3.9. OTHER INFLUENCES ON SAFETY CULTURE

As outlined in Figure 3.1, factors from outside the immediate organisation will have an influence on the safety culture and people's behaviour. Findings from a study by Spangenbergen et al. (2003) seem to indicate that differences in regulatory regimes, training and education exert an effect in countries where the national culture may be anticipated to be similar. Spangenbergen et al. (2003) investigated why Danish workers had approximately four times the lost-time injury rate of Swedish workers during a joint venture to construct the 16 km road/rail link across Øresund (a body of water between Denmark and Sweden). It was possible to compare Danish and Swedish employees working on the same tasks and using the same procedures for reporting occupational injuries. The results were explained at different levels.

At the macro level (national and societal), differences in Swedish and Danish waging practices during sick leave, national educational programmes, work environment legislation and differences in the socio-economic structure of the construction industry in the two countries were proposed as determinants of the differential safety performance. Swedish workers have to pay for their first day of absence from work resulting from illness or injury, whereas in Denmark the employer pays. After the first day, employers in both countries pay workers during the first two weeks of sick leave. As a result of the economic penalty imposed on the first day of absence, Spangenbergen et al. (2003) propose that the Swedish workers are more aware of workplace health and safety. Furthermore, Swedish workers have more formal education than their Danish counterparts, both in terms of public schooling and professional training. Danish construction workers are more likely to be inducted into work practices by their supervisors, whereas Swedish workers undergo long and structured apprenticeships, which impart knowledge of construction methods and techniques and knowledge of health and safety practices and injury prevention.

At the meso level (group and organisation), employment practices and planning and preparation of work were identified as important factors leading to the discrepancy in safety performance. The Swedish contractors were employed on full-time salaries and stayed with the company for a number of years, allowing the employing organisation to enforce their health and safety policy, and thus influence the safety attitudes of the workers. In comparison, Danish workers tended to be employed on a temporary basis and paid according to a piecework system (known to be detrimental to safety), leading to more movement between employers. It is also significant to note that the Swedish contractors spent more time planning and preparing for their work than the Danes, including addressing health and safety issues.

Finally, at the micro (individual) level, training, learning and attitude towards work were believed to be exerting an effect. The focus on formal training and apprenticeships for Swedish construction workers as compared to the practical on-site experiences and the passing down of work practices from supervisor and work team members in the Danish contingent could have contributed to the differences in lost-time injury rate, as could the stronger bonds between the companies and their workers owing to the Swedish employment conditions. In Sweden, it is common for construction workers to return to work immediately after a minor injury, and they will carry out other tasks if they cannot work on-site. In Denmark, it is accepted that injured workers take sick leave (irrespective of the severity of the injury). This situation is no doubt reinforced by the differing waging practices during sick leave that are on offer in the two countries. The important lesson to be learned from this unique study is that even in countries that might be deemed culturally similar according to Hofstede's model – that is, Scandinavian countries in general are identified by high collectivism, low power distance and low masculinity – there are potentially differences in national and company policies and practices that have an influence on work group and individual factors and thereby influence lost-time injury rates.

3.10. SAFETY CULTURE IN AIR TRAFFIC MANAGEMENT (ATM): ARE THERE LESSONS TO BE LEARNED?

The ATM industry has an excellent safety record worldwide and is therefore considered to be an ultra-safe industry (Weick 1987). ATM organisations maintain that safety is their business because accidents involving aircraft greatly undermine the confidence of the travelling public and such disasters are to be averted at all costs. The oil and gas industry also likes to maintain that safety is their business, but this is not entirely true because their real business is to produce oil and gas, which is the lifeblood of industrialised societies. In reality, both ATM and oil and gas industries each have competing priorities. ATM has to cater for an ever-increasing scheduled throughput of traffic, and sometimes this can lead to conflict with safety priorities. On the other hand, holding up or diverting air traffic can also lead to safety problems. As far as the oil and gas industry is concerned, the consequences of major accidents are costly both in terms of human life and finances. For example eleven men died in the 2010 Deepwater Horizon disaster, and BP's costs in the wake of this accident were $42 billion (the major costs are for the oil pollution and clean-up, not the eleven deaths) and are still rising. The only way BP can survive as a company is to keep producing oil and gas to pay for these substantial costs. Thus we come back to the necessity of maintaining this delicate balance between efficiency and thoroughness (Hollnagel 2009). It is ultimately about how these conflicts are managed within the business that leads to success in both production and safety. In the short term, organisations can invest in efficiency at the cost of thoroughness, but ultimately the pendulum swings too far in the direction of efficiency – for example, reducing manning leads to an adverse event, which swings the pendulum back to an increased focus on thoroughness.

An interesting aspect in the ability of ATM to sustain high levels of safety is that up until recently, this industry did not systematically develop and use safety management systems. ATM shared this perspective with the airlines and rail companies until relatively recently because it was assumed that safety was so deeply engrained in these high-reliability organisations they did not have to be explicit about it. The safety and integrity of ATM was maintained through the professionalism of carefully selected and highly trained individuals as air traffic controllers (and associated staff) and engineers. European ATM has now begun to develop and implement safety management systems to maintain an appropriate focus on safety (Perrin, Stroup and Kirwan 2007), although selection and training still plays a key role in sustaining performance.

Part of this impetus for a renewed focus on safety has come from two major accidents involving loss of life attributable to ATM failings: Linate and Überlingen. At Linate airport in Italy, 110 passengers and crew and 4 ground staff were killed when two planes collided on the runway. The Überlingen disaster involved a mid-air collision in Swiss airspace killing 71 passengers and crew. Ultimately poor safety culture was considered to be a contributing factor in both accidents, particularly at Überlingen where air traffic controllers were coping with failures in key systems and did not have sufficient manpower to cope with the situation, necessitating shortcuts and procedural violations. Since 2005, EUROCONTROL (the European Organisation for Safety in Air Traffic Control) has been developing a Safety Culture Measurement Toolkit (SCMT) to measure employee awareness, attitudes, perceptions and beliefs and how these contribute to maintaining the safety and integrity of the ATM system. The SCMT has evolved over many years and has been developed in close collaboration with the European ATM community. The process consists of ATM personnel (controllers, engineers, managers and support staff) completing a fifty-four-item questionnaire measuring seven components, which are believed to represent safety culture (Management Commitment to Safety; Reporting and

Learning; Responsibility for Safety; Communication; Teamwork; Trust and Involvement). The results from the survey are then reported back to the organisation in workshops consisting of six to eight personnel, who are representative of the sample survey. Several workshops are usually run, depending on the size of the organisation. The objective is to discuss the results of the survey to determine their face and content validity. The feedback usually consists of covering the positive aspects of the safety culture before moving on to the more negative aspects, where more than 25 per cent of respondents have provided an 'unfavourable' response to the item under consideration. When discussing unfavourable responses, the objective is to determine why people perceive the situation in this way and what can be done to improve the situation. The EUROCONTROL safety culture team then produce a report for senior management, highlighting the strengths and weaknesses of the safety culture by incorporating the results of the survey with the feedback from the workshops. The report then concludes with a list of recommendations, which management is required to act upon. A number of European Air Traffic Navigation Providers (ANSPs) have acted upon the recommendations provided from their first survey and have conducted follow-up surveys that indicate considerable improvements in their organisation's safety culture. Associated improvements in ANSP safety performance have also been recorded. This willingness to act on recommendations sends a clear signal to staff that management takes safety seriously. Unfortunately, the external validation of such a tool is not possible in an ultra-safe organisation, which is a major shortcoming of the study.

The lessons to be learned from such an approach are that the time and effort expended has been accepted by the ATM industry as an important part of the process, as has the requirement to actually act on recommendations arising from analysis of the data. Too often, in other industries, including oil and gas, safety culture or safety climate measurements are conducted but the results are not presented to the workforce and the workforce is not given the opportunity reflect upon and comment on the results, leading to disenfranchisement and lack of trust in management to take the process seriously and act on the findings. Genuine commitment to safety involves trading off Efficiency and Thoroughness to the extent that Thoroughness is given more emphasis. In the short term this may be seen as damaging the prospects of the organisation, but in the long term major accidents can lead to serious loss of life and long-term injury to the individuals involved, with consequent expensive compensation and clear-up costs and a loss of reputation.

Finally, returning to the conclusions of Spangenbergen et al. (2003), differences between European countries in how ATM is regulated have revealed some interesting differences with regard to reporting and learning from safety incidents. In most European countries, air traffic controllers (ATCOs) can be held responsible and accountable for any errors they make when managing aircraft on the runway and in airspace. Not surprisingly, this leads to a propensity to under-report near-misses. Although there is a mandatory, regulatory requirement to report all losses of aircraft separation in all European countries, in reality the reporting may only occur if someone else witnesses the incident. In 2001, the Danish Parliament passed a new law, mandating the establishment of a compulsory, strictly non-punitive and strictly confidential system for the reporting of aviation incidents.[5] A characterising feature of this legislation was that ATCOs and pilots (or any other employee) were assured of immunity against penalties and disclosure. Furthermore, breaches against the guarantee of non-disclosure became a punishable offence. Loss of separation between aircraft has always been a mandatory reporting requirement for ATM, but with the implementation of the new reporting system under Danish law, personnel were required to report a wider range of safety infringements. This confidential, non-punitive approach led to a marked increase in the number of incidents reported. Reported incidents increased to 980 in the 12 months post implementation, compared to the 15 incidents reported in the previous 12 months. Like-for-like incidents increased from a baseline of fifteen 'loss of separation' incidents to between forty and fifty. Obviously, on the face of it, this might indicate a worsening safety culture (as measured by the outcome of number of incidents), whereas in fact there was an improvement in safety culture with opportunities to investigate more incidents and learn from them, as well as an increase in trust between front-line personnel and management. Another important component is that feedback was given to reporters, thus reinforcing the importance of providing the information.

What can be learned from such an approach? Despite its position as an ultra-safe industry, ATM still strives to sustain and improve its safety performance, through changing the law to encourage reporting and learning to adopting practices such as safety management systems and safety culture measurement approaches. It is important to note that the SCMT developed by EUROCONTROL takes a multi-faceted approach to measuring and managing safety culture, using both quantitative and qualitative methods and delegating responsibility for carrying out recommendations arising from the safety culture assessment of the organisation to management. The work of safety is never done. It requires the focus and the will to prioritise safety, even when it appears to be in conflict with other organisational goals. Managers (particularly senior managers) need to provide time, money and other resources to show that they are genuinely concerned about safety. They must also be willing to engage in dialogue with members of the workforce in order to identify potential problems associated with hazards, risk and safety that may be lying within the system, ready to be realised as harm to people, plant and the environment. This has

5 See paper by Nørberg at http://shemesh.larc.nasa.gov/iria03/p11-norbjerg.pdf

also been found in a Dutch study that investigated a range of other industries (Hale et al. 2010).

3.11. CONCLUSION

A prime assumption in the safety literature is that the workforce's safe or unsafe attitudes and behaviour are a function of the organisation's prevailing safety culture (Guldenmund 2000). Furthermore, it is assumed that organisations that find themselves in a position to discuss the development of a positive safety culture already hold or are beginning to develop 'safety' as a central value, and consequently organisations that have a focus on other beliefs and values are unlikely to achieve a safety culture (Hale 2000). The current chapter has outlined a model of safety culture starting from the underlying values and beliefs, assumptions and expectations and working through the safety climate, attitudes and feelings through to behaviours. The model is very simplistic because these factors do not necessarily unravel in a linear fashion. Worker involvement and engagement from the bottom up can also contribute to establishing a positive safety culture but only if the channels for communication are open and management are ready to accept and act upon workforce contributions. Can we regulate for safety culture? Probably not, but we may be able to affect culture indirectly by focussing our efforts on tackling behaviour and organisational and regulatory practices (Spangenbergen et al. 2003). The experiences of ATM show that if you change the law, you can change reporting practices, which ultimately could lead to improved safety; if you select and train the right people to lead safety-critical organisations, you may be able to influence the safety management policies and practices and by association the safety climate and employees attitudes and feelings in relation to that climate. Finally, we need to ask the question whether it is necessary to regulate for different national cultures. The evidence presented in this chapter would seem to indicate that this is not necessary, as the proximal influence of perceived management commitment to safety is dominant; however, it is important that there is an awareness of the values and norms of particular cultures so that the most appropriate safety management systems are implemented to achieve optimal safety performance.

REFERENCES

Baker Panel (2007) *The Report of the BP U.S. Refineries Independent Safety Review Panel* (January). http://www.bp.com/bakerpanelreport

Burke, M.J., Chan-Serafin, S., Salvador, R., Smith, A. and Sarpy, S.A. (2007) "The role of national culture and organizational climate in safety training effectiveness", *European Journal of Work and Organizational Psychology*, 17: 133–154.

Christian, M.S., Bradley, J.C., Wallace, J.C., & Bruke, M.J. (2009) "Workplace safety: A meta-analysis of the roles of person and situation factors". Journal of Applied Psychology, 94: 1103–1127.
Clarke, S. (2006) "The relationship between safety climate and safety performance: A meta-analytic review." *Journal of Occupational Health Psychology*, 11: 315–327.
Cox, S. and Flin, R. (1998) "Safety culture: Philosopher's stone or man of straw?" *Work & Stress*, 12: 189–201.
Cullen, D. (1990) *The Public Inquiry into the Piper Alpha Disaster, Vols I and II*. HM Stationary Office: London.
Eagly, A.H., and Chaiken, S. (1993) *The Psychology of Attitudes*. Harcourt Brace Jovanovich: Fort Worth.
Flin, R., Mearns, K., O' Connor, P. and Bryden, R. (2000) "Measuring safety climate: Identifying the common features", *Safety Science*, 34: 177–192.
Glendon, I. and Stanton, N. (2000) "Perspectives on safety culture", *Safety Science*, 34 (1–3): 193–214.
Guldenmund, F.W. (2000) "The nature of safety culture: A review of theory and research", *Safety Science*, 34: 215–257.
Hale, A.R. (2000) "Culture's confusions", *Safety Science*, 34: 1–14.
Hale, A.R., Guldenmund, F.W., van Loenhout, P.L.C.H. and Oh, J.I.H. (2010) "Evaluating safety management and culture interventions to improve safety: Effective intervention strategies", *Safety Science*, 48: 1026–1035.
Havold, J.-I. (2005). "Safety culture in a Norwegian shipping company", *Journal of Safety Research*, 36: 441–458.
Hellesøy, O. H. (1985) *Work Environment Statfjord Field: Work Environment, Health and Safety on a North Sea Oil Platform*. Universitetsforlaget: Bergen, Oslo.
Helmreich, R.L. (1984) "Cockpit management attitudes", *Human Factors*, 26: 63–72.
Helmreich, R.L. and Merrit, A.C. (1998) *Culture at Work in Aviation and Medicine: National, Organisational and Professional Influences*. Ashgate: Aldershot.
Hofstede, G. (1984) *Culture's Consequences: International Differences in Work-Related Values* (Abridged edition). Sage: London.
(1991) *Culture and Organisations: Software of the Mind*. McGraw Hill: Maidenhead.
Hofstede, G. and Bond, M.H. (1988) "The Confucius connection: From cultural roots to economic growth", *Organizational Dynamics*, 16: 5–21.
Hollnagel. E. (2009) *The ETTO Principle, Efficiency-Thoroughness Trade-Off*. Ashgate Imprint: Aldershot.
Kopelman, R., Brief, A. and Guzzo, R. (1990) "The role of climate and culture in productivity", in B. Schneider (ed.): *Organisational Climate and Culture*. Jossey Bass: Oxford, pp. 282–318.
Mearns, K. and Flin, R. (1999) "Assessing the state of organizational safety – culture or climate?" *Current Psychology: Developmental, Learning, Personality, Social*, 18 (1): 5–17.
Mearns, K. and Yule, S. (2009) "The role of national culture in determining safety performance: Challenges for the global oil and gas industry", *Safety Science*, 47: 777–785.
Mearns. K., Flin, R., Fleming, M. and Gordon, R. (1997) *Human and Organisational Factors in Offshore Safety*. OTH 87 543. HSE Books: Suffolk.
Mearns, K., Rundmo, T., Gordon, R. and Fleming, M. (2004) "Evaluation of psychosocial and organizational factors in offshore safety: a comparative study", *Journal of Risk Research*, 7 (5): 545–561.

Merrit, A.C. and Helmreich, R.L. (1996) "Human factors on the flight deck: The influences of national culture", *Journal of Cross-Cultural Psychology*, 27: 5–24.

Perrin, E. Stroup, R. and Kirwan, B. (2007) *Future Considerations in ATM Safety R&D. A Summary of the Safety Gap Analysis Report by FAA/Eurocontrol Action Plan 15, Safety Research and Development*. Presentation at the International System Safety Conference, Baltimore, August.

Pidgeon, N. (1998) "Safety culture: Key theoretical issues", *Work and Stress*, 13: 202–216.

Priest, T. (2008) "Wake-up call: Accidents and safety provision in the Gulf of Mexico offshore industry", in U.S. Department of the Interior, Minerals Management Service, Gulf of Mexico (ed): *History of the Offshore Oil and Gas Industry in Southern Louisiana. Volume I: Papers on the Evolving Offshore Industry*. OCS Region: New Orleans, pp. 139–152.

Reason, J. (1997) *Managing the Risks of Organizational Accidents*. Ashgate: Aldershot.

Rundmo, T. (1992) "Risk perception and safety on offshore petroleum platforms. Part II: Perceived risk, job stress and accidents", *Safety Science*, 15: 53–68.

Schwartz, S.H. and Bilsky, W. (1987) "Toward a universal structure of human values", *Journal of Personality and Social Psychology*, 53: 550–562.

Sheen, M.J. (1987) *M.V. Herald of Free Enterprise*. HMSO: Department of Transport: London.

Spangenbergen, S., Baarts, C., Dyreborg, J., Jensen, L., Kines, P. and Mikkelsen, K.L. (2003) "Factors contributing to the differences in work related injury rates between Danish and Swedish construction workers", *Safety Science*, 41: 517–530.

Sunstein, C.R. (1996) "Social norms and social roles", *Columbia Law Review*, 96 (4): 903–968.

Weick, K.E. (1987) "Organizational culture as a source of high reliability", *California Management Review*, 29 (2): 112–127.

Wright, C. (1986) "Routine deaths: Fatal accidents in the oil industry", *The Sociological Review*, 34 (2): 265–289.

(1994) "A fallible safety system: Institutionalised irrationality in the offshore oil and gas industry", *The Sociological Review*, 42: 79–103.

Zohar, D. (1980) "Safety climate in industrial organizations: Theoretical and applied implications", *Journal of Applied Psychology*, 65: 96–102.

(2003) "Safety climate: Conceptual and measurement issues", in J.C. Quick and L.E. Tetrick (eds.): *Handbook of Occupational Health Psychology*. American Psychological Association: Washington, DC, pp. 123–142.

4

Optimising Offshore Health and Safety Inspections

How the Markets Could Help

Emre Üşenmez

4.1. INTRODUCTION

It is commonly understood that enforcement and the rules themselves compose the two primary legs of a regulatory regime. There is, however, a third component, one that performs more of a bridging role between these two branches: the decision-making process of a regulator. It is certainly true that despite the best wills and intentions, performance of a regulatory regime will be in jeopardy without a robust enforcement. This robustness is dependent upon a number of variables including the level of communication between the stakeholders. Both in the United Kingdom and in Norway, the communication regarding health and safety is not confined to those between the regulators and the operators but also include the workforce and, at times, the public. In Norway, the workforce involvement is achieved through what is termed as a "tripartite model" (see Band and Thiestad, chapter 10 in this volume, and Rosness and Forseth, Chapter 12 in this volume) while in the United Kingdom, the involvement is more organic, albeit provided for in the relevant regulations.[1] This reflects the recognition that a sound health and safety regime ought to involve those on the front line and those who may potentially be affected by the outcomes in the decision making. In fact, Ortwin Renn puts forward a linear relationship between the level of stakeholder participation and the level of ambiguity surrounding risks (see Figure 1.3 in Chapter 1 in this volume).

The robustness of enforcement, however, is also dependent upon the decision-making process of an inspecting body in allocating its scarce resources to its inspection services. The regulators tend to evaluate the risks based on quantifiable measures and

[1] Offshore Installations (Safety Case) Regulations 2005 (SI 2005/3117). For more discussion on the UK regulations, see Chapter 6 in this volume and Paterson, J. (2011) "Health and safety at work offshore", in: G. Gordon, J. Paterson and E. Üşenmez (eds.): *Oil and Gas Law – Current Practice and Emerging Trends*. Second Edition. Dundee University: Dundee pp. 1–13 (hereinafter, Paterson, *Health and Safety at Work Offshore* [2011]).

allocate the inspection services accordingly. However, as Renn points out (Chapter 1 in this volume), "risk-related decision-making is neither about risks alone nor usually about a single risk." It can also be added that risk-related decision making tends to be a process not carried out by a single individual, but rather as a group.

This chapter therefore examines the third component of a regulatory regime, namely the decision-making process of a regulator. For the purposes of discussions henceforth it treats a given inspecting body as a collection of individuals grouped together to forecast and decide how and when to inspect and at what cost. It therefore explores the main mechanisms available for optimising forecasting and decision-making processes in a group environment along with their drawbacks. Among these mechanisms the focus will be on the face-to-face (FTF) meetings, nominal group technique, statistical aggregation, Delphi method and prediction markets. It is intended that this chapter contribute to the literature by providing a fresh angle that supports a constructive debate in improving offshore health and safety regimes, and by increasing the awareness of this third component's importance. It is central to this chapter, therefore, to discuss initially the reasons as to why any importance should be attached to this third component in the first place. The chapter identifies the scarce resources of inspectorates and the solutions offered previously before discussing the decision mechanisms.

4.2. SCARCE RESOURCES

Helge Ryggvik (Chapter 15 in this volume) notes that the number of offshore inspections in the Gulf of Mexico has dropped from the peak of 7,000 in 1994 to approximately 5,000 in 2009. Shortly before the fatal Deepwater Horizon accident, Ryggvik had the opportunity to meet with Troy Trosclair, at the time the "leader of all offshore safety inspectors in the Gulf of Mexico," whom Ryggvik found to be "frustrated" because he "felt he was [at the time] under pressure to reduce the number of inspections on offshore installations" (Chapter 15).

Trosclair's frustration was echoed by various notable entities in the United States. In its September 2010 Report,[2] the newly established Outer Continental Shelf Safety Oversight Board (the Board)[3] highlighted both the insufficient numbers of staff in the Gulf of Mexico[4] and the difficulties in recruitment and retention of inspectors.[5] The reasons for these shortcomings, the Board concluded, were not only that the

[2] U.S. Department of the Interior Outer Continental Shelf (OCS) Safety Oversight Board, *Report to Secretary of the Interior Ken Salazar*, 1 September 2010 (hereinafter, *Safety Oversight Board Report*).

[3] U.S. Department of the Interior, The Secretary of the Interior, Secretarial Order No. 3298 (30 April 2010).

[4] *Safety Oversight Board Report*, p. 13.

[5] *Safety Oversight Board Report*, p. 12.

salaries were inadequate[6] but also the promotion opportunities were very limited, with "opportunities to cross-train and move into related positions at higher grades or levels of organization" practically nonexistent.[7] In addition, approximately half of the inspectors the Board surveyed in their report noted that they lacked adequate training, and those "who identified their own training needs often [were] denied that training."[8] Some of them instead had to "rely on industry representatives to explain the technology at a facility."[9]

Parallel views were also expressed by both the National Commission on the BP Deepwater Horizon Oil Spill and Offshore Drilling (the Commission)[10] and the U.S. Government Accountability Office (GAO). In its January 2011 Report,[11] the Commission agreed with the Board's assessment on the need for the inspectors to significantly improve their technical expertise but noted that, regardless of improvements, they would never be able to "possess technical expertise truly commensurate with that of private industry." This is because the "salary differential, combined with the sheer depth of industry expertise on a wide variety of topics critical to understanding and managing offshore drilling operations, would make that goal illusory."[12] As a result, the Commission pointed out, the then-regulator, Minerals Management Service (MMS) of the Department of the Interior,[13] "became an agency systematically lacking the resources, technical training, or experience in petroleum engineering that is absolutely critical to ensuring that offshore drilling is being conducted in a safe and responsible manner."[14]

6 *Safety Oversight Board Report*, p. 11.
7 *Safety Oversight Board Report*, p. 12.
8 *Safety Oversight Board Report*, p. 11.
9 *Safety Oversight Board Report*, p. 11.
10 The White House Office of the Press Secretary, *Executive Order – National Commission on the BP Deepwater Horizon Oil Spill and Offshore Drilling*, 22 May 2010, established the Commission with a mandate to analyse "the root causes of the Deepwater Horizon oil disaster" and to recommend "options" for future.
11 National Commission on the BP Deepwater Horizon Oil Spill and Offshore Drilling (January 2011) *Deep Water: The Gulf Oil Disaster and the Future of Offshore Drilling*. Report to the President. The White House: Washington, DC (hereinafter, *the Commission Report*).
12 *The Commission Report*, p. 240.
13 The responsibilities of the MMS were to be distributed to three separate entities, the Bureau of Ocean Energy Management (BOEM), the Bureau of Safety and Environmental Enforcement (BSEE), and Office of Natural Resources Revenue (ONRR) under the U.S. Department of the Interior, The Secretary of the Interior, Secretarial Order No. 3299 (19 May 2010). Until the establishment of these entities, the MMS was renamed Bureau of Ocean Energy Management, Regulation and Enforcement (BOEMRE). ONRR became operational on 1 October 2010 operating under the Department of Interior's Office of Policy, Management and Budget. Both BOEM and BSEE became operational a year later, on 1 October 2011, replacing BOEMRE completely. See BOEM, *The Reorganization of the Former MMS*, at http://www.boem.gov/About-BOEM/Reorganization/Reorganization.aspx (accessed 3 April 2012). For the establishment of MMS, see *The Commission Report*, chapter 3.
14 *The Commission Report*, p. 57.

The GAO asserted views along the same lines. In his testimonies before the House of Representatives,[15] Frank Rusco, Director of Natural Resources and the Environment at the GAO, stated that the inspectors

> lacked the critical skills because, according to agency officials, Interior 1) has had difficulty in hiring experienced staff, 2) has struggled to retain staff, and 3) has not consistently provided the appropriate training for staff. Interior's challenges in hiring and retaining staff stem, in part, from competition with the oil and gas industry, which generally pays significantly more than the federal government.[16]

As solutions to these issues, the Board has recommended implementing regular and standardised training programmes.[17] It was also recommended that the regulator ought to develop "Individual Development Plans" tailored to each inspector's career progression needs, and one that would "promote sound succession planning and foster employee development and satisfaction."[18] These recommendations were duly adopted three days later in the Regulator's Implementation Plan.[19]

The Commission has also focussed on the remuneration of the inspectors and recommended a more specific salary classification, at similar levels to the employees of the Nuclear Regulatory Commission, the federal regulatory agency for the nuclear industry.[20]

The GAO, on the other hand, voiced its concerns regarding recruitment of new inspectors, stating that it is "unclear whether Interior will be fully successful in hiring, training, and retaining these additional staff."[21] This is presumably owing in large part to the unattractive levels of remuneration and limited career progression options.

[15] GAO, *Oil and Gas Management: Key Elements to Consider for Providing Assurance of Effective Independent Oversight*, Statement of Frank Rusco, Director, Natural Resources and the Environment, Testimony Before the Subcommittee on Energy and Mineral Resources, Committee on Natural Resources, House of Representatives, 17 June 2010, GAO-10–852T (hereinafter, *GAO Subcommittee Testimony*); GAO, *Past Work Offers Insights to Consider in Restructuring Interior's Oversight*, Statement of Frank Rusco, Director, Natural Resources and the Environment, Testimony Before the Committee on Oversight and Government Reform, House of Representatives, 22 July 2010, GAO-10–888T (hereinafter, *GAO 2010 Committee Testimony*); GAO, *Interior's Restructuring Challenges in the Aftermath of the Gulf Oil Spill*, Statement of Frank Rusco, Director, Natural Resources and the Environment, Testimony Before the Committee on Oversight and Government Reform, House of Representatives, 2 June 2011, GAO-11–734T (hereinafter, *GAO 2011 Committee Testimony*).

[16] *GAO Subcommittee Testimony*, p. 4; *GAO 2010 Committee Testimony*, p. 6.

[17] *Safety Oversight Board Report*, p. 12.

[18] *Safety Oversight Board Report*, p. 12.

[19] Michael R. Bromwich, Director of the BOEMRE, *Implementation Plan in Response to the Outer Continental Shelf Safety Oversight Board's September 1, 2010 Report to the Secretary of the Interior*, 4 September 2010.

[20] *The Commission Report*, p. 258.

[21] *GAO 2011 Committee Testimony*, pp. 5–6.

4.3. RECURRING THEME

The GAO's reservation is correct. This is not the first time the regulator was recommended to hire more staff, improve its remuneration and training packages and develop mechanisms to reduce staff turnover rates. During the 1960s, the inspection services of the then-regulator, the U.S. Geological Survey Conservation Division (USGS),[22] were labelled as "lax" and irregular.[23] This was largely attributable to the fact that the regulator was "underfunded and understaffed" and the existing inspectors lacked sufficient training.[24] "Even those inspectors and supervisors who had the appropriate training and competence often did not have the requisite experience in the oil business and grasp of its changing technological capabilities."[25] Following a series of accidents during this period,[26] and as a response to these criticisms, the USGS increased the number of its inspectors and engineers threefold, "introduced a more systematic inspection program, and stopped using industry furnished transportation for inspection purposes."[27]

In 1973, a report by the GAO[28] found that the regulator's own standard for the frequency with which to inspect offshore assets in the Gulf of Mexico was rarely being met,[29] despite the increase in the number of inspection staff.[30] It highlighted the need

[22] Until the leasing and regulatory functions were brought together under Minerals Management Service (MMS) on 19 January 1982, the leases and pipeline rights were issued by the Bureau of Land Management (BLM), while "operational matters, collecting rentals and royalties and policing the operations" were managed by the U.S. Geological Survey Conservation Division (USGS) – both agencies of the DOI. See: U.S. DOI MMS Gulf of Mexico OCS Region (September 2008) *History of the Offshore Oil and Gas Industry in Southern Louisiana*. Volume I: Papers on the Evolving Offshore Industry, OCS Study MMS 2008–042, OCS: New Orleans, pp. 93–95 (hereinafter, *MMS, History of Offshore Vol. I*).

[23] National Commission on the BP Deepwater Horizon Oil Spill and Offshore Drilling (2010) *The History of Offshore Oil and Gas in the United States (Long Version)*. Washington, DC. See also: Staff Working Paper No.22, 2010, p. 13 (hereinafter, *Commission, History of Offshore*).

[24] *Commission, History of Offshore*, p. 13. "In 1969, the Gulf region's lease management office had only 12 people overseeing more than 1,500 platforms."

[25] *Commission, History of Offshore*, p. 13.

[26] These include "blowouts (22 deaths on the *C.P. Baker* catamaran drilling vessel in 1964), helicopter crashes (eleven in one crash in 1966)", a near-miss crash of a Louisiana Air National Guard jet on an *ODECO* rig, Santa Barbara oil spill following a blowout at Union Oil's Platform A-21 in 1969 and another blowout in 1970 at Chevron's Platform C in Main Pass block 41. For more discussion on these incidents, see *MMS, History of Offshore Vol. I*, pp. 145–146.

[27] *MMS, History of Offshore Vol. I*, p. 147.

[28] GAO (1973) *Improved Inspection and Regulation Could Reduce The Possibility of Oil Spills On the Outer Continental Shelf*. Report to the Conservation and Natural Resources Subcommittee on Government Operations, House of Representatives by the Comptroller General of the United States, B-146333, 29 June 1973 (hereinafter, *GAO 1973 Report*). Note that at the time the acronym GAO stood for General Accounting Office and not Government Accountability Office, despite being the same entity.

[29] *GAO 1973 Report*, p. 17.

[30] *GAO 1973 Report*, p. 3.

for staff training and recommended that the regulator should set up a formal training regime.[31] This need for training and lack of offshore experience among the regulator's staff was also highlighted in a 1981 report by the National Research Council.[32] It drew attention to the fact that out of 1,700 staff of the Conservation Division of the USGS only about 17 per cent, or 300 employees, had any experience in the oil and gas industry; and of those who were employed in Gulf of Mexico, "only about half had more than five years' experience."[33] Finally in 1985, the GAO reported[34] that the then-regulator, MMS, was in the process of developing a formalised training programme for its inspectors.[35] It found that the existing "inspection requirements [were] being met for the most part although some problems exist[ed]".[36] Yet it still drew attention to the limited number of inspectors, with 1 inspector for every 102 production facilities.[37] At the time, the Coast Guard was also mandated with regulatory and inspection powers over offshore oil and gas assets.[38] In the same GAO Report, the Coast Guard's inability to inspect all the facilities in the Gulf of Mexico in 1983 was explained by the fact that the Coast Guard "did not have sufficient funding for helicopter transportation necessary to inspect all facilities."[39] It instead had to prioritise its inspection and receive a cash injection of $200,000 to increase its efforts by the latter part of 1983.[40]

This recurring theme, however, was not unique only to the regulators of Gulf of Mexico. In Norway, the safety authority received similar criticisms. Ryggvik points out that despite the denigration it had received after the Deepwater Horizon incident, the MMS

> still conducted far larger number of inspections than the Norwegian Petroleum Directorate (NPD) or after 2004 the Petroleum Safety Authority had ever conducted.... In fact, in [the] Summer [of] 1974 Arne Flikke, [the] then leader of the NPD's safety department, left the organisation in protest because he felt he did not have the necessary resources to visit every platform regularly. The next year, the

[31] GAO 1973 *Report*, pp. 24–25.

[32] Marine Board of the National Research Council (1981) *Safety and Offshore Oil*. National Academy Press: Washington DC, pp. 240, 252–253 (hereinafter, *NRC 1981 Report*).

[33] *NRC 1981 Report*, p. 252.

[34] GAO (1985) *Offshore Oil and Gas: Inspection of Outer Continental Shelf Facilities*. GAO/RCED-86–5, Washington, DC (hereinafter, *GAO, 1985 Inspection Report*).

[35] *GAO, 1985 Inspection Report*, p. 26.

[36] *GAO, 1985 Inspection Report*, chapter 2.

[37] *GAO, 1985 Inspection Report*, p. 21.

[38] The 1978 Amendments to the OCSLA 1953 grants the Coast Guard to "promulgate and enforce ... regulations with respect to ... safety of life and property" on the assets in the Outer Continental Shelf. OCSLA, section 5(d).

[39] *GAO, 1985 Inspection Report*, p. 27.

[40] *GAO, 1985 Inspection Report*, pp. 27–28.

NPD's resources were increased [though] not enough to cover every installations with regular inspections.[41]

Ryggvik (Chapter 15 in this volume) finally notes that "more than 30,000 oil workers on the Norwegian continental shelf will seldom or never meet inspectors from PSA offshore."

In the United Kingdom, a similar theme had emerged in the reports of government inquiries into some of the major offshore incidents. This was certainly true of the independent review of the regulatory regime applicable to the UKCS in the aftermath of the Gulf of Mexico accident (Maitland Review),[42] of the inquiry into the disaster at Occidental Petroleum's Piper Alpha production platform in July 1988 (Cullen Inquiry)[43] and of the Committee that was set up following a major blowout in April 1977 at Phillips Petroleum Company's Bravo production platform in the Ekofisk field in the Norwegian sector of the North Sea (Burgoyne Committee).[44]

In 1977, at the time of Ekofisk Bravo incident, the responsibility for offshore safety in the UKCS was divided among three government agencies: the Department of Trade (DT), the Health and Safety Executive (HSE) and the Petroleum Engineering Division of the Department of Energy (PED).[45] The Committee set up under the chairmanship of Dr. J.H. Burgoyne following this incident in the Norwegian sector had argued for consolidation of these fragmented responsibilities under a single agency.[46] Yet it also highlighted that the staff of the HSE had a "lack of expertise in certain areas unique to the offshore industry" while the PED, despite having "qualified and experienced personnel," was nonetheless "under-staffed."[47] For the proposed single agency Burgoyne recommended "further recruitment" of "similarly highly qualified and experienced specialists, so that the inspectorate retain[ed] the confidence of the industry that problems [would] be discussed between equally qualified people."[48]

In his report in 1990, the Honorable Lord Cullen (the Cullen Report) drew attention to the inadequacy of the training, guidance and leadership the inspectors received, and

41 See Chapter 15 in this volume.

42 Maitland, G. (2011) *Offshore Oil and Gas in the UK: An Independent Review of the Regulatory Regime*. Report by Chairman Geoffrey Maitland, FREng, December 2011 (hereinafter, *Maitland Review*).

43 Lord Cullen (1990) *The Public Inquiry into the Piper Alpha Disaster*. Report by the Chairman The Hon. Lord Cullen, Cm. 1310, 1990. HMSO: London (hereinafter, *Cullen Report*). The events and responses surrounding the Piper Alpha disaster are beyond the scope of this work. For further discussion on this, see Chapter 6 in this volume.

44 Burgoyne, J.H. (1980) *Offshore Safety: Report of the* Committee. Report by Chairman Dr. J. H. Burgoyne, Cmnd. 7866, HMSO: London (hereinafter, *Burgoyne Committee*). The events and responses surrounding the Ekofisk Bravo blowout are beyond the scope of this work. For further discussion on this, see Chapter 6 in this volume.

45 *Burgoyne Committee*, para. 4.8.

46 *Burgoyne Committee*, para. 4.10.

47 *Burgoyne Committee*, para. 4.13.

48 *Burgoyne Committee*, para. 4.14.

highlighted the "persistent under-manning" at the PED.[49] The then-director of safety at PED, J.R. Petrie, explained to the Cullen Inquiry that remuneration packages were partly to blame in the difficulty of attracting new staff. He was in agreement that

> [the offshore] industry provided opportunities for higher salaries and promotion, along with other attractions such as foreign travel. Within the PED the prospects of career development were limited for inspectors because of the departmental grading which they were in and because of the comparatively small size of the PED.... The loss of inspectors to industry was not an annual event but it was not infrequent.[50]

Lord Cullen concluded that the regulatory control as well as the responsibility to provide inspection services ought to transfer to the HSE.[51] Yet he still drew attention to the difficulties the HSE had in recruitment. Despite the pay increase, the HSE was the target for the private sector, which was "poaching" its experts.[52] Regardless, he pointed out that the HSE had a better system in place for career progression, which had a positive impact on recruitment and lowering turnover rates.[53]

Following the Cullen Report, the health and safety inspection services have been brought under the HSE. The environmental inspections, on the other hand, are conducted by what is currently called the Department of Energy and Climate Change (DECC).[54] Approximately two months after the Gulf of Mexico incident, the DECC has announced its plans for recruiting more inspectors.[55] In its September 2010 submission to the House of Commons Energy and Climate Change Committee, the DECC said the recruitment would be for three new environmental inspectors in order to double the mobile drilling rig inspections from eight to sixteen a year.[56] This increase in the frequency of inspections was deemed "appropriate."[57] Yet the Committee still highlighted the turnover of inspectors as they move to the industry, arguing that this "may make it difficult for DECC to recruit and maintain high-quality inspectors in the future."[58]

49 *Cullen Report*, paras. 15.49 and 21.23.
50 *Cullen Report*, para. 15.46.
51 *Cullen Report*, chapter 22. For a detailed discussion, see Chapter 6 in this volume.
52 *Cullen Report*, para. 22.25.
53 *Cullen Report*, para. 22.26.
54 The regulator, now the DECC, was previously the Department for Business, Enterprise and Regulatory Reform, which itself previously was the Department of Trade and Industry.
55 DECC (2010) *UK Increases North Sea Rig Inspections*, Press Release 10/067, 8 June 2010.
56 House of Commons Energy and Climate Change Committee (2010) *UK Deepwater Drilling – Implications of the Gulf of Mexico Oil Spill*. Second Report of Session 2010–11, Vol. I, HC 450-I, 14 December 2010 (hereinafter, *Energy and Climate Change Committee Report Vol. I*). Written Evidence of the Department of Energy and Climate Change, Health and Safety Executive, and Maritime Coastguard Agency, Ev 109, para. 59.
57 *Maitland Review*, p. 85.
58 *Energy and Climate Change Committee Report Vol. I*, p. 12.

The number of health and safety inspectors, however, did not increase after the Gulf of Mexico incident, as HSE had "ongoing recruitment difficulties" with 15 per cent vacancy rate.[59] The Maitland Review argued that these difficulties were attributable to "significantly higher remuneration levels" offered in the private sector, as well as to the issues of "career progression and advancement opportunities."[60] As a solution, the Review Panel recommended that the regulators should

> develop strategies to ensure that each authority is in a position to recruit and retain inspectors and managers of the right number, quality, experience and range of specialities. The strategies should also consider issues around age profile, plans for career progression through both technical and managerial routes and commit to an ongoing programme to market-test remuneration rates amongst relevant, specialist staff.[61]

As it is apparent from the past four decades, even when adopted, these recommendations provide short-term fixes at best. Perhaps it is time to change the approach to this issue and seek a longer-term solution. Increasing the salaries will not in and of itself reduce offshore health and safety risk – the aim of health and safety inspections and enforcement. In fact, suggesting, for decades, that the remuneration differential is corresponding to the technical expertise of the inspectors may be construed as a back-handed comment on the competence of those who *remain* as inspectors that they are not on par as those in the private sector. In other words, it may be construed that some inspectors are in their profession not by choice but rather by lack thereof. Incidentally, this has been stated during the evidential hearing of House of Commons Climate Change Committee by one of the Committee members: "Presumably anybody half decent is employed by the operator, not the regulator."[62] This is, and should be, unacceptable. Despite this remark by one of its members, the Committee – or the Maitland Review, for that matter – has not found any evidence giving weight to the questionability of the quality of inspections.

What these four decades ought to have demonstrated instead is that inspecting all the offshore assets as frequently as a regulator would like is an elusive goal. The regulators and policy makers may have devoted time and resources more efficiently had this fact been taken as a starting point for an inspection regime instead. Given the ubiquitous resource constraints of the regulator – both in terms of human and financial capital – the goal of minimising offshore health and safety risk could perhaps be better served if the inspection regime instead focusses on allocating these scarce resources optimally.[63]

59 *Maitland Review*, p. 86.
60 *Maitland Review*, p. 86.
61 *Maitland Review*, Recommendation 9.1.
62 *Energy and Climate Change Committee Report Vol. I*, Ev 50, Q288.
63 A clarification is needed here. It is not being argued that inspections by themselves would minimise the health and safety risk. Without a robust regulatory regime, transparency, clear and frequent communication among stakeholders and other variables discussed elsewhere in this book, health and

4.4. TRANSFERRING RESPONSIBILITY

One possible way of optimising the regulator's scarce resource allocation is through transferring the responsibility for providing inspection services to the private sector. The underlying assumptions for this view are that the governmental entities tend to always "be inherently less efficient than private firms,"[64] that allocation of resources can happen more efficiently through competition[65] and that "it may be substantially cheaper for government to delegate power to private actors than to undertake an activity itself."[66]

In this approach the transferee of this responsibility can be one or both of the two private parties: either the party that is the subject of inspections or a third party. In the case of the former, the private party would self-inspect through internal audits, checks and evaluations, whereas in the latter, the private party would procure the services of an external private party to carry out inspection on its facilities.

In 2009, this transfer of responsibility is exactly what the Conservative Party in the United Kingdom has announced as one of their key policies: "The power of Government inspectors will be drastically curbed by allowing firms to arrange their own, externally audited inspections and, providing they pass, to refuse entry to official inspectors."[67] This policy is effectively a combination of the two approaches in transferring the inspection responsibility to a private party, envisaging that the private parties would self-police in the first instance but complement it with third-party inspections.

As novel as it may seem from the outset, this idea was not unique to the UK's Conservative Party. In 1979, the Supreme Court of Iowa in the United States put forward a similar point when deliberating on the role of municipalities in building inspections.[68] In the event that the municipality withdrew from its inspection role, the Supreme Court noted, "the void might be filled by private agencies whose certificates could be relied on by persons risking their lives and property in multiple dwelling

safety risks would not reduce dramatically. The focus here is limited to the extent inspections can contribute to attaining this goal.

64 Shirley, M.M. and Walsh, P. (2011) *Public versus Private Ownership: The Current State of the Debate*. World Bank Policy Research Working Paper No. 2420. World Bank: Washington, DC, p. 3 (hereinafter, *World Bank Policy Research Working Paper No. 2420*), citing Alchian A. (1965) "Some economics of property rights", *Politico*, 30 (4): 816–824.

65 *World Bank Policy Research Working Paper No. 2420*, p. 5. For a concise and good literature survey on privatisation, see Megginson, W.L. and Netter, J.M. (2001) "From state to market: A survey of empirical studies on privatisation", *Journal of Economic Literature*, 39: 321–389 (hereinafter *A Survey of Empirical Studies on Privatisation*).

66 Lawrence, D.M. (1985–1986) "Private exercise of governmental power", *Indiana Law Journal*, 61: 647–695, here 657.

67 *Conservative Policy Paper*, p. 4. Note, however, that this was not referring to offshore inspections specifically.

68 *Wilson v. Nepstad*, 282 N.W.2d, 664–677 (1979).

apartments."[69] A decade later, this idea was discussed as a "desirable alternative" to the existing regime.[70] The rationale for this was fourfold. First, private parties would carry out the inspection function more cheaply. Second, remuneration packages would not have "limitations of civil service. Third, private programs [would] have a greater capacity for flexible responses to new conditions. Finally, the profit incentive [would] increase the possibility of service improvements."[71] This final rationale, when coupled with the threat of liability, would go a long way in improving the quality of inspections because together they would act as motivators.[72] This is why it is concluded that the "legislatures should place the inspection function in the hands of the private sector."[73]

A similar programme that was adopted in Florida in 2000 saw the responsibility of inspection of elevators transferring to private actors. Eight years later, it was reported that this transfer was largely a success, although certain discrepancies between the government agency and the private actors were noted.[74] In Europe, a similar trend had evolved in building and construction inspection services as well. A recent study[75] had found that despite a number of countries such as Belgium, England, France, Germany, Norway, Sweden and Wales having a varying degree of private inspection services, there was no evidence of diminished quality in the services provided.[76] In fact, it was highlighted that the reason these countries had allowed private inspections in the first place was to "improve the quality of building control." This was why the "inspection need" was envisaged to be supplied by private inspectors.[77]

Application of this approach to inspection of offshore installations may not be as straightforward, however. The theoretical argument that transferring the inspections

[69] *Wilson* v. *Nepstad*, at 674. A discussion on this case is beyond the scope of this work. For a case note, see Carla A. Scholten, C.A. (1979–1980) "Torts – a municipality is liable for tortious commissions and omissions when authority and control over the fire safety of apartments have been delegated to it by statue and breach of that duty involves a foreseeable risk of injury to an identifiable class to which the plaintiff belongs – Wilson v. Nepstad (Iowa 1979)", *Drake Law Review*, 29: 207.

[70] Levowitz, J.E. (1988) "Privatization of the building inspection function: An alternative to municipal liability", *Washington University Journal of Urban and Contemporary Law*, 34: 267–295 (hereinafter *Privatization of Municipal Inspection*).

[71] *Privatization of Municipal Inspection*, pp. 293–294.

[72] *Privatization of Municipal Inspection*, p. 294.

[73] *Privatization of Municipal Inspection*, pp. 294 and 295. Also interesting to note: "Negligent inspection problems arise because municipalities are unable to perform their responsibility. Municipalities claim that insufficient funding and staffing cause negligence."

[74] The Florida Legislature Office of Program Policy Analysis and Government Accountability (April 2008) *Privatization Has Helped Improve Elevator Safety; Additional State Oversight Is Needed*. Report No. 08–18.

[75] Meijer, F. and Visscher, H. (2006) "Deregulation and privatisation of European building-control systems?", *Environment and Planning B: Planning and Design*, 33: 491–501 (hereinafter *European Inspection Privatisation*).

[76] *European Inspection Privatisation*, Abstract.

[77] *European Inspection Privatisation*, p. 500.

services to private actors rests on the assumption that there exists a competitive market in which the transferee would operate. "If the change is simply from one monopoly supplier to another, then neither cost nor performance is likely to change very much."[78] The level of competitiveness, therefore, has a direct impact on the degree of efficiency gained from transferring the responsibility to private actors.[79]

Yet the inspection of offshore industry has to be monopolistic in structure. This is because a private entity mandated with inspection of offshore assets ought to receive its remuneration from the regulator, not from the inspected party – at least not directly – in order to avoid any conflicts of interest. If there are multiple private agents competing in offshore inspection, the regulator would inevitably face bills higher than its constrained budget could afford. This inevitability would be a result of the incentives the private agents would have in inspecting in parallel or, worse, in duplicity. In the event that the regulator has a mechanism whereby the parties would compete per inspection and get remunerated accordingly, the private parties would have an incentive to maximise the frequency of inspections in order to maximise their remuneration. This, however, would run the risk of incurring expenses over and above the given budget of the regulator. Even if it could be done within the budget of the regulator, it does not ensure, or even create incentives for, high-quality inspections.

Of course, the regulator could tender out the inspection services clearly specifying the scope, liabilities and duration in the winning contract. This would avoid the regulator paying more than it intends to and would create a desired competitive environment at the bidding process. This contracting regime, in turn, would imply a hierarchical structure – that is, the regulator and the private contractor would not have an equal, horizontal relationship – with distinct risks.[80] These risks stem from two separate but related asymmetries of information the parties would have: as an endogenous factor, the contractor would have more accurate information on its own level of competence and skills than the regulator could determine, for example.[81] This is commonly referred to as moral hazard.[82] As an exogenous

[78] Kolderie, T. (1986) "The two different concepts of privatization", *Public Administration Review*, 46 (4): 285–291.

[79] *A Survey of Empirical Studies on Privatisation*, p. 8 at 3.1.1., and p. 10 at 3.1.6.

[80] de Palma, A., Leruth, L. and Prunier, G. (2009) *Towards a Principal-Agent Based Typology of Risks in Public-Private Partnerships*. International Monetary Fund, IMF Working Paper, WP/09/177, Washington, DC, p. 9 (hereinafter, *IMF WP/09/177*). Note that although this IMF Working Paper specifically refers to Public-Private Partnerships, its discussion on risks distinct to hierarchical structure can also be applied to contracting inspection services.

[81] *IMF WP/09/177*, p. 9. Even if the regulator implemented a costly measure – perhaps through independent auditing of contractor – to obtain this information, there still exists "conflicting incentives for transparency: the [contractor] tries to blur the signal, whereas the [regulator] tries to clarify it." *IMF WP/09/177*, fn 24.

[82] Laffont, J.-J. and Triole, J. (1993) *A Theory of Incentives in Procurement and Regulation*. MIT Press: Cambridge, MA, p. 1 (hereinafter, *A Theory of Incentives*).

factor, the contractor would also have better information than the regulator on the conduct of inspections it carries out and on the state of the inspected parties, as the regulator would only "observe the output."[83] This is commonly referred to as adverse selection.[84] When this "output" is lower than expected, the regulator would not be able to "determine whether it corresponds to a low level of effort on the part of the private company"[85] or to a high level of compliance by the inspected parties with the safety regime.

As this structure is hierarchical, it may also create a potential principal-agent problem[86] where not only information constraints such as moral hazard and adverse selection are present, but also the private actor, as an agent, may prioritise its own interests, which may not be in alignment with the regulator, the principal. For the regulator there is a real problem of identifying "an inefficient agent operating in a good environment from an efficient one operating in an adverse environment."[87] The solutions do exist, such as insurance,[88] but at a much higher cost.

Even if the transfer of responsibility for offshore inspections to private agents was a feasible and a cost-efficient outcome, it still does not ensure a robust inspection regime. Apart from the real potential for the presence of moral hazard and adverse selection issues, the private inspection service would still have to work within the budgetary constraints of the regulator. Although this means not every offshore asset will be inspected, this in and of itself does not mean the inspection regime is not robust. What it does mean, however, is that there still exists an optimisation problem. Regardless of who carries out the inspection, private or public, the duty holder will still have to optimise its allocation of human capital within the constraints of its scarce budget. An improvement of the inspection regime, therefore, may not lie in the identity of the responsible entity for conducting the service, but rather in how the decision for allocation of its resources for carrying out the inspection service is done.

83 *IMF WP/09/177*, p. 10.
84 *A Theory of Incentives*, p. 1.
85 *IMF WP/09/177*, p. 10.
86 A detailed discussion of principal-agent and incentives theory is beyond the scope of this work. For an application of incentive theory within the framework of regulations and government procurement, see *A Theory of Incentives*, which is also the theoretical foundation for the discussion on principal-agent problem in Public-Private Partnerships in *IMF WP/09/177*.
87 *IMF WP/09/177*, p. 15.
88 This would be a suggestion given by the theory of incentives. See *IMF WP/09/177*, p. 16 for a specific discussion. More generally, also see *Theory of Incentives*, cited earlier. This may also be considered within the context of "third party policing": see Hsuan, C.-T. (2011) 'The research on "Application of the Third Party Policing to Social Security"', *Asia Pacific Business Innovation and Technology Management* (APBITM). 2011 IEEE International Summer Conference of Asia Pacific, 10–12 July 2011, pp. 126–130.

4.5. OPTIMISATION OF DECISION MAKING

Both in Norway and in the United Kingdom, those on the front line of hydrocarbon exploitation activity, namely the workforce, take part to a varying degree in each of the stages in shaping the health and safety regime. In Norway, what is termed Norwegian (Nordic) tripartite system[89] allows for the three predominant stakeholders – the regulator, the employers and the workforce – to communicate and cooperate. This cooperative environment is provided for under Safety Forum, a platform established in July 2000 under the administration of then Norwegian Petroleum Directorate (NPD) and later Petroleum Safety Authority (PSA).[90] Safety Forum is where the three stakeholders carry out continual dialogue regarding health and safety issues and policies.

Similarly, in the United Kingdom, the workforce involvement is quite an important aspect of the health and safety regime. However, unlike in Norway, the regulator is not a direct party to dialogues between the workforce and the employers insofar as health and safety of offshore work is concerned. The dialogue is between the employees and the operators, and usually takes place in preparation of the safety cases and, since 2006,[91] in the periodic reviews of, or in the revisions of, the safety cases.[92]

In both countries, however, the involvement of the workforce concerns predominantly the processes and activities of the employer. In Norway, there is an element of influencing future policies of the regulator via the Safety Forum,[93] but outside of this the involvement not just of the workforce but any or all of the stakeholders is limited both in Norway and the United Kingdom. This is especially true when it comes to the decision-making process of the regulator in carrying out its inspection services.

Currently in the United Kingdom, the HSE is deemed to have a "sophisticated inspection regime" with independent third-party verification system for well design and "safety-critical elements offshore."[94] Interestingly, these verifications are conducted not by another governmental agency but by entities in the industry such as

[89] Chapter 11 (Kringen), p. 6.

[90] See Chapters 10 (Bang and Thuastad) and 12 (Rosness and Forseth) in this volume for a discussion on the events surrounding, and leading to, the establishment of Safety Forum.

[91] Offshore Installations (Safety Case) Regulations 2005 (SI 2005/3117) Schedule 2, para. 3 amending the Offshore Installations (Safety Representatives and Safety Committees) Regulations 1989 (SI 1989/971).

[92] See Chapter 6 (Paterson) for discussions on the United Kingdom's regulatory regime and also see Paterson, J. (2011) "Health and safety at work offshore", in J. Paterson, G. Gordon and E. Üşenmez (eds.): *Oil and Gas Law: Current Practice and Emerging Trends*. Second edition, Dundee University Press: Dundee, pp. 187–230.

[93] See Chapter 12 (Rosness and Forseth) in this volume.

[94] *Energy and Climate Change Committee Report Vol. I*, Ev 29, Q189. The third-party independent verification system is a legal requirement under The Offshore Installations and Wells (Design and Construction, etc.) Regulations 1996 (SI 1996/913).

Lloyds and DNV.[95] One can argue, therefore, that there is already a degree of third-party involvement in the offshore health and safety inspection regime in the United Kingdom. The DECC, on the other hand, prioritises its environmental inspections based on the ranking of risks it calculates. Following three streams of inspection – auditing of records and management systems in place, carrying out interviews and observing activities and practices on location – the DECC calculates a risk profile for a given activity.[96] Its environmental strategy, then, "takes into account all the activities being conducted, the companies undertaking those activities, risk to the environment and regulatory controls" to determine the allocation of its inspectors and other resources.[97]

Current decision making on allocating the regulator's resources, therefore, relies to a large extent on measurable variables that are quantified, ranked and co-ordinated through a centralised mechanism. The centralised mechanism, in this context, is a group – or groups – of specialists who are attempting to forecast those offshore assets that are least likely to be compliant with the existing regulations and/or that carry the highest health and safety risks. Consequently, the accuracy of these forecasts is very important in that the inspectorate would be relying on them when deciding to allocate its inspectors and other resources to where they are most needed. This is why it is important to look at the different methods of forecasting processes in order to identify those that may aid a heavily constrained inspectorate.

4.6. GROUP-BASED FORECASTING METHODS

Although there are virtually endless ways in which group members can pool their information together for forecasting, there are five relatively more common categories of group-based forecasting that may be of assistance for the purposes of this chapter. These are face-to-face (FTF) meetings, the nominal group technique (NGT), the Delphi method, statistical aggregation and prediction markets.[98]

Sometimes referred to as unstructured FTF meetings,[99] or interacting group processes,[100] the FTF meetings are the most basic form of forecasting. It is essentially what it says on the tin. Most commonly through meetings – though increasingly

[95] *Energy and Climate Change Committee Report Vol. I*, Ev 29, Q189.

[96] See Chapter 1 (Renn) in this volume for a discussion on charcaterizing risk.

[97] *Maitland Review*, Appendix F, section 2.4.1.

[98] Graefe, A. and Armstrong, J.S. (2011) "Comparing face-to-face meetings, nominal groups, Delphi and prediction markets on an estimation task", *International Journal of Forecasting*, 27: 183–195 (hereinafter, Graefe and Armstrong, *Four methods of prediction*).

[99] Graefe and Armstrong, *Four methods of prediction*, p. 184.

[100] van de Ven, A.H. and Delbecq, A.L. (1971) "Nominal versus interacting group processes for committee decision – making effectiveness", *Academy of Management Journal*, 6: 203–211 (hereinafter, Van de Ven and Delbecq, *Nominal vs. Interacting*).

through conference calls and other technological means as well – members of a group discuss and deliberate on each of their view points and their justifications in order to converge upon a decision or a forecast.[101]

Developed in 1968,[102] the NGT, on the other hand, introduces a structure to these unstructured FTF meetings. The difference of nominal group from the interacting group lies in the degree of communication that takes place between the group members. While in the latter all, or almost all, of the communication takes place between the group members, in the former there are no, or very limited, verbal communication among the group members despite working in the presence of the group.[103] The developers of the technique, Andrew H. Van de Ven and André L. Delbecq, structure the meetings thus:

> (a) Individual members first silently and independently generate their ideas on a problem or task in writing [and in presence of the other group members]. (b) This period of silent writing is followed by a recorded round-robin procedure in which each group member (one at a time, in turn, around the table) presents one of his ideas to the group without discussion. The ideas are summarized in a terse phrase and written on a blackboard or sheet of paper on the wall. (c) After all the individuals have presented their ideas, there is a discussion of the recorded ideas for the purposes of clarification and evaluation. (d) The meeting concludes with a silent independent voting on priorities by individuals through a rank ordering or rating procedure, depending upon the group's decision rule. The "group decision" is the pooled outcome of individual votes.[104]

Contrary to FTF meetings and NGT, under the third method, the Delphi method (also referred to as Delphi process), the group members do not have to be all in the same location, even though they continue to communicate with each other.[105] Developed by Norman Dalkey and his colleagues at the RAND Corporation in the

[101] Kerr, N.L. and Tindale, R.S. (2011) "Group-based forecasting?: A social psychological analysis", *International Journal of Forecasting*, 27: 14–40 (hereinafter, Kerr and Tindale, *Group-based forecasting?*).

[102] van de Ven, A.H. and Delbecq, A.L. (1974) "The effectiveness of nominal, Delphi, and interacting group decision making processes", *Academy of Management Journal*, 12: 605–621 (hereinafter, Van de Ven and Delbecq, *The Effectiveness*). Fn. 2 of this work states the following: "NGT was developed by André L. Delbecq and Andrew H. Van de Ven in 1968 from social-psychological studies of decision conferences, studies of industrial engineering problems of program design in the NASA aerospace field, and social work studies of citizen participation in program planning. Since that time, NGT has gained extensive use and recognition in health, social service, education, industry, and public administration organizations."

[103] Van de Ven and Delbecq, *Nominal vs. Interacting*.

[104] Van de Ven and Delbecq, *The Effectiveness*, p. 606.

[105] Van de Ven and Delbecq, *The Effectiveness*. Note, however, that it is also considered that Delphi is a variant of FTF meetings. See Kerr and Tindale, *Group-based forecasting?*, p. 18.

early 1960s,[106] today the method has a number of variations including Imen-Delphi, Wideband Delphi, Policy Delphi, Group Delphi[107] and SOON, among others.[108] In its simplest version, the method requires a "facilitator" who sends a questionnaire to a group of experts individually asking for their assessments on a given issue. For example, a questionnaire may ask the probability of platform X in the Northern North Sea having a fatal accident within the next twelve months. The questionnaire would also ask the reasoning and methods used in reaching that probability. The process would then continue as follows:

> The facilitator ... produces a summary of the panel's forecasts, along with their justifications for the forecasts. The panel members are anonymous in this summary (and any subsequent summaries.) The summary is then returned to the panel members, and they are allowed to update their forecasts (if they choose) and send a new forecast (along with justifications, counterarguments, etc.) to the facilitator. This process may be repeated for more rounds. Usually, the panels' forecasts converge over the rounds. A compound decision rule (one rule for when to stop, and another for how to statistically aggregate the final set of forecasts) ends the process.[109]

Statistical aggregation, as a fourth method, on the other hand, is distinct from the foregoing three processes in that it does away with all the communication or interaction between the members of a group. In its basic form it involves pooling together all the forecasts of the group members and applying an aggregation function to reach an outcome. Under this method, for example, the inspectorate can take the arithmetic mean of all the probability forecasts received from the group members in determining the likelihood of a fatal accident happening on platform X within the next twelve months. Alternatively, the inspectorate can request from the group members a priority ranking of infrastructure to be inspected within the next twelve months. The inspectorate then can simply take the mean rank order from the responses.

The fifth and final method involves decentralising the decision-making process in aggregating the collective wisdom of a (much larger) group. A decision, commonly referred to as prediction, market is a platform the participants use to purchase and

[106] Dalkey, N.C. and Helmer, O. (1963) "An experimental application of the Delphi method to the use of experts", *Management Science*, 9: 458–467. Note, however, that it is cited as 1950s elsewhere. See, for example, Rowe, G. and Wright, G. (1999) "The Delphi technique as a forecasting tool: issues and analysis", *International Journal of Forecasting*, 15: 353–375 (hereinafter, Rowe and Wright, *Delphi as a forecasting tool*) where it is stated: "The Delphi technique was developed during the 1950s by workers at the RAND Corporation while involved on a U.S. Air Force sponsored project."

[107] Webler, T., Levine, D., Rakel, H., Renn, O. (1991) "The group Delphi: A novel attempt at reducing uncertainty", *Technological Forecasting and Social Change*, 39: 253–263.

[108] Kerr and Tindale, *Group-based forecasting?*, p. 18.

[109] Ibid.

sell futures contracts depending on what each participant thinks the outcome of an event will be. Although it works along the same principles as any other futures market,[110] prediction markets instead focus on the likelihood of real-world events taking place. One of the earliest examples of this is a platform set up in 1988 that focusses on political elections: the Iowa Electronics Market (IEM) operated by Henry B. Tippie College of Business at University of Iowa,[111] which has been more accurate in predicting the election outcomes than the opinion polls.[112] Currently, there are a number of different markets operating predicting not only elections and other political events but also "movie box receipts, corporate earnings, returns, stock prices, incidences of influenza, hurricane landfalls, etc."[113]

None of the aforementioned five methods ought to be mutually exclusive. A regulatory body can incorporate more than one of the techniques and/or apply different processes to different and specific issues. It is, therefore, important to highlight the limitations and drawbacks of each before drawing any conclusions.

4.7. DRAWBACKS

Group dynamics tend to give rise to a number of issues. In a group environment there exists a pressure to avoid contradicting the points of views of the majority, albeit not always explicitly.[114] As a result, not all of the members of a group can contribute to a robust and creative solution-oriented process. One of the reasons for what Van de Ven and Delbecq call "inhibitory influences" – those that prevent a dissenting member from contributing – comes from the fact that the other members

[110] Angrist, S.W. (1995) "Iowa market takes stock of presidential candidates", *The Wall Street Journal* (28 August). Angrist then quotes Steven Feinstein, assistant professor of finance at Boston University's School of Management.

[111] Iowa Electronics Market, The University of Iowa Henry B. Tippie College of Business, available at http://tippie.uiowa.edu/iem/ (accessed 30 March 2012). Note, however, that IEM is not the only one. There are a number of other prediction markets, including The Foresight Exchange Prediction Market, available at http://www.ideosphere.com/, Hollywood Stock Exchange, available at http:// www.hsx.com/, NewsFutures Prediction Markets, available at http://www.newsfutures.com/, and, of course, perhaps the most controversial of them all FutureMAP and PAM, funded by the research arm of the U.S. Defense Department, Defense Advanced Research Projects Agency (DARPA). More information on these can be found at http://hanson.gmu.edu/policyanalysismarket.html.

[112] Berg, J.E.; Nelson, F.D. and Rietz, T.A. (2008) "Prediction market accuracy in the long run", *International Journal of Forecasting*, 24: 285–300. The authors gathered "national polls for the 1998 through 2004 U.S. Presidential elections and ask[ed] whether either the poll or a contemporaneous Iowa Electronic Markets vote-share market prediction [was] closer to the eventual outcome for the two-major-party vote split. [The authors compared] market predictions to 964 polls over the five Presidential elections since 1988. The market [was] closer to the eventual outcome 74% of the time. Further, the market significantly outperform[ed] the polls in every election when forecasting more than 100 days in advance."

[113] Ibid, p. 288.

[114] Kerr and Tindale, *Group-based forecasting?*, p. 28.

of a group may make "covert judgments" even if they are "not expressed as overt criticisms."[115] Of course, an individual may not contribute simply because of the so-called self-weighting effect whereby he or she may not feel as competent as the other group members.[116] Similarly, presence of dominant participants, or high-status individual(s), may prevent those with lower status or less dominant position to conform to the contrary viewpoints, even if they believe in their own dissenting opinion.[117] Group environments also tend to prioritise minimisation of the time spent on decision making.[118] In seeking "speedy decisions," groups may also create an environment where a contrarian view may receive "implied threat of sanctions from the more knowledgeable members" or from the group as whole.[119] These group pressures can even be more pronounced if either the members of a group are stressed, the "non-normative stance" is aligned with the "normative position of an outgroup," or it is perceived as a threat to the identity of the group.[120]

Consequently, latent knowledge of individuals in a group will not contribute to the decision(s) of that group.[121] Corollary to this is the tendency of groups to disproportionately weigh the information or the ideas that are common to the majority of the members. Although this may expedite consensus reaching, it may do so at the expense of discarding information that may otherwise be central to optimal outcome. As a result, it may inevitably lead to the "exacerbation of any biases that are shared among the members."[122]

In addition to the inhibitory influences, motivation and confidence may also impact the performance of a group. Kerr and Tindale note three separate processes that lead to loss of motivation among the group members.[123] The first process, they note, is when a contributing member cannot be identified. This may lead to the lessening of the motivation as the individual is not receiving recognition for the contribution. The second process involves the opportunity for a free ride where a member is under the impression that the group decision would be the same with or without the participation of that member. The final process that may lead to loss of

[115] Van de Ven and Delbecq, *Nominal vs. Interacting*, p. 206.

[116] Van de Ven and Delbecq, *Nominal vs. Interacting*, p. 206.

[117] Gustafson, D.H., Shukla, R.K., Delbecq, A. and G. Walster, G.W. (1973) "A comparative study of differences in subjective likelihood estimates made by individuals, interacting groups, Delphi groups, and nominal groups", *Organizational Behavior and Human Performance*, 9: 280–291 (hereinafter, Gustafson, Shukla, Delbecq and Walster, *A Comparative Study of Differences*). Also see Van de Ven and Delbecq, *Nominal vs. Interacting*, p. 206.

[118] Graefe and Armstrong, *Four methods of prediction*, p. 184.

[119] Van de Ven and Delbecq, *Nominal vs. Interacting*, p. 206, and Graefe and Armstrong, *Four methods of prediction*, p. 184.

[120] Kerr and Tindale, *Group-based forecasting?*, p. 28.

[121] Van de Ven and Delbecq, *Nominal vs. Interacting*, p. 206.

[122] Kerr and Tindale, *Group-based forecasting?*, p. 29.

[123] Kerr and Tindale, *Group-based forecasting?*, p. 31.

motivation is essentially the opposite of the second process where a member feels that the others in the group are not contributing in an equal and fair manner.[124]

The issue of group confidence arises from the overconfidence of the individuals that make up that group in the accuracy of their own views or forecasts. Interestingly, "groups tend to be more confident in their accuracy of their collective judgment than their (already overconfident) individual members are. This seems to be particularly true when the decision task is complex or difficult, or when the accuracy is actually poor."[125]

All of these drawbacks are present to a varying degree in the five categories of group-based forecasting. In FTF meetings, all of the foregoing issues are strongly present bar the motivational loss. FTF meeting setting is particularly vulnerable to the inhibitory influences and to the group overconfidence.[126] Motivational loss is less common in FTF meetings because the contributors in the group are easily identified, and as each member tends to have an equal opportunity to participate, it is less likely that free riding will occur.[127]

One of the five methods that is less vulnerable to the foregoing issues, on the other hand, is thought to be the NGT. Because of the way it is structured, it noticeably slows down the tendency to reach "speedy decisions" as it provides participants time to consider and record their views. This recording also avoids to a large extent any dominance of high status or influential individuals in the group, and allows for dissenting and contrary views to be expressed. Yet the process assumes the problem is clearly defined. If the issue to be addressed is defined vaguely, or if group members are not in agreement as to the particulars of the issue to be addressed, the input from the participants may "overlap, or worse yet, widely vary in their specificity."[128] In fact, Van de Ven and Delbecq, the developers of the technique, recommend the use of NGT when the group is tasked with "fact-finding or information generation," whereas the FTF meetings, they note, may be better suited for "information synthesis, evaluation, or working toward group consensus."[129]

The drawbacks of the Delphi method are thought to lie between the FTF and the NGT. It is not as vulnerable to the inhibitory influences because the group members are anonymous, and the process allows for minority views to have equal voice as the majority. This anonymity of group members may also contribute to avoidance

[124] Kerr and Tindale, *Group-based forecasting?*, p. 31.

[125] Kerr and Tindale, *Group-based forecasting?*, p. 31.

[126] See Kerr and Tindale, *Group-based forecasting?*, Van de Ven and Delbecq, *Nominal vs. Interacting*, and Graefe and Armstrong, *Four methods of prediction*.

[127] Kerr and Tindale, *Group-based forecasting?*, p. 32.

[128] Bartunek, J.M. and Murninghan, J.K. (1984) "The nominal group technique: Expanding the basic procedure and underlying assumptions", *Group & Organization Studies*, 9 (3): 417–482 (hereinafter, Bartunek and Murninghan, *The Nominal Group Technique*).

[129] Van de Ven and Delbecq, *Nominal vs. Interacting*, p. 210.

of motivational loss.[130] However, inhibitory influences are not fully absent because the "dissidents in early rounds may drop out or comply with the majority position."[131] Similarly, the Delphi method is designed to eliminate the "status incongruities"[132] through anonymity. Yet the facilitator, through its central position, has disproportionate levels of influence over the outcome. It not only chooses the participants but also singlehandedly summarises and redistributes the contributions and decides when to stop.[133] Accordingly, the problem of latent information could also be present in the Delphi method because the process is subject to "sampling advantage of shared information."[134] As the perception of being an expert would contribute to individual's overconfidence, this in turn may translate into overconfidence of the panel of experts reflected in the outcome of the Delphi method.[135]

As communication among the group members is nonexistent under statistical aggregation method, it is less impacted by inhibitory influences of group dynamics. Yet the aggregation itself may suffer from certain underlying assumptions of the methodology. Because of the function used in calculation, a minority view may be weighted less than the forecasts of majority. Related to this, arithmetic mean may dramatically change with a single outlier forecast. Assumption of normal distribution of expected forecasts may be problematic in this regard. A possible solution may be to readjust the distribution model, and consequently the weight of outliers, to incorporate high-impact – low-probability events. Yet the readers of Nassim Nicholas Taleb[136] or Benoit B. Mandelbrot[137] will know the difficulty of incorporating a statistical layout (black swan) that can consistently provide accurate forecasts a priori.

Similar to the aggregation methods, communication levels between the participants in prediction markets tend to be near zero. Therefore, the prediction markets also tend to be less affected by inhibitory influences. It is possible that prediction markets may be vulnerable to herd behaviour whereby individual participants may follow the actions of the majority – the herd – lessening the degree of accuracy of the market's forecasts. This may also be further exacerbated by the overconfidence of individual participants. Yet such negative influences would also create opportunities

[130] Kerr and Tindale, *Group-based forecasting?*, p. 31.
[131] Kerr and Tindale, *Group-based forecasting?*, p. 29.
[132] Van de Ven and Delbecq, *Nominal vs. Interacting*, p. 206.
[133] Kerr and Tindale, *Group-based forecasting?*, p. 29.
[134] Kerr and Tindale, *Group-based forecasting?*, p. 30.
[135] Kerr and Tindale, *Group-based forecasting?*, p. 32.
[136] Taleb, N.N. (2010) *Black Swan*. Second Edition. Random House: New York. See also Taleb, N.N. (2007) *Fooled by Randomness: The Hidden Role of Chance in Life and in the Markets*. Penguin: London.
[137] Mandelbrot, B.B. and Hudson, R.L. (2008) *The (Mis)Behaviour of Markets: A Fractal View of Risk, Ruin and Reward*. Profile Books: London. See also Mandelbrot, B.B. (1982) *The Fractal Geometry of Nature*. W.H. Freeman: New York.

for increasing benefits to contrarian, but more accurate, views. This may induce more individuals to join in the market, or to convince the existing participants to reallocate the resources to these opportunities. One final drawback with this method may lie in its application. The operators of the inspected platforms as a result of the prediction market may object the regulator's decision to inspect their assets, as the decision to do so would not be based on "expert analysis."[138] In more extreme circumstances, it may lead to serious loss of trust between the regulator and the industry.

4.8. THE FIVE METHODS AND RISK MANAGEMENT

Ortwin Renn (Chapter 1 in this volume) highlights the importance of widening the net of stakeholder involvement as the risks that require managing get more ambiguous. Given that neither the FTF nore the NGT method limits the membership to a group in any way, it is possible to structure groups whereby the identity of the participants can reflect the levels of knowledge associated with risks. This is also true to an extent for the statistical aggregation method. However, this latter method lacks the key elements that, as Renn argues, "provide a better basis for responsible risk management." Specifically, the interactions that reflect the individual group member's "knowledge, experiences, interpretations, concerns, and perspectives"[139] are not provided for in the statistical aggregation method.

Similarly, prediction markets method also incorporates wide stakeholder participation. However, it falls outside Renn's linear relationship. This is because the participation level is not a function of the level of knowledge possessed on a given risk, but rather is open to anyone at all times. In fact, rather counter-intuitively, the participation does not require *any* knowledge. The only requirements in prediction markets are participants' willingness to bet on a future event and the ability to pay for that bet. Therefore, one can say the participants use their intuition to forecast a future event. This is true. It is also true that the participants can use their knowledge, experiences, interpretations, concerns and perspectives.

On the other hand, the Delphi method is the most restrictive of all. Like the prediction markets, it falls outside Renn's linear relationship because the participation is again not a function of knowledge, but the same for all levels of ambiguity surrounding risks. This time, however, it is restricted only to experts. Yet, unlike the statistical aggregation, the participants are in a position to communicate their reasoning as the questionnaires they receive request their rationale for their position.

[138] Regardless of how that term is defined.
[139] See Chapter 1 in this volume, Section 1.6.

4.9. FINAL COMMENTS

The offshore regulators in the United States, the United Kingdom and Norway always had to manage their duties within highly constrained environments. When an accident took place in their offshore jurisdictions, the blame was repeatedly and consistently laid, among other factors, on the inadequate budgets and staff of the regulators. In fact, the rhetoric at times reached the unacceptable levels that implied the staff of the inspectorates were inferior to their colleagues elsewhere. The solution had equally been repetitive and consistent: allocation of more money for the budgets of the regulators and the increase in staffing levels. Yet, with the next offshore accident, this blame-solution cycle would continue to take its undeserved seat in the inquiries that followed.

This chapter has investigated possible solutions that would take the constraining factors available to the regulators not as a source of blame but as a starting point for a solution. One possible solution could be for the private sector to carry out the inspection duties of a regulator, as it is assumed that it could generally be cheaper for a government entity to delegate a function than to run it. However, the monopolistic structure of offshore inspections may not provide the cost savings that could be envisaged. In fact, because of this structure, the incentives could be misplaced. Finally, as the services of the private entity would be purchased by the regulator, the former would still have to operate within the budgetary constraints of the latter.

The solution, then, may not lie in the reshuffling of the enforcement of regulations, but may instead lie in the third branch of a regulatory regime: the decision-making process of the regulator that bridges the regulations and enforcement. Prior to inspecting an offshore infrastructure, a regulator has to decide on how to allocate its scarce resources, two of which are its budget and its staff. A regulator effectively forecasts those infrastructures that require to be inspected the most in order to send its few inspection staff available in the cheapest manner possible. It is, therefore, important to look at the main forecasting methodologies available to ascertain whether they can be of help in optimising the decision-making process.

The discussions cover the five main techniques employed in forecasting, ranging from unstructured FTF meetings to structured NGT, from statistical aggregation to the Delphi method and to prediction markets. They are, by all means, not exhaustive.[140] The discussion of these five techniques does not intend to champion one technique over the others, but instead is intended to inform the reader of the possibilities with the hope that further discussion can ensue on improving this third branch of a regulatory regime.

[140] For the classification of "all possible types of forecasting methods," see "Forecasting principles: Evidence-based forecasting", *Methodology Tree*, 24 April 2010, at www.forecastingprinciples.com (accessed 22 August 2012).

PART II

Regulatory Regimes: Norway, United Kingdom, United States, and Australia

INTRODUCTION TO PART II

The most fully realized manifestation of risk governance is the modern regulatory regime. That is not to say that such a regime is a complete and final realization of risk governance, nor does it infer that regimes have the same structure, procedures, substantive features, decision-making processes, and other essential ingredients.

The reality is that each regime is unique and incomplete, and although its legal mandate may remain unchanged, it evolves as it develops its own internal culture and norms. It also evolves in response or as a reaction to many dynamic external factors and changing conditions in its national and global contexts. These contextual factors and conditions include, for example, the stream of new knowledge arising from experience and research, changes in values and norms and expectations, changes in economic circumstances and societal priorities, and the unfolding consequences of its own actions.

This section examines four regulatory regimes that are assigned the role of regulating the safety of offshore oil and gas operations, their main features and modes of regulation, the premises and concepts on which they act, their interactive relationship with industry and stakeholders, and other matters including their accomplishments and ability to cope with change and other challenges.

Three regimes – Norway, the United Kingdom, and Australia – represent the "new governance" co-regulatory model with performance-based rules in which the government regulator supervises and mentors self-regulatory efforts by industry to fulfill the rules, and intervenes to enforce compliance as a last resort for dealing with uncooperative companies. Yet subtle and notable differences are revealed: Norway holds allegiance to a tripartite policy that promises inclusion of labor in virtually all aspects of regime activities and remains ambiguous about enforcement of rules; the United Kingdom features the safety case approach, and is taking special initiatives needed for its invigoration and ability to detect systemic risks; and Australia, which employs a safety case approach moored to engineering expertise, is seen to be in need of a greater infusion of social and behavioral science knowledge.

The fourth regime, the United States, stands apart with its highly prescriptive, command-and-control approach and commitment to hard law and strict compliance. Shaken by the Macondo accident and spill and knowing of other incidents causing unacceptable levels of harm to workers, it is taking some still modest steps to improve its regime. Capping the discussion of these regimes is Chapter 9 on the potential value of safety performance indicators to government regulators and industry, difficulties in developing sufficient and comparable data, current indicators, and how derived data is being put to use.

Regimes are works in progress. Achieving and maintaining robust regulation under changing conditions is their goal. The chapters in this part of the book provide a basis for comparison. Although comparison can only be qualitative at this time and differing contextual factors must be taken into account, there is much to be learned about robust regulation from the information presented.

5

Safety Regulation on the Norwegian Continental Shelf

Knut Kaasen

5.1. INTRODUCTION

5.1.1. *The Role of Law in Safety*

Improving the level of safety in the petroleum activities implies trying to avoid accidents that may cause damage to health, environment and investments. At first glance this is primarily a technological challenge. Structures, equipment and components have to be constructed in such a way that accidents do not happen when the installations are exposed to the physical stresses of operation and nature.

A closer look reveals, however, that technology alone will not suffice. 'Soft issues' come into play as well, owing to the fact that the installations have to be constructed and operated by humans. This simple fact entails the need for adequate organisational structures, individual competence and adequate ad hoc discretionary decisions – all leading to safety as one of the end results. If these factors are disregarded, no technological achievement can guarantee that accidents are avoided.[1]

Of course, such guarantees cannot be given under any circumstance. Any activity will unavoidably involve 'risk' that damage is caused to life/health, environment or installations. This 'risk' may be defined as the product of 'probability of occurrence

[1] The Piper Alpha disaster on the UK sector of the continental shelf on the evening of 6 July 1988, which killed 165 people, may serve as an illustration. The immediate cause of the disastrous fire was ignition of condensate flooding from a blind flange that could not withstand the pressure. The blind flange was replacing a valve that had been removed for repair. At that time there was no condensate in the pipe, and the flange was not intended to withstand the pressure of condensate. However, the operators in the control room on the night shift were not properly informed by the day shift that the valve had been taken out for repair. That night operational irregularities occurred. The operators took the natural action of leading the condensate into the alternative pipe, and the tragedy was a fact – resulting from inadequate communication rather than technological challenges. See the Lord Cullen report (The Public Inquiry into the Piper Alpha Disaster, Cm 1310, November 1990) at pp. 1 and 119–122.

of an undesirable event' times 'the probable consequences if it occurs'. This implies that the risk can be reduced by reducing the probability that the event occurs at all, and/or by reducing the likely consequences if it nevertheless does.[2]

The endeavours to improve safety imply that all these elements have to be taken into consideration: The technological aspects, the 'soft issues' and the choice between reducing likelihood of occurrence and/or consequences. Various types of expertise are needed for this purpose. Understanding of technological, psychological, economic, organisational and decision-making aspects is vital. Nor can the rather cynical cost-benefit analysis be avoided – which in turn also calls for political considerations, such as whether it is considered acceptable that environment is put at risk if the cost of avoiding it exceeds a given amount, or if the alternative simply is that the activity in question cannot take place.

So where does law come into this?

Law in this sector, like elsewhere, is a general tool for enforcing decisions that result from considerations based on such other types of expertise. But the legal tool is not a given element: it takes legal expertise to design the legal tool in such a way that it provides the best means for reaching the goals that have been defined in other arenas. The legal aspect also brings in some additional parameters of its own, for example, certain limitations as to how and when 'the rules of the game' may be changed.

The following examination of the Norwegian regulatory regime for offshore safety discusses how legal tools are applied to reduce risk of damages caused by petroleum activities on the Norwegian continental shelf. It is also intended to highlight elements of safety regulation which are to be found in most jurisdictions governing offshore safety. The details[3] of Norwegian regulations are largely left aside in this chapter but are discussed in the chapters comprising Part III of this book. Rather, emphasis here is put on the general themes and issues, the relationship between the different players and the various legal techniques that are used to promote safety in the Norwegian context.

5.1.2. *The Emergence of Norwegian Offshore Safety Regulation*

(a) Ever since the very start of petroleum activities offshore Norway in the 1960s, there have been regulations aiming at preventing accidents from happening.

[2] Thus, the general health hazard ('risk') offered by crossing the street of Karl Johan in Oslo may be in the same order as the risk offered by a nuclear power plant: the likelihood of an accident happening on the street is far greater than the likelihood of the power plant blowing up, but the consequences of an incident are far greater in the latter case.

[3] The details of the safety regulation are numerous, and they have a tendency of changing ever so often.

The *first set of rules*[4] was designed in the tradition of industry safeguarding regulations, concentrating on rather practical 'do's and don'ts' directed to the industry carrying out the activities. Some ten years later, the same path was followed in a more detailed manner, distinguishing between drilling, production and working environment issues.[5]

(b) At this stage, in the mid 1970s, a new concept was introduced in addition to the traditional approach: the *internal control* was made a part of the safety regulation.[6] The basic idea was that the industry should not limit itself to complying with straightforward regulatory obligations and prohibitions directly aiming at an acceptable level of safety. The industry was also – as a separate obligation – required to establish a system for identifying relevant requirements, checking that these were adhered to, implementing corrective measures if needed and reporting all these activities to state authorities supervising offshore safety. This was to take place within the framework of general requirements imposed by the state in the form of 'functional requirements': see (c) below.

The internal control concept was far from fully developed back in the 1970s. In the following years the concept has been further refined and structured, a key element being that the industry is expected to supplement the often vague requirements of the regulations by defining more specific norms to be applied internally. The (so far) last stage was introduced by the regulations issued in 2010.[7] But it is fair to say that since 2000 the internal control system is a fundamental element in Norwegian safety regulation, and that the impact on petroleum activities imposed by the safety regulations cannot be fully understood without this fact being appreciated.

(c) Another line of development in the safety regulations is the move from detailed rules specifying methods to be applied to rules that simply state which results are to be achieved. While the Decrees of the 1960s and 1970s would give specific details on, for example, the construction of cranes or load-carrying structures, the 2010 regulations restrict themselves to stating that all facilities, including cranes and load-bearing structures, should be constructed in such

4 The 1967 Royal Decree (25 August 1967) on safety in exploring and drilling for subsea petroleum deposits.

5 Royal Decrees 3 October 1975 replacing the 1967 Decree, 9 July 1976 on safety in offshore petroleum production, and 24 June 1977 on working environment in offshore petroleum exploration and production.

6 The Internal control system is by no means a Norwegian invention. It has been developed inter alia in the U.S. car and aviation industries over several decades. But Norwegian regulations in the last twenty-five to thirty years have taken the system a step further by more explicitly making it a distinct formal part of the state safety regulation regime, not just a matter for industry's internal organising of activities.

7 For further discussion, see Section 5.4.

a way that they are able to function safely under the assumed operating conditions. The lack of engineering guidelines in such *function requirements* is often remedied by detailed 'cookbooks' contained in guidelines and recommendations issued pursuant to the regulations, or issued by (national or international) standardisation organisations and subsequently endorsed in such official guidelines or recommendations. But these details do not constitute legally binding requirements, just indications on possible (but not necessarily sufficient) ways to fulfil the legally binding requirements.

(d) Consequently, in parallel with the emergence of function requirements, there has been a *move from Decrees to guidelines* and the like as the primary basis for identifying in detail which obligations are actually placed on the industry and individuals employed in the petroleum activities. It is no longer possible – if ever it was – to derive specific directions on how to safely build and operate offshore installations from reading formal regulations. Accordingly, any such specific directions to be found will not be legally binding, as they will emerge from documents that do not possess such standing. This observation of course carries some legal implications that we will return to in Subsection 5.4.2 (c).

(e) The regime governing offshore safety does not, however, operate solely on the general level that we have now described. The individual company (or, in principle, person) that is subject to the safety regulations will also have an individual link to relevant state authorities: the regulations may call for governmental approvals or exemptions at certain stages, or the authorities may want to intervene in the activities on a separate basis – for example, because checks reveal that the company is not complying with what the authority thinks is a safe practice. In such instances the *general* provisions will be supplemented by *individual* administrative decisions, in Norway as elsewhere. In other words, in addition to requirements that are applicable to all subjects operating in the sector, this specific subject is also obliged to comply with individual requirements that in principle apply to him or her only. Such administrative decisions will normally constitute *legally binding* requirements that are *more specified* and (often) precise than the general requirements, as previously discussed in Chapter 2 (Baram and Lindøe) of this volume. Thus, the regulator to a large degree abstains from expressing legally binding detailed requirements on a general level, but is not as reluctant on an individual level.[8]

[8] There may be several reasons for this, the obvious one being that relevant facts and considerations are easier to identify in individual cases than on a general level. On the other hand, the authorities generally emphasise that rather than making individual 'approvals' to the industry a condition for certain activities to commence, the preferred term is 'consent'. This change of terminology aims at reducing the legal implications of the authorities' 'fingers in the pie'.

(f) *In conclusion*, what we are left with is a legally binding framework defining administrative and organisational systems to be established[9] and some general function requirements to be complied with in the activities. But the details as to how safety in construction and operation is to be achieved are to be found either in non-legally-binding documents (guidelines etc.) or in individual administrative (legally binding) decisions issued within the general framework. This shift away from regulatory details to rather abstract regulations is a characteristic feature in the development of Norwegian offshore safety regulation.

5.1.3. *Different Categories of Safety Regulations*

(a) Norwegian regulation deals with two distinctly different issues.

First and basically, it establishes requirements that potentially have a *direct* effect on the risk level. In this category one finds, for example, rules on how blowout preventers for exploratory drilling should be constructed, and which qualifications welders should hold. These may be labelled requirements as to the *state of matters*. But the direct safety regulations also include requirements regarding *actions or occurrences*. Most important in this latter category are rules on procedures for operations, for example, rules on how to weld certain critical elements of a structure or how to run risk analyses on the structure. The use of both methods, requiring the qualified welder to weld in the specified way, is likely to improve the level of safety. If direct safety requirements are not complied with, the risk for accidents is likely to increase.

Second, it deals with matters that are likely to have an *indirect* effect on risk. For example, the operator may be obliged to apply for approval to run a drilling program, specifying the equipment he intends to use for the purpose. Or he may be ordered to report logs from his running of maintenance programs for the production facilities. The non-compliance with this kind of regulations will not in itself increase risk offered by the operations, as opposed to the situation if the operator employs unsafe drilling equipment or refrains from carrying out maintenance activities. But failure to report or apply for approvals may cause the direct safety rules (on maintenance or equipment) to become less effective, and may thus indirectly affect the risk.

(b) From a legal perspective, the *direct safety rules* are rather simple. They constitute traditional examples of straightforward legal obligations, and non-compliance is likely to have the traditional consequences – admittedly with some 'petroleum flavour' to them. To the lawyer or the engineer they may

[9] I.e. the internal control; see Subsection 5.6.2.

pose a challenge when determining their exact meaning,[10] but they do not constitute complex systems that are difficult to see or understand.

(c) *The indirect safety rules* are more complex. They form the legal basis for *safety control*. This includes both a passive element – monitoring that direct safety rules are observed in the operations – and an active element if such monitoring reveals a need for corrective measures. The latter may involve the prescription of individual and more specified and/or precise direct safety norms (by means of administrative decisions) to provide further guidance (Subsection 5.4.4). The need for corrections may also necessitate the use of legal means of enforcement, such as coercive fines (Subsection 5.5.2).

Therefore, a prerequisite for an effective safety control is that relevant state authorities are given the right of insight into the activities that are subject to control and to interfere with these activities should the monitoring reveal a need for that. The industry must be obliged to submit applications, plans, reports and so forth, and the authorities must be empowered to employ adequate corrective measures.

The indirect safety rules are, however, not restricted to establishing a necessary legal basis for the *state*'s safety control. Safety control is also a matter for the industry itself. An important type of indirect safety rules are those dealing with the operator's internal control system: by requiring the operator to establish and maintain a defined system for managing his compliance with the safety regulation in general, his operations are not made totally safe, but the risk for errors causing accidents and damage is likely to decrease.

(d) Based on the above, two important observations may be made – both of them important to get the grips on safety regulation.

First, from a legal perspective, the endeavour to improve offshore safety involves two major players: the state and the industry – or the operator in the individual case.

Second, the regulatory regime in this area requires that each of them run a system for *safety management* comprising two elements: they should prescribe safety norms – that is, define norms that should be met in order for risk for accidents to be reduced – and they should *establish a system* for safeguarding that these norms actually are met. The further details of this 'safety management' of course vary between the state and the operator. But it is helpful to realise that both types basically amount to the same system.

The petroleum industry is of course engaged in safety aspects of its activities for several other reasons than compulsory state requirements: the level of compensation for damages caused by safety failures may call for increased efforts laid into

[10] We will return to that in Subsection 5.4.2.

technological innovation,[11] the general public image of the company involved may suffer or gain by its safety record, damage to equipment and installations may put them out of use and thus result in great losses, and so on. However, in the following we concentrate on the compulsory elements – 'the state's finger in the pie'.

(e) In *summing up* this rather complex picture, safety norms are twofold: the direct safety norms and the indirect safety norms. An important type of indirect safety norms establish the legal basis for safety control. The prescription of safety norms and the performance of safety control together constitute safety management. The safety management is a matter for both state authorities and industry. From both a legal and a practical perspective, these two players run separate systems for safety management. But of course they are also interlinked.

In the following we take a closer look first at the state's safety management, then at the operator's and finally at the interaction between the two. But first we look briefly into the issue of jurisdiction: What is the jurisdictional competence of the coastal state outside of its own territory?

5.2. THE PROBLEM OF JURISDICTION

The continental shelf is located outside the boundaries of national territory and territorial waters. Consequently, the coastal state needs a specific basis in rules of international law in order to exercise jurisdictional powers[12] over the continental shelf and the activities taking place there. We will not delve into these issues in general, but illustrate their link to safety regulation by looking briefly into the issue of jurisdiction over floating devices employed in offshore petroleum activities.

Whereas some of these devices are operating on a fixed location their whole life, others are moving to and from. The latter are often registered in a national registry of ships. If this state of registry is not Norway, the fact that the vessel operates on the Norwegian continental shelf introduces a problem of international law: How is the power of jurisdiction over the vessel distributed between the 'flag state' and the 'coastal state'?

This question is relevant for several issues other than safety regulation, for example, employment, insurance, tax, liability for tort, and so forth. In most of these other matters, the coastal state (here: Norway) cannot exercise any jurisdiction over a foreign vessel operating in connection with petroleum exploration and exploitation off its

[11] This aspect is also influenced by the state; ref. Petroleum Act Sect. 10–9 on liability for independent contractors, and Petroleum Act Ch. 7 on liability for pollution damage.

[12] Generally, these powers include the power to regulate, adjudicate and enforce.

coast – the vessel is more or less in the same jurisdictional position as any vessel on the high seas.[13] On safety issues, the situation is more complex. As long as the safety of the vessel, its operation and its personnel does not pose a threat to its surroundings, there is little need for the coastal state to intervene in any way; such matters can be left for the vessel's flag state to handle. But we can easily foresee situations where the interests of the coastal state are put in jeopardy by the presence and operation of the foreign vessel. This is specifically the case when the vessel is employed in such state's petroleum activities: the safety of the coastal state's installations can be put at risk by the vessel's faulty manoeuvring or construction activities, the environment may be harmed by spills, and so on. For this reason, the coastal state would like to have the authority under international law to regulate (and enforce) certain aspects of the vessel and its operation. On the other hand, as seen from the state of registry, such interference would constitute competing jurisdiction, either removing said issues from flag state jurisdiction or complicating matters severely (specifically because the vessel could – and in practice would – be subject to differing coastal state requirements depending on its geographical area of operation).

The balance is set by the UNCLOS art. 77:[14] 'The coastal state exercises over the continental shelf sovereign rights for the purpose of exploring it and exploiting its natural resources.' This clause can hardly be said to clarify all issues related to continental shelf jurisdiction. But it emphasises that in our case the crucial point is to which extent the regulatory issue in question is likely to have a direct effect on safety aspects of exploration/exploitation, that is, the core petroleum activities. On this basis, there is likely to be a jurisdictional difference between, for example, general working environment and crane operation procedures on board a foreign construction barge engaged in heavy lift operations in an offshore development project. Also, the argument for coastal state jurisdiction grows stronger the closer the activities on board the floating device – or the function of the device in general – resembles that of fixed installations. On this basis, there will be a general jurisdictional difference between ordinary ships (e.g., supply ships and shuttle tankers) and floating production platforms.

> In between these extremes there are numerous examples of floating devices that have more or less in common with ordinary shipping. One group comprises vessels specifically designed for standby, supply services, anchor handling, seismic or geological exploration, subsea work and so forth. These vessels may both look like

[13] Note that these jurisdictional issues are complex: the fact that the vessel is employed in petroleum activities distinguishes it from ordinary seagoing vessels because separate rules of international law are then brought into play. Also, the international law rules (and even the internal regulations) on safety zones establish a separate basis for a certain coastal sate jurisdiction over foreign vessels. However, we need not look into the general implications of this.

[14] United Nations Convention of the Law of the Sea (Montego Bay, 1982). The same clause is contained in the Convention on the Continental Shelf (Geneva, 1958) art. 2 (1).

ordinary ships and operate more or less like them. The same goes for shuttle tankers, transporting crude from the offshore field to refineries onshore. All of these vessels at the outset would seem closer to the flag state than to the coastal state. At the other end of the scale we find vessels carrying out construction, pipe-laying or maintenance services. These vessels are often constructed exclusively for petroleum-related operations, their activity and appearance typically deviate strongly from those of ordinary vessels, and their link to 'core petroleum activities' is close. Safety aspects of activities on board such vessels perhaps hold stronger resemblance to that on fixed platforms than to that on board ships. Consequently, it may be argued that these vessels should be subject of coastal state jurisdiction.

In practice, the alternatives are not 'no' or 'total' coastal state jurisdiction. Rather, the issue is which aspects (activities, operations, physical arrangements, qualifications etc.) on board the mobile facility should be subject to what kind of jurisdiction. This follows from the key wording of UNCLOS art. 77: '…sovereign rights for the purpose of exploring it and exploiting…'.

Quite another matter is whether the coastal state sees fit to exercise the jurisdictional power vested in it by international law. The competing considerations may be illustrated by the Norwegian position, which resembles that of most states hosting both its own international maritime activities and its own offshore petroleum activities. In general, Norway's flag state jurisdiction in the internal legal system takes the form of maritime legislation, while the coastal state jurisdiction is exercised by means of petroleum legislation (comprising several acts; see Subsection 5.3.1). A vessel registered in Norway would in principle be subject to both of these sets of regulations when it operates on the Norwegian continental shelf. As no problem of international law exists in this situation, it is for Norwegian authorities alone to decide which set of rules are to apply. The choice is based on considerations quite parallel to those relevant under the international law perspective, that is, mainly the strength of the link between (1) the vessel and its operations and (2) the specific features of petroleum activities as opposed to ordinary shipping. The result generally is that the vessel is not made subject to petroleum safety regulation only – as opposed to maritime safety regulation. On the contrary, it is for most purposes made subject to maritime regulation only, unless the vessel is very closely connected to core petroleum activities. This is not just a result of considerations parallel to those applied under international law. It is also a fear that giving the 'petroleum perspective' the lead would in turn result in other coastal states doing the same, which would result in practical – and therefore commercial – restrictions being imposed on Norwegian vessels operating worldwide.[15]

[15] Similar arguments can be seen as basis for the efforts made for international standardisation of inter alia safety regulations within the shipping industry.

In other words, even if a state takes the position of both a flag state and a coastal state in relation to the same vessel, the choice between 'jurisdictional hats' may pose challenges.

5.3. STATE SAFETY MANAGEMENT: THE STRUCTURE OF THE SAFETY REGULATION

The state's regulatory safety management comprises two main elements: the prescription of safety norms (directly and indirectly affecting the safety level) and the activities to see to it that the norms are complied with. There are strong interactions between the two. As a third element we find the means of enforcing compliance with the safety norms.

On the formal level, the offshore safety regulations comprise (1) statutes and (2) decrees and regulations issued pursuant to each of the statutes (or pursuant to several of them in combination).

5.3.1. *Norwegian Statutes*

Petroleum activities combine elements from various other activities. There are elements of industrial as well as maritime operations, and there are aspects of working environment as well as external environmental issues of pollution, to take just a few examples. Each of these aspects often carries with it a separate piece of legislation. For this reason, there is no single statute that exclusively regulates safety aspects of the petroleum activities. At least four distinctly different statutes are relevant for offshore safety:

Naturally, the most general statute applicable to offshore petroleum activities is the *Petroleum Act 1996*. This act contains several sections on various safety aspects of the activities, covering the whole range from safety for personnel to availability of installations (see further on this in Subsection 5.4.2 (a)).

The 2005 *Working Environment Act*[16] is also relevant. This act contains several sections with a bearing on safety aspects of the activity.

If, on the other hand, the personnel work onboard a *floating* device engaged in petroleum activities, the 2005 act will not come into play at the outset. After all, it is designed for onshore (inter alia) working environment issues, which may differ greatly from those involved in shipping. Instead, the 2007 *Ship Safety and Security Act*[17] and the 1977 *Seamen's Act*[18] will take care of inter alia safety aspects of the working environment onboard.

[16] Act of 17 June 2005 No. 62 relating to working environment, working hours and job protection, etc. (the Working Environment Act).

[17] Act of 16 February 2007 No. 09 relating to Ship Safety and Security (The Ship Safety and Security Act).

[18] Seamen's Act of 30 May 1975 no. 18.

Specific risk factors relevant also in the petroleum industry are covered by separate pieces of legislation, such as the 1929 *Electrical Supervision Act*[19] and the 2002 *Fire and Explosion Protection Act*.[20] Also, because health issues may well have an impact on safety level, the specific health legislation is relevant to offshore safety. A total of six health-related acts thus form part of the basis for the detailed safety regulations (see further in Subsection 5.4.1).[21]

Petroleum activities may pose a risk to the external environment. Therefore, the 1981 *Act on Protection against Pollution* and on waste[22] is obviously relevant to this type of activity. So is the 1976 *Act on Control of Products and Consumer Services*.[23]

We may safely conclude that it is a challenge to keep track of all acts relevant to petroleum safety.

5.3.2. *The Role of Decrees and Regulations*

While the Petroleum Act by its very purpose is directly applicable to the petroleum activities, the other statutes are applied to these activities by specific provisions to that effect.[24] The details are complex and to some extent difficult to interpret. Also, the acts partly overlap, both in substance and in scope of application. The total picture thus becomes somewhat complex, to put it mildly.

In practice, however, the regulatory regime is simpler than this. First, at the level of formal acts, there are few substantive provisions on safety to be found anyway. Second, most of the acts leave it to the king (i.e., the government by means of royal decree) to provide the details of how and under which circumstances the act shall be applied to offshore activities. Consequently, we can turn to the decrees (and

[19] Act of 24 May 1929 No. 4 relating to supervision of electrical installations and equipment (the Electrical Supervision Act) (amended several times, last time in 2009).

[20] Act of 14 June 2002 No. 20 relating to protection against fire, explosion and accidents involving dangerous substances and relating to the fire department's rescue tasks (the Fire and Explosion Protection Act).

[21] The six acts are the Act of 2 July 1999 No. 64 relating to health personnel, etc. (the Health Personnel Act), the Act of 2 July 1999 No. 63 relating to patients' rights (the Patients' Rights Act), the Act of 5 August 1994 No. 55 relating to protection against contagious illnesses, the Act of 23 June 2000 No. 56 relating to health-related and social preparedness, the Act of 19 November 1982 No. 66 relating to the municipal health service, and the Act of 19 December 2003 No. 124 relating to food production and food safety, etc. (The Food Safety Act)

[22] Act of 13 March 1981 No. 6 relating to protection against pollution and relating to waste (the Pollution Act).

[23] Act of 11 June 1976 No. 79 relating to the control of products and consumer services (the Product Control Act).

[24] At the outset, the geographical scope of the acts is restricted to Norwegian territory. (As to the basis in international law for extending the application to the continental shelf, see Section 5.2.) Specific provisions are needed to make the acts applicable to offshore activities. All the acts mentioned here contain such specific provisions.

the regulations issued pursuant to them) in order to establish the total picture of the offshore safety regulation. At the outset these decrees (and regulations) can be understood and implemented without regard to the act or acts constituting their formal basis.

This also provides help in relation to the numerous governmental bodies involved:[25] by means of the central decree (see further in Subsection 5.4), the Petroleum Safety Authority Norway (PSA) is appointed the main state body in the field of offshore safety, co-ordinating the activities and responsibilities of the other state bodies administering the various pieces of legislation in the field. Its role is extensively discussed in the chapters comprising Part III of this book.

5.3.3. *Simplified Approach*

The conclusion so far is that although the legal and administrative basis for the state's offshore safety management is very complex at the outset, it boils down to a few components when we turn to the practical aspects of state engagement. To get a general understanding of the main issues we can concentrate on one set of Norwegian regulations and one governmental body: the 2010 Framework Regulations and the Petroleum Safety Authority.

5.4. THE 2010 FRAMEWORK REGULATIONS AND THE PURSUANT REGULATIONS

5.4.1. *The Structure of Regulations*

The basic legal framework for offshore safety is laid down in the Royal Decree 12 February 2010 relating to health, safety and environment in the petroleum activities and at certain onshore facilities (the *Framework Regulations*).[26] Pursuant to

[25] The various acts, decrees and regulations relevant to offshore safety are administered by different ministries, directorates and other governmental bodies. Parallel to the heterogeneous set of acts etc., we therefore find a similar heterogeneous set of legal entities administering the sector. This complex picture is simplified by the Petroleum Safety Authority being made a coordinating body for all other governmental bodies involved in HSE aspects of offshore – and to some extent onshore – petroleum activities; ref. item 2 and 3 of the Royal Decree of 19 December 2003 No. 1592 establishing the Petroleum Safety Authority.

[26] The inclusion of 'certain onshore facilities' in the scope of the decree – and consequently in that of the adjacent regulations – is the main reason that the whole set of decrees and regulations was amended in 2010 (from the previous version of 2001) following a lengthy process of deciding the extension of 'offshore' HSE to onshore processing plants closely linked to the offshore production facilities, and the related consequences for the powers of the enforcement agencies. The means of delimitation is simply that of explicitly referring to named existing onshore facilities (such as Kårstø, Sture, etc.); see the Framework Regulations section 6 litera f.

this decree,[27] four subordinate regulations have been issued by the competent directorates.[28] Each of these regulations covers a separate aspect of safety issues – Management, Facilities, Activities and Technical and Operational Matters – and each of them except the latter one is laid down jointly by all the four involved directorates. While the Management and Technical/Operational regulations contain provisions relevant in all phases and all aspects of the operations, the scope of the Activities and the Facilities regulations is defined according to the distinction between operations as opposed to state of matters.

The Framework Regulations provides exactly that: a framework for the offshore safety regulation. It defines the common scope of application for all the regulations, their common purpose and definitions, who is to be responsible for complying with all the regulations, and the common main principles for health, safety and environment, including what is labelled 'health, safety and environment culture' (Sect. 15). Within this framework, the four regulations spell out in some detail what is required in each of the specific areas.

The common *scope of application* for all the regulations is generally defined by reference to Sect. 1–4 of the Petroleum Act, with some adjustments following from the parallel provisions contained in the other acts upon which the decree is based (see Subsection 5.3.1).[29] This implies that the basic criterion is whether the matter in question is 'petroleum activit', defined in the Petroleum Act as 'all activities associated with subsea petroleum deposits', and further defined by examples of such activities (Sect. 1–6 (c) and (e)–(i)). We need not go further into this; for our purpose it suffices to note that the decree covers all aspects of activities that reasonably can be considered to have such a link to offshore petroleum activities that they may be relevant for safety in that activity,[30] including activities related to certain facilities onshore.

27 Formally speaking this is not quite correct, as the Framework Regulations do not constitute the legal basis for the four subordinate regulations. For undisclosed reasons, they are all legally rooted directly in (most of) the statutory provisions that also form the legal basis for the Framework Regulations – as well as in the provisions of the Framework Regulations that empower the relevant directorates to issue regulations (Section 68, first subsection, litera b). From a legal perspective the latter basis would suffice, and it would also underline the hierarchical structure that is intended between the decree and the regulations.

28 They are The Petroleum Safety Authority Norway, the Climate and Pollution Agency, Norwegian Directorate of Health and the Norwegian Food Safety Authority. This reflects the fact that the regulations are based on the abovementioned acts, the enforcement of which is vested in the four different state agencies.

29 This method of defining the scope of application is rather complex. It may result in the various parts of the decree (and hence of the four regulations) having different scope of application depending on which act must be considered to form the formal basis for the provision in question. On the other hand, this can hardly be avoided, assuming that (1) the decree has to base itself on several acts, and that (2) these acts define their respective application on offshore matters differently.

30 Again, the various types of vessels engaged in petroleum activities create problems (ref. 2 above). The Framework Regulations Sect. 4 second para. state that 'The following are exempt from the

The *purpose* of the Decree and the four regulations is threefold (Sect. 1 of the Decree): To further 'a high standard' for safety, achieve 'a systematic implementation' of measures to fulfil safety requirements and objectives, and 'further develop and improve' safety standards. While the first and last of these objectives aim directly at the fundamental objective of any safety legislation, the second objective points in a slightly different direction: in order to be able to 'systematically implement' relevant measures to comply with safety requirements, the responsible party has to establish an administrative and organisational structure for this purpose. This objective thus constitutes the inception of the internal control system (Subsection 5.1.2 (b)).

Note should also be taken of the separate objective of developing and improving safety standards. This explicitly stated dynamic element of the regulations is natural in the sense that society's views on acceptable safety standards are indeed changing over time. Therefore, the regulations should reflect that the required safety level is not static. But the dynamic objective also introduces some legal challenges, at least if it is supplemented by operational requirements to the same effect – which it actually is.[31] At a given point in time it may be difficult to establish the relevant requirement, and consequently establish a basis for enforcement, ultimately in the form of criminal sanctions.

Finally, the *responsible parties* are defined. The Framework Regulations state that 'the operator and others participating in the activities are responsible pursuant to these regulations' (Sect. 7, 1st para.). The term 'others' covers both companies and individuals, implying that anyone engaged in 'petroleum activities' has to observe the requirements of the decree and the four regulations. But among this lot, the operator carries a special responsibility: it shall 'ensure that anyone who carries out work on his behalf, either personally, through employees, contractors or subcontractors, complies with the requirements stipulated in the health, safety and environment legislation' (Sect. 7, 2nd para.). By this clause, the operator is defined as the central actor in the play, and its internal control system is defined as a key factor in making the safety legislation effective (see further in Subsection 5.6.3 and Section 5.7).

Working Environment Act and provisions in these regulations, which are laid down in pursuance of the Working Environment Act: a) supply, standby … with vessels … and other comparable activities which are considered shipping, b) vessels carrying out construction … in the petroleum activities'. As explicitly stated, these exemptions only apply to the provisions of the decree which are founded in the Working Environment Act, the reason being that the Act calls for such exemptions. However, the Petroleum Act does not, implying that those parts of the regulations that are laid down on the basis of that act may be applied to 'activities which are considered shipping', to the extent that is allowed under Sect. 1–6 litera d defining 'facility' under the Petroleum Act. This exemption thus illustrates the complexity added by the fact that the Framework Regulations are founded in several acts having slightly different scopes of application. The situation is somewhat confusing, as it has been in this area ever since the first safety regulations were issued in 1967.

[31] See e.g. Sect. 10, 2nd para. of the decree: 'A high standard of health, safety and environment shall be established, maintained and further developed'.

5.4.2. *Direct Safety Requirements*

Requirements specifically prescribing 'do's and don'ts' to achieve an acceptable level of risk are not easily found in Norwegian legislation pertaining to the offshore sector. Rather than prescribing methods to be employed, the legislation defines what is to be achieved – the results that the legislation and the administering authorities are aiming at. Such *function requirements* leave it for the responsible party to choose a method that is likely to achieve the required result.

> Therefore, the preference for function requirements is closely linked to the emergence of the internal control system: it becomes vital that the responsible party is in a position to be able to make the correct choices within the function framework defined by safety legislation. This internal prescription of norms is one element of internal control, the other being the act of checking that these norms are complied with and the introduction of possible corrective measures; see further in Subsection 5.6.1.

(a) The safety legislation provides numerous examples of function requirements. On the very top level, the general provision on offshore safety is contained in the Petroleum Act: 'The petroleum activities shall be conducted in such manner as to enable a high level of safety to be maintained and further developed in accordance with the technological development' (Sect. 9–1). The chapter containing general provisions states that 'Petroleum activities according to this Act shall be conducted in a prudent manner and in accordance with applicable legislation for such petroleum activities. The petroleum activities shall take due account of the safety of personnel, the environment and of the financial values which the facilities and vessels represent, including also operational availability' (Sect. 10–1, 1st para.).

Three observations may be made regarding these provisions. First, the level of safety is not well defined. There is not very much help to be found in knowing that 'a high level of safety [shall] be maintained' and that 'due account of the safety of personnel' shall be taken. Admittedly, the subordinate regulations offer some specification, as we shall see later, but the basic problem remains: The required level of safety is by no means precisely defined – and can hardly be. Second, whatever the required level may be, it shall be 'further developed' to keep up with technological development. This dynamic aspect is essential, but it makes it even more challenging to identify the required level at any given point in time. Finally, the Act explicitly makes it a distinct end to secure 'operational availability'. In this way, the traditional concept of safety has been given a broadened meaning: even if no harm is done to persons, environment or installations, it would contravene the industry's *safety* obligations under the Petroleum Act if an unplanned occurrence renders the installation unable to fulfil its purpose, potentially entailing all consequences of contravening safety requirements.

(b) On a general level, the Framework Regulations offer some further guidance on the standards contained in the Act: the petroleum activities shall be carried out in a safe and prudent manner 'based both on an individual and an overall assessment of all factors of relevance for planning and implementation of the activities as regards health, safety and the environment', and 'Consideration shall also be given to the specific nature of the activities, local conditions and operational assumptions' (Sect. 10, 1st para.). Further, Sect. 11 provides, inter alia, that safety assessments 'shall be carried out during all phases of the petroleum activities' and that if there is 'insufficient knowledge concerning the effects that the use of technical, operational or organisational solutions can have on health, safety or the environment, solutions that will reduce this uncertainty, shall be chosen'. But these provisions are obviously far too general to be of much help in the day-to-day business of planning, constructing and running offshore installations in a safe manner.

Some help can be found in the more specific provisions of the Framework Regulations and the subordinate regulations. However, most of them are function requirements, albeit on a more detailed level. One illustration is a provision on loads and resistance: 'Accidental loads and natural loads with an annual probability greater than or equal to $1\mathrm{x}10^{-4}$ shall not cause the loss of a main safety function' defined as, inter alia, 'maintaining the main load carrying capacity in load-bearing structures until the facility has been evacuated'.[32] There are no legally binding directions as to how this end shall be reached (but there are guidelines directing the user to accepted standards; see further (c) below).

This is also the general picture: the regulations scarcely offer any detailed guidance in the form of *legally binding* and *generally applicable* rules on how the required safety standards shall be met. In the *individual* case, however, the regulations may be supplemented by administrative decisions which may contain further details and even specific methods to be applied in order to achieve the prescribed functions. This detailing in the form of individual norms is an aspect of the state's safety control, see further in Subsection 5.4.4.

(c) In the legally *non-binding* form, extensive help is given to the responsible party in the form of guidelines and recommendations attached to the regulations.[33]

[32] Ref. the Facilities Regulations Sect. 11, 1st para. cf. Sect. 7, 2nd para. (b).

[33] The aforementioned provision on accidental loads is accompanied by an official guideline recommending the use of inter alia several specified NORSOK standards for the purpose of designing the structure in such a way that the function requirement is met.

The Facilities Regulations alone make reference to and recommend the use of a total of some 100 standards issued by some 12 different institutions in Norway and worldwide; the similar figures from the guidelines for the other regulations are somewhat lower. Each of the standards generally offer comprehensive and detailed guidelines and recommendations within their specific scope, usually describing methods to be used to achieve the results prescribed by the function standards rather than just detailing the results. In other words, there is no lack of detailed norms, not even in the form of detailed methods to be applied. But they are all *non-binding*.

This does not imply that the recommended standards are irrelevant from a purely legal perspective (from a practical perspective – e.g. to the engineer designing the facilities – they are of course highly relevant). The Framework Regulations state that 'When the responsible party makes use of a standard recommended in the guidelines to a provision of the regulations, as a means of complying with the requirements of the regulations in the area of health, safety and the environment, the responsible party can normally assume that the regulatory requirements have been met' (Sect. 24, 1st para.). The wording may seem less stringent than what is normally found in royal decrees. But there is little doubt that the provision implies that by implementing the recommended standards, the party responsible is in full compliance with the relevant regulation, unless specific circumstances strongly indicate otherwise. In this sense, the standard takes the role of a regulation. But not in the opposite meaning: the party responsible is in principle free to choose another way to achieve the prescribed results than the method given by the standard – such attitude will not necessarily amount to a breach of statutory obligations inherent in the function requirement of the regulation. But in this instance, the party responsible 'shall be able to document that the chosen solution fulfils the regulatory requirements' (Sect. 24, 2nd para.). An important indication when determining whether the chosen solution actually fulfils the requirements is of course the general impression left by the non-implemented standards.

(d) In summing up, most of the direct safety norms contained in the Norwegian regulations are function requirements, leaving it to the party responsible to decide how the described results are to be achieved. But this freedom of choice is in practice limited because the official comments to the regulations recommend certain standards to be applied. The party responsible carries the burden of proof that this method is as good as the recommended one if he elects to deviate from it.

An important effect of this system is that there are no exhaustive regulatory requirements that relieve the industry from employing their best know-how in trying to achieve an acceptable safety level. This in turn means that safety requirements are not becoming static in the same way as if detailed do's and don'ts had been spelled out in the regulations. It also tends to place the responsibility for safety where it belongs – with the industry itself. But the system is totally dependent on the industry being able to undertake such a central and – to some extent – independent role in safety management. And the state would not do its part of the job properly if it did not see to it that the industry (and any party responsible) actually was in a position to play its role properly.

Finally, it should be noted that other regulatory regimes differ in their regard for and reliance on private standards and recommended practices, such as the U.S. regime discussed by Baram in Chapter 7. This leads us to the indirect safety norms.

5.4.3. *Indirect Safety Requirements*

While the direct safety requirements are directly aiming at improving the level of safety, by function requirements accompanied by recommended standards, the indirect safety requirements are dealing with matters that in turn may have this effect, but which are not aiming at having a direct impact on safety. If drilling operations require approval of a plan presented by the operator, the likely effect is that drilling becomes safer: in preparing the plan, alternative equipment and operational procedures have to be considered, decided and described, increasing the probability that the best choices are made and that they are subsequently properly implemented when drilling. Thus, the risk of damage is likely to decrease. But the direct approach specifying how the drilling actually should be performed is likely to have a stronger impact on safety level than just requiring plans to be approved. Similarly, an obligation to report on maintenance work may have an indirect safety effect, whereas an obligation to perform certain maintenance work would potentially have a direct impact on safety.

The indirect safety norms may well be used in parallel with the direct norms. And there is little reason not to do that: although the indirect norms may seem less potent than the direct norms as a means to achieve safety, they widen the range of tools available for the purpose. Rather than putting all bets on one horse – the detailed 'do's and don'ts' that are likely to affect safety directly – the safety regulation also requires the industry to establish procedures and expertise for *handling and complying with* the direct safety requirements.

These indirect norms are also 'do's and don'ts': the operator shall set up his organisation to certain standards, employ qualified personnel, check and report his activities, plan and apply for approvals and so on. But although the legal tool basically is the same, the different object of regulation implies that the indirect norms operate on another level. They impose two important obligations on the industry that do not follow from the direct norms. First, the industry shall establish a system that enables it to comply with the safety requirements in all its operations. It is not left entirely to the industry's discretion how this should be done – it has to carry out its own safety management in the required way. This requirement results in the internal control system (see further in Subsection 5.6.3). Second, there shall be a formalised link between this internal control system and the state safety management. The industry shall, by legally binding obligation, establish a system for effectively relating to the state authorities that enforce safety regulations. This constitutes an important element in state safety control, in that various types of input from the industry itself in turn form a basis for the authorities' check that safety regulations are complied with. For more on this, see Section 5.7.

In this way, the indirect safety norms play a role in safety management in two important ways: they establish the legal basis for the internal control system, and they constitute the legal link between this and state safety control. The safety regulations contain numerous examples of indirect safety requirements. Some illustrations: 'The responsible party shall prepare and retain material and information necessary to ensure and document that the activities are planned and carried out in a prudent manner. The responsible party shall ensure that documentation demonstrating compliance with requirements stipulated in or pursuant to these regulations, can be provided'.[34] There are several provisions obliging the party responsible to obtain consents and approvals at various stages of operations, and also a provision that generally authorises the Petroleum Safety Authority to decide by regulations or individual decisions that the operator 'shall obtain consent from the Petroleum Safety Authority Norway before certain activities are initiated'.[35] A less general, but not less important, example is the provision that 'In connection with shift and crew changes, the responsible party shall ensure necessary transfer of information on the status of safety systems and on-going work'.[36]

The most general indirect safety requirement is the provision that obliges the party responsible to 'establish, follow up and further develop a management system designed to ensure compliance with requirements in the health, safety and environment legislation'.[37] This is the very basis for the internal control system.

[34] The Framework Regulations Sect. 23, 1st para.

[35] The Framework Regulations Sect. 29, 1st para. Detailed provisions are given in the Management Regulations Sect. 25, which lists ten different activities that need prior consent, e.g. manned underwater operations, major modifications and disposal of a facility.

[36] The Activities Regulations Sect. 32. Failure to transfer such information was the direct cause of the Piper Alpha disaster in 1988; see note 1 to this chapter.

[37] The Framework Regulations Sect. 17, 1st para. Section 12, 2nd para. reads: 'The responsible party shall ensure that everyone who carries out work on its behalf in activities covered by these regulations, has the competence necessary to carry out such work in a prudent manner'. Together, these sections imply that both the organisational structure and the individuals operating it shall be capable of identifying safety norms, complying with them, seeing to it that they are complied with and performing necessary corrective measures. For further discussion, see Subsection 5.6.3.

5.4.4. *General Regulations and Individual Administrative Decisions*

Both direct and indirect safety norms are general – they are prescribed in the form of statutes, royal decrees or regulations which are applicable to any party 'participating in activities covered by these regulations'.[38] We have also seen that the norms are general in the sense that they normally do not offer detailed guidelines directly applicable to a given situation.

There are two types of legal tools that can transform the general provisions into individual requirements tailored to a specific situation and party. First is *exemptions*. The relevant ministries and their supervisory bodies 'can grant exemptions from the provisions stipulated in or in pursuance of' the Framework Regulations, subject to taking into account the enforcing powers vested in other bodies, provided only that 'special circumstances exist' and that a statement from the elected representative of the employees shall be enclosed with the application for exemption if the exemption 'could impact safety and the working environment'.[39] Under Norwegian administrative law, this implies that the supervisory bodies are entrusted with wide discretionary power in deciding whether exemptions should be granted, and if so, under which conditions.

Second is *individual decisions*. The supervisory bodies 'can make the administrative decisions necessary to enforce the provisions stipulated in' the Framework Regulations.[40] Generally speaking, under general administrative law principles such decisions cannot amount to more burdensome requirements to the industry than those contained in the regulations; they can only specify and give details within the borderline of the general provisions of the regulations. Nor can they modify requirements contained in the regulations – such decisions have to take the form of 'exemptions' and comply with any restrictions applied to such decisions. But in practice, these types of administrative decisions occur simultaneously, and then the scope may be wider. A typical example is approvals and consents for which the operator has to apply at certain stages of its operations. In deciding whether approval shall be granted, the supervisory body may consider the option of granting an exemption from a regulatory provision, in combination with a condition that tightens the requirements that already follow from another regulatory provision.[41]

38 Ref. the Framework Regulations Sect. 6 litera a defining 'the responsible party' in relation to the regulations.

39 Ref. the Framework Regulations Sect. 70.

40 Ref. the Framework Regulations Sect. 69.

41 If and to which extent the granting of exemptions may widen the authority to issue related individual decisions that exceed the requirements contained in regulations is a complex and difficult question of general administrative law.

Together, the provisions on exemptions and individual decisions result in the regulations becoming flexible: The requirements may be adjusted to individual circumstances, and experience gained may be reflected in the operational requirements without necessarily having to engage in the demanding process of amending regulations. This flexibility is also an important prerequisite in the correlation between the state's safety management and the industry's, which is discussed further in Section 5.7.

5.5. STATE SAFETY CONTROL

5.5.1. *The Objects of Control and the Means of Controlling Them*

Norway does not restrict itself to influencing safety by laying down legal requirements in the form of direct and indirect safety norms. State safety management also includes checking that the party responsible actually complies with the norm.

The safety control depends on which type of requirements comprise the basis for the control.[42] Checking compliance with direct safety requirements differs fundamentally from monitoring compliance with the indirect norms, simply because the factual objects of the exercise are so dissimilar. We can draw a distinction between direct verification and indirect supervision.

(a) The *direct verification* deals with the industry's actual adherence to specific safety requirements: Are the valves tight, the level of corrosion acceptable, the load carrying structures sufficiently strong, the subsea operations safe, the crane operator qualified? This type of supervision is a form of 'hands on' check of the activities and state of matters that are likely to have a direct impact on the risk for damages inherent in the industry. It follows the tradition of state involvement in industrial activities ever since the industrial revolution, and it also used to form the major part of state offshore safety control in the early days – prior to the emergence of the internal control system.

(b) The *indirect supervision* monitors and audits the industry's system for controlling its own activities, that is, the internal control system. As we shall see in Subsection 5.6.3, the industry is required to document its internal control system. This documentation gives a strong indication of whether the

[42] The term 'control' is not precise. It may mean just the passive checking, or it may denominate the active steering – taking control. When describing state safety control, it is preferable to make a distinction between these two. The passive control implies to verify or audit the industry's operations, whereas the active steering requires that the passive control is followed by administrative decisions directed towards the industry. See further Subsection 5.4.4 on administrative decisions linked to control.

required system is in place and working, which makes it a relevant object of safety control.[43]

But having a structured system for internal safety management in place is one thing; quite another is having this system actually lead to the activities being performed in compliance with relevant requirements. In order to verify this, the state control will have to look into the primary activities themselves: Does the internal control system work in that the valves actually are tight, and so forth? In order for the indirect supervisory type of control to be an effective tool to ensure safety standards, it therefore has to be supplemented by the direct verification of the technical details that the internal control system is designed to handle – simply to safeguard that the system is actually working as intended.[44]

Consequently, there is a close link between the two types of state control – once the internal control system has been introduced and thus forms a natural object of state control itself. But the balance between the two has shifted: the development is towards increased weight being attached to the indirect supervision. This is explained by the emergence of function requirements. Along with a changed method for prescribing safety norms – from specific methods to general goals – the state's system for supervising safety has changed. The fundamental need is to ensure that the industry is able to make adequate choices within the wide boundaries defined by the function requirements and that these choices are actually implemented in the activities – that is, that the internal safety management works.

(c) Both direct verification and indirect supervision may in principle take two forms: The control could be restricted to *just monitoring* what is happening, or it could be a part of a *go-stop-go system*, implying that certain milestones in the operations cannot be passed unless the controlling authority – for example, the Petroleum Safety Authority – has positively concluded that relevant requirements are met. Usually this is combined with a requirement that the industry at such milestones shall apply for approval or consent to proceed, and that the application shall be supported by plans, information, documentation and the like allowing the Petroleum Safety Authority full insight into present situation as well as planned activities to the extent relevant to safety issues. Obviously, such approval system is more effective

[43] See the Framework Regulations Sect. 67 2nd para.: the Petroleum Safety Authority 'will carry out supervision of the management systems established pursuant to these regulations and will make the decisions necessary to implement provisions regarding the requirements for the administrative parts of the management systems'.

[44] An illustrative example of these alternatives is expressly stated in the Framework Regulations Sect. 67 last sentence: 'Within their respective areas of authority, the supervisory authorities can order the operator to carry out verifications itself, or to have such verifications performed by others.'

in the sense that the 'burden of proof' is placed on the industry, while the monitoring system implies that the Petroleum Safety Authority itself has to pick up all relevant information and decide to act on that basis. On the other hand, the go-stop-go system unavoidably means that there will be 'stops' awaiting approvals and other actions by the monitoring authority, which could be most disturbing to a rational progress of operations. In this respect the 'looking over the shoulder' monitoring system has its benefits as it allows industry to continue operations until positively stopped by the control authority.

(d) The actual physical control work *need not be performed by state employees*. State control may be based on input provided by others, for example, classification societies like Lloyd's or Det norske Veritas, following *their* physical control. Although this system is much more developed in the area of maritime safety, it is also important in offshore safety.[45] Far more important, however, is the basis for state control that is provided by the industry itself, based on the work of its own employees or classification societies engaged by industry. The internal control system naturally includes that physical checks have to be performed and reported by the operator's own personnel (e.g. on the status of a high pressure pipeline in the process module). State control may ensure that this is properly done by checking the same item itself. Or it may limit itself to checking that the internal control system is established (i.e. indirect supervision), and then use the output of the system – the reports form the operator's own checks – as a basis for state control of the physical items. The fact that this is in practice a frequent basis for state control makes it even more crucial that the internal system works properly – and consequently calls for an intensive indirect supervision.

(e) *Summing up*: Along with the emergence of function requirements, the state control has moved into supervising the quality of the industry's internal control system, including spot-checking that this system actually picks up and rectifies non-compliance. State control is not concentrating on directly verifying that safety requirements regarding physical aspects of the operations are complied with. Instead, this kind of control is based on the output of the internal control system – further emphasising the importance of this system being adequate and operational. The risk inherent in this approach is to some extent reduced by the implementation of go-stop-go approval systems, leaving it for the industry to convince the state control authorities at certain milestones that operations are and will remain safe.

[45] On the UK continental shelf, the use of classification societies has been formalised in the concept of the 'Safety Case' and the 'Certifying Authority' certifying that relevant requirements have been met.

5.5.2. *Enforcing Safety*

Neither prescribing safety norms nor controlling that they are complied with will alone result in reduced risk for damages if the industry does not comply with the requirements. There is also a need for means for enforcing the requirements.

The Petroleum Act (and the other relevant acts) and the various regulations provide two types of such means of enforcement: the administrative and the criminal sanctions. The imposition of sanctions is based on the positive provisions of the safety legislation combined with the general principles contained in administrative and criminal law.

Under general administrative law, a licence may be revoked if the licensee commits a major breach of conditions upon which the licence is based. The violated condition need not be express for a breach to have this effect, but it is of course a prerequisite that the rule violated can be fairly and objectively attributed to the licence. In relation to safety aspects, this implies that even if the compliance with safety requirements has not specifically been made a condition for an approval, licence, exemption or other type of individual decision, the non-compliance with safety requirements may result in the revocation of the beneficial administrative decision. In this way, general principles and rules of administrative law supplements shortcomings that may exist in the rules on revocation that are found in the safety legislation. Similarly, the Criminal Code is applied to the offshore petroleum activities by general reference,[46] which implies that the specific criminal sanctions contained in the safety legislation are supplemented by general rules.

Neither the Framework Regulations nor the detailed regulations issued pursuant to it contain any specific provision on sanctions. The Framework Regulations limits itself to making a general reference to the acts that form the legal basis for the regulations: 'Provisions with regard to penalties and other sanctions contained in the legislation relating to health, environment and safety are applicable to violation of provisions stipulated in and pursuant to these regulations' (Sect. 62). As the Petroleum Act contains the most general provisions in this respect, we disregard the other acts in the discussion which follows.

Approvals, licences and so forth may be revoked if provisions stipulated in or pursuant to the Petroleum Act are violated seriously (in which case a single violation is in principle sufficient) or repeatedly (in which case neither of the violations in principle need be serious).[47] The power vested in the supervising governmental bodies by this provision is restricted by the general administrative law 'principle of proportionality' applied to an evaluation of the nature of the violation, what will

[46] Criminal Code Sect. 12 (1) (a)-(c), making the act generally applicable to activities taking place on installations engaged in petroleum exploration or production and located on the Norwegian continental shelf, or activities within the safety zones established around such installations.

[47] Petroleum Act Sect. 10–13.

be gained by a revocation and what will be the negative consequence of it to the licensee.

Revoking a licence – or even an approval – may under the circumstances be a rather drastic reaction. A more flexible alternative, introduced by the Petroleum Act Sect. 10–16, is therefore using 'coercive fines': the issuance of an administrative order (e.g. that the operator should take or refrain from taking certain actions for safety reasons) may be linked to a threat that violation of the order will result in a daily fine, payable for as long as the violation lasts. This does not constitute a criminal sanction, just an administrative strong pressure on the operator.

Finally, the Petroleum Act Sect. 10–16 also empowers the supervisory bodies to order a halt in operations that impose a safety hazard – that is, those that violate general or specific provisions of regulations or individual decisions. This, however, must be considered to be a direct means of ensuring an acceptable level of safety rather than a means of enforcing compliance with safety provisions that merely indirectly aim at reducing risk.[48]

The Petroleum Act authorises criminal sanctions in the form of penalties or imprisonment for wilful or negligent violations of provisions stipulated in or pursuant to the act. Imprisonment is naturally not relevant in relation to corporate bodies, but they may, on the other hand, be subject to very substantial fines even if no individual physical person acting on the company's behalf can be fined, for example, because they cannot be proven to have acted negligently (or wilfully).[49]

5.6. INDUSTRY SAFETY MANAGEMENT

5.6.1. *Overview*

Like the state's safety management, the industry's safety management consists of two elements: the prescription of safety norms, and the safety control checking that the norms are complied with and instigating corrective measures if need be.

Obviously, the safety norms that the industry itself defines and implements in its own activities have to observe the requirements that are laid down in the acts and regulations pertaining to the activities, as well as in the individual administrative decisions issued on that basis. It may seem less obvious, however, that the control activities also should be governed by statutory requirements. In relation to the state, it might be assumed that the industry's primary obligation is to comply with direct safety norms, potentially having a direct effect on the safety level in the activities,

[48] The same goes for the Safety Representative's right to halt operations that he considers pose an immediate danger to life and health of employees; ref. the Working Environment Act Sect. 6–3.

[49] See the Criminal Code Sect. 48a.

and that it would be for the industry to decide which means – in the form of safety control or other – would seem feasible to secure such compliance. But as we have seen from the preceding discussion, an important element in the state's safety management is indeed the safety control carried out by the industry itself – the internal control. Consequently, there is a need to direct the content of the internal control by stipulating legal requirements to that effect.

Thus, both elements of the industry's safety management have a legal compulsory basis. But they differ, both in terms of structure and detailed content.

5.6.2. *Internal Prescription of Safety Norms*

The direct safety norms prescribed in acts, regulations and individual administrative decisions normally leave room for several alternative lines of action that will all be in compliance with the requirements. This is not just a necessary consequence of rapidly changing technology and the like; it is a deliberate means of forcing the industry to implement safety aspects into all daily activities without being tempted to simply lean on predefined regulatory solutions. The extensive use of function requirements is a consequence of this approach, and so is the effect of placing the burden of activity on the industry by means of implying an approval system rather than a monitoring system (Subsection 5.5.1 (c)).

This system implies that the industry has to identify the limits of freedom, and then to define its preferred alternatives within these limits. This amounts to internal prescription of norms. Often, material elements consist of mere reference to various standards issued by (in most cases) private institutions and referred to in the guidelines that are attached to the regulations (see Subsection 5.4.2 (c)).

The internal norms may surface in internal procedures, technical project specifications, operating manuals and the like. These documents will in turn constitute relevant objects for state safety control: state supervisory bodies evaluate the solutions and may react by requiring amendments to be made. More likely, the choices made by industry will be reviewed by the supervisory bodies in the context of the industry's applications for the various approvals, permits and licences needed to conduct the petroleum activities.

5.6.3. *Internal Safety Control*

While there are no specific provisions requiring the industry to perform 'internal prescription of norms' (the need for this activity simply follows from the fact that the legally binding safety norms are not precise), there are several provisions requiring the industry to establish a system for internal control.

The general obligation follows from the Petroleum Act Sect. 10–6:[50] 'The licensee and other persons engaged in petroleum activities comprised by this Act are obliged to comply with the Act, regulations and individual administrative decisions issued by virtue of the Act *through the implementation of systematic measures*'.[51]

The single purpose of the required activities is to ensure that relevant requirements contained in the act and other legislation are complied with. The establishing of this system is a requirement of its own: the obligations contained in Petroleum Act Sect. 10–6 are not fulfilled by complying with prescribed safety norms. And conversely, the obligation derived from this provision is not necessarily fulfilled by complying with all direct safety norms contained in acts and regulations.

More detailed requirements on the system for internal safety control are given in the Framework Regulations. The party responsible 'shall establish, follow up and further develop a management system in order to ensure compliance with' the safety legislation.[52] On verifications, which are important elements of safety control, the Framework Regulations Sect. 19 provides that 'The responsible party shall determine the need for and scope of verifications, as well as the verification method and its degree of independence, to document compliance with requirements in the health, safety and environment legislation'. The actual verification 'shall be carried out according to a comprehensive and unambiguous verification programme and verification basis'.

The Framework Regulations Sect. 23 add to this by requiring that the party responsible 'shall ensure that documentation demonstrating compliance with requirements stipulated in or pursuant to these regulations can be provided'. Compliance on this point means that both internal prescription of norms, the checking that they are complied with and any corrective measures following such check are well documented – which in turn implies that the internal control system must constitute a fully developed administrative structure within the organisation of the party responsible, though not necessarily (and in practice not) a separate part of the organisation. Sect. 18 points in the same direction, by obliging the industry to ensure that parties to contracts have the qualifications necessary to fulfil *their* obligations under the safety legislation (as such parties are themselves a 'responsible party'; see Subsection 5.4.1), and further that the contractual parties actually comply with these obligations during their performance of work under the contract. Again, these requirements mean that the internal control must constitute a systematic administrative approach to safety.

[50] Similar provisions are not contained in the pollution act, the working environment act or in the products control act, but all of them provide legal basis for issuing regulations requiring internal control systems to be implemented.

[51] Emphasis added.

[52] Framework Regulations Sect. 17, 1st para.

Albeit that these requirements do not constitute a detailed cookbook, they definitely provide guidance beyond the general requirement that there shall be internal control.

5.7. THE LINK BETWEEN STATE AND INDUSTRY SAFETY MANAGEMENT

In the preceding description of safety regulation in Norwegian petroleum activities, it will have emerged that there are two players – state and industry – and that they both run a system for 'safety management' that comprise two main elements: the prescription of safety norms and the control that these norms are complied with. It will also have shown that these two systems for safety management are not operating independently of each other. In this concluding discussion, we shall look closer at the interaction between these elements by placing them into a four square matrix (see Figure 5.1).

The simple interrelation is that the state safety requirements (1) define the limits within which the industry's choices have to be made, and that state safety control (2) checks that these choices are kept within these limits – and that they also are desirable from the supervisory body's point of view. This is rather banal. The interesting aspect is what happens if the control reveals that changes should be made to the industry's approach. As a matter of principle, this happens in two fundamentally different situations. One is that the state control reveals that industry has not complied with the regulatory requirements. This may lead to sanctions being invoked in order to enforce the safety norms (see Subsection 5.5.2). The other situation is that the control reveals that the industry's choices are legally indisputable, but still not desirable. In this case, the control is likely to result in the norms being modified (3). This could take different forms: the regulations could be amended, or – more likely – individual decisions could be made to the effect that the operator in question is left with a more narrow room for manoeuvring without changing the general regulatory provisions. Such detailing or specifying of the content of the regulations in turn means that the operator has to make new choices within the boundaries of legal requirements – the applicable safety norm has become more precise and the previous choice is no longer within its limits. So the operator chooses (1), the state controls (2), the norms that may be further refined (3) by new administrative decisions based on the findings and evaluations of the control, and so on – until the supervisory body is satisfied with the operator's choices or time has run out because of the progress of activities. This circle of interrelated activities can be labelled '*prescription of norms through control*'. It has the obvious potential effect of ensuring that an acceptable level of safety is achieved. The means is to restrict the area of flexibility left to industry by, for example, function requirements, based on insight gained by control.

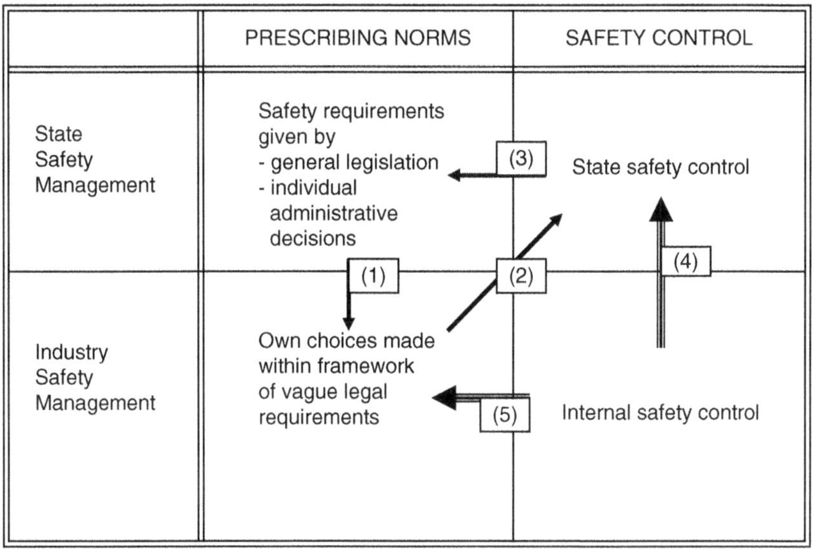

FIGURE 5.1 State and industry safety management.

So far, the circle does not involve the industry's internal control. This is by no means a sign that this box in the matrix is less important. On the contrary, the internal control is a crucial part of safety management in both regimes.

In relation to *state* safety management, the internal control provides an important input and basis for state safety control (4) in that reports and the like flowing from the industry's own control activities in practice quantitatively constitute the most central input – more central than supervision and verification carried out directly by state inspectors. This means that the quality of the internal control and its output have to be secured. Therefore, the internal control also constitutes a separate object of state control in order to ensure that it can serve the intended purpose. The internal control is made both a basis and an object for state safety control.

In relation to the *industry* safety management, the request for an internal control system means that the industry is forced to establish a structured system for managing safety issues. Inherent in this system are also the technical and organisational prerequisites for making the right choices within the flexible framework defined by, inter alia, function requirements – and indeed also the prerequisites for identifying such flexibility. Consequently, the internal control as it is required under the safety legislation is a crucial factor in the industry's internal prescription of norms (5).

Rather than falling outside the scheme of safety management, we can conclude that the internal control system is a major contributor to the scheme, both on state and industry sides.

6

Health and Safety Regulation on the UK Continental Shelf

Evolution and Future Prospects

John Paterson

6.1. INTRODUCTION

Health and safety at work in the offshore oil and gas industry has moved up the political agenda once again, this time not only because of immediate concerns for the well-being of the workforce but also because of a growing realisation of the connection between safe operations and environmental protection. In the aftermath of the Macondo disaster in the Gulf of Mexico in 2010 in particular, the political debate has in part been about the appropriate *orientation* of such safety regulation. This is because the United States, perhaps uniquely among developed hydrocarbon producers, had maintained its faith in a detailed prescriptive regulatory approach, whereas most others had shifted to a goal-setting safety case approach following the Piper Alpha disaster on the UK continental shelf in 1988.

In such circumstances, it is timely to revisit the rationale for the United Kingdom's shift in regulatory orientation and to consider how the safety case approach has performed after a decade and half of operation. Accordingly, this chapter first of all traces the evolution of offshore safety regulation in the United Kingdom with particular emphasis on the nature of the shift from the approach under the Mineral Workings (Offshore Installations) Act 1971 and its related regulations to that heralded by the Offshore Safety Act 1992 and implemented by subsequent regulations. Thereafter, it examines experience with the safety case regime, especially the reasons for the 2005 reforms and the implications of the critical 2007 asset integrity report from the Health and Safety Executive's Offshore Safety Division. The chapter also considers the implications of the October 2011 proposals for offshore safety regulation from the European Commission. Insofar as these appear not to threaten the United Kingdom's approach in the way that some had feared given the Commission's initial Communication and especially the European Parliament's Resolution, both published in October 2010, the political debate in Europe might be perceived to have been resolved in favour of the status quo.

This chapter argues, however, that both the 2007 asset integrity report and the findings of the UK Parliament's Energy and Climate Change Select Committee suggest that while the safety case approach may remain the best option among the alternatives (and especially detailed prescription), questions remain as to whether it is as yet being *implemented* as well as it might. The chapter concludes less with definitive answers to the question of implementation and more with indications of the direction in which those answers might lie and of the multidisciplinary approach that would appear to be required.

6.2. THE EVOLUTION OF OFFSHORE HEALTH AND SAFETY REGULATION ON THE UNITED KINGDOM CONTINENTAL SHELF

If there is a recurring theme in the discussion of health, safety and environmental regulation in the context of the global oil and gas industry, it is that the effectiveness of such regulation is compromised by an inevitable tension between safety and cost which is commonly resolved in favour of cost. The argument to the effect that industry managers under pressure to bring projects in on time may be prone to cut corners in relation to safety is not one that is difficult to understand and is one that appears again in the context of the Macondo disaster (Bergin 2011). Less obvious can be the influence of the regulatory orientation itself: to what extent does the way in which health, safety and environmental regulation is set up either allow or even encourage this kind of behaviour? The fact is that when hydrocarbons are discovered in any given jurisdiction, there can be a desire on the part of political actors to ensure rapid progress with a view towards electoral advantage, which is not necessarily consistent with safe and clean operations. While this last point may appear to be a reference to what has happened in developing economies (McNeish and Logan 2012), there is an argument that no country is entirely immune from this political dimension to the problem of safe oil and gas operations. There is certainly an argument that this issue was a factor in the early stages of hydrocarbon activities on the UKCS (Carson 1980).

6.2.1. *1964–1971: De Facto Self-Regulation*

The possibility that there might be oil and gas under the North Sea was first recognised when the full extent of the massive Groningen gas field became apparent in the early 1960s. Insofar as it extended offshore from the Netherlands, there was an obvious possibility that other hydrocarbon formations were present there also. The less than spectacular economic performance of the United Kingdom in the 1960s meant that this possibility was greeted with some enthusiasm in government circles. There was accordingly a need to put legislation in place quickly to allow the necessary exploration (and hopefully production) work to go ahead. As such, the Continental Shelf Act 1964 was passed, which vested all oil and gas under the UKCS in the Crown and required anyone who wished to explore for and produce these resources to obtain a licence from the Secretary of State.

The continental shelf was, however, in many respects uncharted legal territory at this time. The United States had been quick to spot the potential for the development of subsea resources and had in the Truman Proclamation of 1947 laid claim to the continental shelf as a natural prolongation of its territory. Faced with the possibility of a multitude of (not necessarily compatible) claims in this regard, the United Nations Convention on the Continental Shelf of 1958 was passed, which set out the rights of coastal states in relation to their continental shelves. The 1964 Act was thus the United Kingdom's effort to lay down the legal regime that would prevail over its continental shelf. In this regard, while the identity of the Secretary of State and the fact that a licence would be required were relatively straightforward matters to resolve, more complicated would be, firstly, the extent to which UK law generally could be applied to the area and, secondly, the content of the licence itself.

With regard to the application of existing UK law, the government discovered that its efforts to achieve this by means of a simple statement to that effect in section 3 of the 1964 Act fell foul of a canon of statutory interpretation to the effect that only laws explicitly stated to extend beyond the territorial waters of the United Kingdom could be applied there (Daintith et al. 1984). This was an early, but significant, indication of the extent to which political ambition was running ahead of a detailed understanding of the novelty of the exercise being embarked upon. As regards the terms of the licence which would be issued to those interested in exploring for and producing hydrocarbons, there was yet another indication of an approach predicated more on speed than on a thoughtful consideration of what was actually required. Here, in the Petroleum (Production)(Continental Shelf and Territorial Sea) Regulations 1964, the United Kingdom simply adapted an existing onshore model petroleum licence that dated from the 1930s in a move that has later resonances in the actions of the governments of developing countries who make minimal adaptations to onshore legal instruments when they are confronted with the need to produce offshore equivalents (Havemann 2011).

That there was reason both to be optimistic about the United Kingdom's prospects as a producer of natural gas and to be pessimistic about its ability to regulate safe operations came in the following year. The summer of 1965 saw the Sea Gem jack-up drilling rig record the first commercial discovery of natural gas on the UKCS. The winter of that same year saw the same rig collapse, with the loss of thirteen lives. The accident was of course a serious blow to the United Kingdom's hydrocarbon ambitions, but quite how serious only became apparent when the Minister of Power attempted to establish an inquiry to examine the causes and to recommend any necessary actions. Insofar as the Sea Gem as a jack-up drilling rig did not fall into any category of vessel recognised by the law, it was in fact impossible for the minister to establish an inquiry. Insofar as an inquiry did actually go ahead, this was only on the basis of the voluntary compliance of all concerned.

This embarrassing situation was only compounded when the inquiry proceeded to examine the adequacy of the regulation of safety under the initial approach constituted by the 1964 Act and the 1964 Regulations. The whole issue of safety was dealt with in a single term of the licence. In this regard the relevant clause required the licensee to comply with such instructions as may be issued in writing from time to time by the minister. For his part, the minister had not been long detained by the question of what his instructions should be. With absolutely no experience of offshore oil and gas operations and with practically no resources for the development or implementation of regulations, his instructions had taken the form of a brief letter telling licensees to comply with a Code of Practice for safe operations which had been recently prepared by the Institute of Petroleum, an industry body (Ministry of Power 1967).

This approach was problematical on a number of levels. First of all, depending so heavily on the licence as the vehicle for the transmission of safety instructions restricted very severely the range of actors the Minister could actually engage with. Insofar as the UK petroleum licence is recognised as possessing a hybrid quality, it is important to recognise not only its regulatory dimension, in which the minister retains the right to exercise certain powers, but also its contractual dimension, in which both the minister and licensee accept obligations (Gordon 2011). It is this latter dimension which caused problems as far as the Sea Gem inquiry was concerned, noting that only the licensee as a contracting party could be subject to the control of the minister. Given that the oil and gas industry was then and is now characterised by a high degree of subcontracting, this meant that the minister would have no ability to control or sanction subcontractors who were failing to comply with his instructions. Secondly, in relation to the sanctions available to the minister in the event that he detected breaches of his instructions, these appeared to be limited to revocation of the licence, an option surely only of relevance in the context of the most egregious acts or omissions on the part of the licensee.

Thirdly, and moving from the licence to the industry Code of Practice upon which the minister placed such reliance, the inquiry was not at all impressed with the quality of its drafting from a legal point of view, describing it as a recipe for unlimited litigation. In fairness to the Code's drafters, they had never set out to produce something that was to be subjected to detailed forensic examination by lawyers, but rather a guide to operations in a new and emerging field. But in the context of an inquiry into a serious accident involving multiple fatalities, the shortcomings stood out sharply. Most serious was the fact that where the personnel on the rig had followed the clear terms of the guidance by mustering on the helideck to await evacuation once the emergency was under way, this action actually led to many of the fatalities insofar as in the context of the collapse of the installation the appropriate response was to take to the life rafts, of which there were sufficient on board for all concerned (Ministry of Power 1967).

Problems such as these led the Sea Gem Inquiry to recommend a significant shift in the regulatory orientation away from the hands-off approach that had characterised the government's involvement under the 1964 Act and 1964 Regulations and towards a code of statutory authority with credible sanctions. In other words, if regulatory options were to be considered in relation to their position on a spectrum, the initial approach would lie close to the end representing complete self-regulation while what the Inquiry was now proposing would lie close to the opposite end representing detailed prescriptive regulation. While this shift is understandable in the context and appreciation of the manifold weaknesses of the initial approach, it is nevertheless interesting to reflect on the inquiry's own apparent recognition of the difficulty, if not indeed the impossibility, of what it was asking the government to do: the inquiry explicitly recognised that, given the novelty of the operations and their ongoing rapid development, generalisations could be inapt and even dangerous (Ministry of Power 1967). And yet insofar as the inquiry was asking for a detailed prescriptive code, generalisations were exactly what it was asking the government to produce.

6.2.2. 1971–1992: *Detailed Prescription*

That the problem identified at the end of the preceding section was by no means only theoretical can perhaps be detected in the length of time it took for the legislation that had been urgently called for by the Sea Gem Inquiry to be passed by Parliament. Whereas the Inquiry's Report had been published in 1967, the legislation intended to implement its recommendations did not appear until 1971 and was not in force until the following year. When Parliament came to debate the Bill that would in due course become the Mineral Workings (Offshore Installations) Act 1971, there was cross-party support for the initiative. The analysis produced by the inquiry was not questioned, but rather wholeheartedly endorsed. Insofar as there was any dissent, it took the form of expressions of concern that more detail was not included in the Bill. The government's response to this was to point to the evolving nature of the industry and the fact that it would be difficult and inappropriate to provide too much detail at the level of primary legislation. The Bill was, therefore, envisaged very much as a framework, enabling the Secretary of State in due course to fill in the detail at the level of secondary legislation. The benefit of this approach would be that when, as would inevitably happen, technology moved on and regulations required to be updated, this could be done more quickly and efficiently insofar as they were contained in statutory instruments rather than in an Act of Parliament.

If it was anticipated, however, that the necessary detail would be filled in relatively soon after the coming into force of the Act, then this expectation was to be disappointed. The full set of regulations required to provide the comprehensive code called for by the Sea Gem Inquiry was not in place until 1980.[1] Furthermore, substantive health and safety regulations did not appear until 1976 and, perhaps surprisingly, the final regulations produced were those dealing with well control. The significance of this substantial delay in producing a detailed prescriptive approach to offshore health and safety becomes fully apparent when it is considered that during this period some of the United Kingdom's major first-generation production platforms were designed, constructed and commenced production. The Forties field began production in September 1975, Brent in November 1976, Piper in December 1976 and Ninian in December 1978.

It is not difficult to accept that those who claim that political and economic imperatives dominate safety concerns in the oil and gas industry may have a point (Carson 1981). It is also important to consider who precisely had been given the task of producing these detailed regulations. While responsibility shifted during the years as government departments were renamed and realigned, it was always the case that the department responsible for health and safety offshore was also responsible for petroleum licensing. For the majority of the period during which the 1971 Act and its subsidiary regulations were in force, these responsibilities fell to the Petroleum Engineering Division of the Department of Energy. Given this collocation of responsibilities, it is again not difficult to accept arguments which point to a problematic balancing of safety and cost.

As if problems such as these were not enough to call into question the wisdom of the detailed prescriptive approach which had been requested by the Sea Gem Inquiry and enthusiastically endorsed by Parliament, there is in fact one more issue that appears to hammer a final nail into the coffin. Recognising that health and safety at work was a general problem in all industries, a government-sponsored inquiry had been established under the chairmanship of Lord Robens to consider the matter in its entirety. Reporting in 1972, after the passing of the Mineral Workings (Offshore Installations) Act but before it had entered into force, the

[1] The regulations introduced under the 1971 Act in this period were: Offshore Installations (Registration) Regulations 1972 (SI 1972/702); Offshore Installations (Managers) Regulations 1972 (SI 1972/703); Offshore Installations (Logbooks and Registration of Death) Regulations 1972 (SI 1972/1542); Offshore Installations (Inspectors and Casualties) Regulations 1973 (SI 1973/1842); Offshore Installations (Construction and Survey) Regulations 1974 (SI 1974/289); Offshore Installations (Public Inquiries) Regulations 1974 (SI 1974/338); Offshore Installations (Operational Safety, Health and Welfare) Regulations 1976 (SI 1976/1019); Offshore Installations (Emergency Procedures) Regulations 1976 (SI 1976/1542); Offshore Installations (Life-saving Appliances) Regulations 1977 (SI 1977/486); Offshore Installations (Fire-Fighting Equipment) Regulations 1978 (SI 1978/611); Offshore Installations (Well Control) Regulations 1980 (SI 1980/1759).

Robens Committee effectively called into question all the assumptions that had underpinned the Sea Gem Inquiry's, and subsequently Parliament's, analysis of the problem of and solutions to occupational health and safety. Rejecting a detailed prescriptive approach, the Robens Committee concluded that this was really part of the problem rather than the solution. Robens thus suggested that there was simply too much such regulation: a whole series of industries were subject to lengthy statutes which attempted to provide in detail how each should operate safely. As a result of the progress of technology, much of this law was simply out of date and irrelevant. Furthermore, many industries were looked after by their own regulator, leading both to a duplication of effort and also inconsistencies in approach. Finally, the fact that the law purported to provide such detail in statutes and regulations served in effect to persuade employers and employees alike that health and safety were actually the government's responsibility, rather than something they should worry about (Robens 1972).

In view of this very comprehensive critique, it is not surprising that Robens recommended a radically different approach. His idea was to do away both with detailed, industry-specific statutes and regulations and with industry-specific regulators and to replace them with a simple set of broadly expressed duties and a general regulator. His belief was that such an approach would help to place responsibility for health and safety where it properly lay – with those who created and managed workplace risk. Employers would thus bear the responsibility for identifying the hazards affecting their workplace, assessing the risks and taking such steps as would reduce those risks to the lowest practicable level. Legislation, regulation and indeed the regulator itself would all exist essentially to facilitate an effective self-regulatory approach (Robens 1972).

Notwithstanding the radical nature of the proposals, these were rapidly accepted by the government and found expression in a comparatively short time in the Health and Safety at Work, etc. Act 1974. Employers (and indeed employees) were subject to new general duties and to the oversight of a new dedicated regulator, the Health and Safety Executive. While this set in motion the reform of the regulation of occupational health and safety in onshore industries, the question remained of how, if at all, this new Act would affect the offshore industry which had only so recently been made the subject of precisely the sort of legislative and regulatory arrangement of which Robens had been so critical. Robens had not addressed the 1971 Act comprehensively, merely indicating that there was in principle no reason why in due course the offshore industry should not be brought within the ambit of the new approach. Perhaps because of the newness of the 1971 Act regime and the novelty of the industry, there was no enthusiasm for wholesale change, but in retrospect the approach adopted looks less than ideal. What actually happened was that most, albeit not all, of the 1974 Act's provisions were extended offshore and were implemented alongside those of the 1971 Act and its subsequent regulations by the dedicated offshore regulator.[2]

The problems for the Petroleum Engineering Division (PED) of the Department of Energy in managing the implementation of two diametrically different regulatory regimes should have been obvious from the outset, but even if that were not the case, an opportunity to revisit this arrangement occurred in the late 1970s in the context of the Burgoyne Committee (Burgoyne 1980). This had been set up in the aftermath of the Ekofisk blowout in the Norwegian sector, in order to consider whether this event raised any questions about the United Kingdom's regulatory approach. This committee was certainly presented with evidence of the problems that the PED was facing, for example, the difficulty in meeting the Sea Gem Inquiry's recommendation that there should be a code of statutory authority: while the Inquiry had bemoaned the inadequacy of the language of the industry Code of Practice from a legal point of view, the PED now complained of the difficulty of making regulations understandable to the industry. In addition, the complexity of the task the PED had been handed of producing a comprehensive and detailed set of regulations was discussed. For example, despite the considerable length of time between the passing of the 1971 Act and the point at which the full set of regulations was in place, the PED was still able to describe one key regulation as having been introduced too hastily and without sufficient consultation.

Despite these problems, the Burgoyne Committee did not recommend any significant changes to the status quo. As regards the specific problem of the PED trying simultaneously to implement two different regulatory approaches, it merely recommended that communication be improved between the PED and the HSE in order that the former had access to the expertise of the latter in relation to occupational health and safety, it being recognised that the PED's strengths lay in the engineering dimension (Burgoyne 1980). And so an opportunity to sort out what was undoubtedly a very difficult if not indeed impossible regulatory situation was missed. Perhaps Burgoyne was impressed by the evidence he received from the industry that it was confident that it had arrangements in place sufficient to prevent disaster in the UKCS and to deal with any emergency should it ever arise.

Matters remained like this until the Piper Alpha disaster in 1988 in which 167 men were killed, making this the worst accident in the history of the offshore industry globally in terms of lives lost. The unprecedented scale of this disaster was a profound shock to industry and regulator alike and prompted the establishment by the government of a further inquiry into its causes under the chairmanship of the senior

2 Health and Safety at Work, etc. Act 1974 (Application Outside Great Britain) Order 1977 (SI 1977/1232). The PED carried out the HSE's inspection function under an agency agreement between the HSC and the DEn.

Scottish judge Lord Cullen. The inquiry conducted by Cullen was at the time the lengthiest and most comprehensive in UK history. It produced a comprehensive two-volume report which detailed the causes of the accident and made no fewer than 106 recommendations designed to ensure that there would be no repetition (Cullen 1990).

Cullen's findings in relation to the causes of the disaster were not only a damning indictment of the way in which safety had been dealt with on the platform and by the operator, Occidental, but also of the regulatory approach and indeed of the regulator. Thus, Cullen found that the immediate cause of the accident was a failure of the permit to work system which ensured that maintenance of safety critical elements on the platform was only conducted on the basis of explicit permission, which in turn was known to those responsible for the operation of the equipment in question. On this occasion incomplete maintenance on pipework was not communicated to operations staff at a change of shift, and thus hydrocarbons were allowed to flow into pipework that was not secure. The ensuing leak led to explosions and fires. The safety systems designed to respond to such emergencies were actually destroyed by the explosions. The Offshore Installation Manager was found to have failed to take the actions required to save lives. The operator was found to have an inadequate approach to risk, to have inadequate safety procedures and even where such procedures existed it was found that they were not followed. Both the installation's stand-by vessel and a dedicated fire-fighting vessel which happened to be nearby proved ineffective when confronted with the sort of emergency it had been assumed they would be able to respond to. In view of this litany of inadequacy, it is perhaps not surprising to discover that Lord Cullen also found that the regulator had conducted only superficial inspections of the platform which were not well adapted as a test of safety.

Faced with such comprehensive evidence that what had been happening on the UKCS was very far from the reassuring picture that had been presented to the Burgoyne Committee, Cullen did not hesitate to recommend wholesale reform of the regulatory regime. Out would go the prescriptive regulations developed under the 1971 Act, which Cullen believed (in an echo of Robens) actually served to persuade people that safety had been taken care of by government rather than thinking for themselves. In would come an approach which would finally see the ethos of the 1974 Act applied in an unalloyed way to the UKCS. Noting the undoubted complexity and ongoing evolution of the technology involved, Cullen essentially proposed that the precise means by which the law would facilitate and support the fulfilment of the general duties of the 1974 Act would be by setting broad goals in dedicated offshore regulations and then requiring operators to demonstrate how they were going to achieve those goals by means of a Safety Case – that is, a document which literally made the case that the design, construction and operation of the installation was safe, so far as was reasonably practicable. In this regard he was influenced by existing practice onshore in the nuclear industry (see Tromans 2010) and in relation to the regulation of industrial major accident hazards, itself inspired by European law.[3] Such a demonstration would first of all involve a Formal Safety Assessment to identify the hazards to which the installation and those on it were exposed. It would then be necessary to indicate how the risks (that is, the probability of those hazards occurring and their impacts) had been calculated, using Quantitative Risk Assessment (QRA) if necessary. Finally, it would be necessary to explain how those risks had been or would be avoided or mitigated. While the preparation of the Safety Case would be the responsibility of the operator, Cullen recognised that those involved in the actual processes and working with the actual equipment on the installation would often have the most relevant information regarding its safe operation. He, therefore, also recommended workforce involvement in the preparation of the Safety Case. Insofar as the regulator was satisfied that the case that the operation of the installation was safe, then it would be accepted and the installation would be permitted to operate on the UKCS. As regards the identity of the regulator, given Cullen's findings in relation to the status quo at the time of the Piper Alpha disaster, it is not surprising that he finally recommended the ending of the special arrangements which the offshore industry had enjoyed heretofore and sought the transfer of responsibility from the PED to the HSE (Cullen 1990).

6.2.3. *1992–Present: The Safety Case Approach*

Nor was the government in any mood to question any of these radical suggestions. All 106 recommendations contained in Cullen's report were accepted following its publication in 1990, and by the following year the legislation[4] required to repeal the majority of the 1971 Act and the regulations required to establish the safety case regime were in place, marking a very stark contrast with the considerable delay that followed the recommendations of the Sea Gem Inquiry in 1967. The goal-setting regulations called for by Cullen were also quickly in place.[5] The Offshore Installations (Safety Case) Regulations 1992 required all installations to

3 SI 1984/1902 implementing the so-called Seveso Directive (Council directive 82/501/EEC of 24 June 1982 on the major-accident hazards of certain industrial activities) and now replaced by the Control of Major Accident Hazard (COMAH) Regulations 1999 (SI 1999/743) which in turn implement the Seveso II Directive (Council Directive 96/82/EC of 9 December 1996 on the control of major-accident hazards involving dangerous substances, as amended by Directive 2003/105/EC).

4 Offshore Safety Act 1992.

5 Offshore Installations (Management and Administration) Regulations 1995 (SI 1995/738) as amended by the Offshore Safety (Miscellaneous Amendments) Regulations 2002 (SI 2002/2175); Offshore Installations (Prevention of Fire and Explosion, and Emergency Response) Regulations 1995 (SI 1995/743); Offshore Installations and Wells (Design and Construction, etc.) Regulations 1996 (SI

have accepted Safety Cases by November 1995 if continued operations were to be permitted. This deadline was achieved, again signalling a much greater degree of urgency than was evident at the time of the establishment of the 1971 Act regime. The speed with which a considerable amount of work was done can be read, on one hand, as demonstrating the significantly lower cognitive burden placed on the regulator by the need to draft only goal-setting as opposed to detailed prescriptive regulations and, on the other, as an indication of the comparative ease with which operators accepted the responsibility which they always had under the 1974 Act once the confusion of the contradictory 1971 Act was removed and the structure of the Safety Case was in place.

This relatively rosy picture does, however, need to be nuanced. Firstly, there could be an extent to which the alacrity with which the industry took to the risk-based regulatory approach to safety might be explained by the fact that it dovetailed with the then-current industry initiative to reduce costs (CRINE 1994). In some respects the ability of the regulatory approach to benefit from a more proactive attitude to costs can be seen as a good thing, but it also raises the possibility that the balance between cost and safety can be struck in an inappropriate way. Secondly, while the industry during the Cullen Inquiry had been enthusiastic supporters of the idea of utilising QRA (perhaps because of the cost-saving opportunities that it presented), when the time came to utilise this tool in the preparation of the first Safety Cases, it transpired that even major companies lacked the necessary in-house expertise and had to contract this element of the work out to specialists. As a consequence, questions could be asked as to whether the crucial aspects of the key document in the new regulatory approach were actually owned by those with whom the responsibility actually lay. Thirdly, while the workforce might be involved in the preparation of the Safety Case, there was no requirement for them to be involved with its ongoing review. Cullen had explicitly stressed that the Safety Case had to be understood as a living document, that is, one which was updated on a continuing basis to reflect the specific circumstances of the installation. Thus, if those in the front line of the installation's operation were not part of that ongoing process, there must surely be questions as to the efficacy and indeed relevance of the whole approach (Woolfson et al. 1996).

That the Health and Safety Executive was alert to these issues is evident both from the fact that it was willing to repeal and replace the original Safety Case Regulations in 2005 and from the fact that it has carried out a series of specific programmes of inspection and review when it perceives the emergence of problems which may have industry wide significance.

1996/913); Pipelines Safety Regulations 1996 (SI 1996/825); the Diving at Work Regulations 1997 (SI 1997/2776); the Lifting Operations and Lifting Equipment Regulations 1998 (SI 1998/2307).

In relation to the changes brought about by the Offshore Installations (Safety Case) Regulations 2005, these first of all sought to reduce the level of bureaucracy associated with the Safety Case process – not least because the HSE had discovered that the process of three-yearly resubmission was subject to the law of diminishing returns (HSC 2004a, 2004b). The three-yearly resubmission obligation was thus replaced with a less demanding five-yearly "thorough review".[6] The new regulations also removed the need for separate safety cases for combined operations, design, or decommissioning. In the case of the first two, the separate safety case is replaced by a simpler notification procedure,[7] while in the third it is now only necessary to modify the existing safety case – although this must still be accepted by the HSE.[8] The 2005 regulations also sought to deal with the question of the adequacy of workforce involvement in the safety case by requiring that the document should summarise consultation with the workforce both in relation to its preparation and its subsequent revision and review.[9]

Concerns in relation to the industry's ability to make appropriate use of quantitative risk assessment were dealt with not in the new regulations, but rather in guidance from the HSE that appeared shortly afterwards (HSE 2006a). The regulator observed that the original safety case regulations focused attention on QRA, often requiring the involvement of specialist consultants. While the HSE saw this as having been useful in the immediate aftermath of the Piper Alpha disaster, it believed that by 2006 the understanding of offshore risks was mature. It was, therefore, of the view that risk assessment should focus on adding value and be management-owned rather than consultant-owned. To achieve this objective, the guidance pointed out that risk assessment should be proportionate to the complexity of the problem in hand and the magnitude of risk. Accordingly, QRA should only be used where the risk level and the complexity of a problem are high, while qualitative and semi-quantitative approaches are more appropriate for situations involving lower risk and complexity. In case this should be misunderstood as indicating a reduction in the level of responsibility rather than a reduction of the regulatory burden, the guidance goes on to make clear, firstly, that the main purpose of risk assessment is to decide whether more needs to be done to reduce risk and, secondly, that the duty holder must demonstrate that risks are controlled and are not intolerable (HSE 2006a).

6 2005 Regulations, regulation 13. Note that there is an exception to this rule where there are "material changes" which will still require to be accepted (regulation 14).

7 Regulation 10 (Notification of combined operations), regulation 6 (Design and relocation notification for production installation) and regulation 9(1) (Design notification in respect of a non-production installation). Note that the terms "fixed" and "mobile" used in the 1992 regulations are thus replaced in the 2005 regulations by "production" and "non-production" respectively.

8 Regulation 11 (Safety case for dismantling fixed installation).

9 Schedule 2, para. 3. The Offshore Installations (Safety Representatives and Safety Committees Regulations 1989 (SI 1989/971) are consequently amended.

The extent to which the detail of offshore operations is a matter for the operator rather than the regulator under the safety case regime could raise questions as to the role which the latter now plays. Even if safety cases are not public documents (for reasons associated with their commercial confidentiality) (see Steinzor 2011), an impression of the scope that remains for the regulator to influence the way in which the industry operates may be gained from the detailed procedures utilised by the HSE for their assessment and for dealing with thorough reviews. These are published online as the Safety Case Handling and Assessment Manual[10] and include aims, objectives and principles, performance standards, roles and responsibilities and assessment procedures. While it may be the case that the burden of producing detailed regulations is lifted from the regulator by the safety case regime, the manual suggests that the burden of assessing and reviewing what the operators now produce is by no means negligible. As just one example, one may note that the personnel involved in assessing a safety case include the senior management of the Offshore Safety Division, managers of inspections teams, managers of topic teams, a case manager and deputy case manager, a topic assessment manager and topic assessors, safety case coordinators, legal and operational strategy teams, to name only some of those concerned. It may also be asked how the new less formalised approach works in practice in terms of what constitutes a "material change" outside the normal five year period. Here the HSE guidance states that such an alteration "is likely to be one that changes the basis on which the original safety case was accepted" (HSE 2006b). It should also be noted that the HSE can direct a revision of the safety case if it considers that a material change should be made.

As regards the enforcement of the regulations by the HSE, it is clear that insofar as so much detail is now produced by the duty holder rather than the regulator, it is not so much in the position of holding the regulated to standards that it has set, but rather of holding them to "the procedures and arrangements described in the current safety case which may affect health or safety".[11]

6.3. EXPERIENCE WITH THE SAFETY CASE APPROACH

Given the degree of development of offshore health and safety regulation on the UKCS by this point, it could have been assumed that there was in place a robust and mature regime that would not require much in the way of further attention from the regulator, not least because of the fundamental idea that the safety case was to be understood as a living document that would be updated as required throughout the

[10] http://www.hse.gov.uk/offshore/scham/index.htm

[11] 2005 Regulations, reg. 16. The sets out a risk-based approach for assisting the regulator in determining what sort of enforcement action is appropriate in given circumstances, including in the context of a permissioning regime such as the offshore Safety Case http://www.hse.gov.uk/enforce/emm.pdf

lifetime of the installation. The HSE has, however, remained very closely engaged in monitoring the industry's performance, a decision which would appear to be entirely justified given its findings over the past decade during a series of so-called Key Programmes, most notably Key Programme 3.

During a significant operation between 2000 and 2004 to reduce hydrocarbon releases in the offshore industry, the HSE "became increasingly concerned about an apparent general decline in the condition of fabric and plant on installations". As a consequence it established a further initiative focused on the issue of asset integrity, designated Key Programme 3 (KP3), which was to run from 2004 to 2007. KP3 involved the inspection of some 100 installations (or 40 per cent of the total on the UKCS), with the regulator concentrating in particular on the maintenance management of safety critical elements. The HSE defined *asset integrity* as "the ability of an asset to perform its required function effectively and efficiently whilst protecting health, safety and the environment" and *safety critical elements* as "the parts of an installation and its plant … whose purpose is to prevent, control or mitigate major accident hazards … and the failure of which could cause or contribute substantially to a major accident", whilst *maintenance management* in relation to safety critical elements was understood to be "the management systems and processes which should ensure that [such elements] would be available when required" (HSE 2007).

Insofar as there is obviously a very close connection between these issues and the very idea of the safety case as a "living document" designed to ensure the ongoing safe operation of an installation, the regulator's findings during this programme raised some very troubling questions. Regarding maintenance management, there was found to be considerable variation in performance across the industry and even within the same company. Where performance was poor, this was attributable to problems in monitoring which equipment was defective or overdue for maintenance. The HSE found that there was "a poor understanding across the industry of [the] potential impact of degraded, non-safety-critical plant and utility systems on safety-critical elements in the event of a major accident", and that "the role of asset integrity and [the] concept of barriers in major hazard risk control" was "not well understood" – extremely worrying findings given the advances claimed by the industry in the years following the Piper Alpha disaster. One conclusion of the KP3 report that might have been well received by the industry's critics in the 1970s – that monitoring by management tended to focus on occupational safety – was nevertheless seen as problematical by the HSE insofar as this focus served to mask the significance of "major accident precursors" (HSE 2007).

Regarding the overall condition of the infrastructure, the regulator was able to report more positively that structural integrity was "well controlled" and that the main hydrocarbon boundary was "reasonably well controlled", but it remained concerned that other parts of the hydrocarbon infrastructure such as pipes and valves were in decline. Given the facts of the Piper Alpha disaster, this last finding is particularly troubling. It further appeared that periods of reduced oil prices had encouraged deferrals in maintenance that had not been reversed when prices had increased, especially where there were plans to sell on assets in due course. This situation was perceived by the regulator as having had an adverse effect on workforce morale. Finally, but again raising questions as to whether the safety case could really be regarded as a "living document", the HSE also found that there was insufficient testing of safety critical elements leading to diminished reliability (HSE 2007).

In an effort to account for these problems, the regulator highlighted three underlying issues relating to learning, the engineering function and leadership. In relation to learning, there was seen to be a problem both of inadequate auditing and monitoring and of a lack of processes to allow learning to be embedded within the organisations concerned. This is a problematical conclusion given that auditing and monitoring are supposed to be integral parts of the safety management system prioritised in the setting of the safety case. As regards the engineering function, the issue here was its relative strength within companies which was perceived to have declined "to a worrying level". The KP3 report did not specify which other functions engineering had lost out to, but it may be inferred that these are related to the companies' financial operations – a conclusion supported by the third underlying problem identified earlier.

In relation to leadership, the regulator observed that in setting priorities for spending senior management had to balance safety and financial risks, but did not always properly understand the impact on these risks of operating with "degraded [safety critical elements] and safety-related equipment" (HSE 2007). While the United Kingdom's ALARP approach has always explicitly envisaged cost-benefit analysis as a feature of decision making in relation to statutory health and safety duties, this finding would nevertheless appear to indicate that cost factors were able to dominate decision making to such an extent that ALARP could no longer be understood to be the end achieved. That this could happen in the context of the safety case is, of course, even more troubling given its emphasis on risk and safety assessment on an ongoing basis.[12] In light of these problems, it might have been expected that there would be moves to adjust the underlying regulatory approach. That no such changes were forthcoming may be attributable to the fact that a later (more limited) follow-up survey suggested improvements had been made to the maintenance management of safety critical elements in the aftermath of the original KP3 report (HSE

[12] That such an interpretation is by no means unreasonable may be inferred from the comments of the Chief Executive of Petrofac, Ayman Asfari, at the Oil and Money Conference in October 2008, where he indicated that "his company had seen installations which were in bad need of repair", that he feared "firms will fail to spend enough in improvements" and that he was "concerned that the industry would end up in a situation where budgets were curtailed, leading to more risk of accidents." Oil and

2009), and to the fact that a new initiative has been launched which will, among other things, follow up the issues identified until the end of 2013.[13]

6.4. MACONDO AND THE POSSIBILITY OF A NEW EUROPEAN DIRECTIVE

Since the Macondo disaster, however, and the questions which this has raised for existing regulatory approaches globally, it may turn out that the future of offshore health and safety regulation on the UKCS will no longer lie in the hands of the British authorities. It is certainly the case that following the disaster, there were efforts on the part of the UK authorities to reassure the public and the EU that there were no problems with offshore safety regulation. A review announced by the Secretary of State for Energy and Climate Change rapidly concluded that the current regime was "fit for purpose".[14] But it was also clear that there was no desire to appear complacent. The HSE established a Deepwater Horizon Review Group which is reviewing the findings of the investigation into the accident and which will, among other things, "make recommendations as necessary with regard to the control of wells and the safety of the exploitation of offshore oil and gas in the UK".[15] On the side of the industry, Oil and Gas UK set up the Oil Spill Prevention and Response Advisory Group (OSPRAG), bringing together industry, regulators and trade unions "to provide a focal point for the sector's review of the industry's practices in the UK, in advance of the conclusion of investigations into the Gulf of Mexico incident."[16] Irrespective of the conclusions of these initiatives, however, it is interesting to note that the House of Commons Energy and Climate Change Select Committee essentially endorsed the current approach (House of Commons Energy and Climate Change Committee 2011). It may nevertheless be wondered whether its finding that "the UK has high offshore regulatory standards, as exemplified by the Safety Case Regime", and that the UK approach is "superior" to that in force in the Gulf of Mexico at the time of the Deepwater Horizon disaster, is compatible with the finding that the "offshore oil and gas industry is responding

Gas UK disagreed with this sentiment whereas the offshore arm of the RMT Union indicated that this served to confirm their warnings in this regard. See BBC news online at http://news.bbc.co.uk/1/hi/scotland/north_east/7696232.stm. The fact that the offshore industry was criticised by the Chair of the HSE at this time in relation to its accident statistics for problems relating to the "control of potential major incident risks" would tend to suggest that the Chairman of Petrofac and the RMT had a point. See HSE Press Release E039:08, 13 August 2008.

[13] KP4: The Ageing and Life Extension Inspection Programme (2010–2013). See http://www.hse.gov.uk/offshore/ageing/kp4-programme.htm (visited 18 March 2011).

[14] DECC Press Release: PN10/067, UK increases North Sea rig inspections, 8 June 2010.

[15] Details online at http://www.hse.gov.uk/offshore/deepwater.htm (visited 18 March 2011).

[16] Details online at http://www.oilandgasuk.co.uk/knowledgecentre/OSPRAG.cfm (visited 18 March 2011).

to disasters, rather than anticipating worst-case scenarios and planning for high-consequence, low-probability events" (House of Commons Energy and Climate Change Committee 2011) insofar as the safety case approach was specifically supposed to do precisely that.

Similar concerns arise when one reads in the OSPRAG Final Report that its recommendations include those "leading to the creation of the Well Life Cycle Practices and the Oil Spill Response forums as permanent mechanisms to drive the industry forward on a path of continuous improvement" (OSPRAG 2011); the question may be asked, however, whether that sort of continuous improvement was not supposed to be an inherent feature of the Safety Case approach. Even more problematical for the industry and the UK regulatory approach is another of the conclusions drawn by OSPRAG in its Final Report to the effect that "OSPRAG is confident that the UK stands in good stead to ensure that high standards in well design, construction and management will continue to be enforced on the UKCS and that, in the unlikely event that a major uncontrolled well incident occurs, the industry's strengthened contingency plans will allow an effective and robust response." The uncontrolled gas leak at the Elgin field in 2011, which took almost two months to fix, notably did not involve OSPRAG's much-vaunted capping device developed in the aftermath of the Macondo disaster (Oil and Gas UK 2011).

Given such problems, it may be wondered whether the United Kingdom's regulatory regime is really well placed to survive the current attentions of the European Commission, which has (along with the European Parliament) become much more interested in offshore safety since the Macondo disaster. In October 2010, both the Parliament[17] and the Commission[18] indicated that legislative action was necessary at the European level. Whereas there was initially a suggestion that such action might be restricted to an amendment of the Extractive Industries Directive,[19] the eventual proposal from the Commission is that there should be a new regulation.[20] This document has caused consternation in both the United Kingdom and Norway which, though not a member of the EU, would be affected by the Regulation through its membership of the European Economic Area. This situation is very fluid, with indications emerging from discussions with industry and regulators that there have been very strong representations to the Commission about perceived

[17] European Parliament resolution of 7 October 2010 on EU action on oil exploration and extraction in Europe (hereafter "Parliament Resolution").

[18] European Commission, Communication from the Commission to the European Parliament and the Council: Facing the challenge of the safety of offshore oil and gas activities, SEC(2010) 1193 final (hereafter "Commission Communication").

[19] Directive 92/91/EEC.

[20] Proposal for a Regulation of the European Parliament and of the Council on safety of offshore oil and gas prospection, exploration and production activities, COM(2011) 688 final, 27 October 2011.

problems in the regulation, and more recent indications that the Commission is willing to consider legislating by way of a directive.[21]

Key features of the European Commission's initial proposal for a Regulation and the likely impact on the UK

The operator would prepare a Major Hazard Report (MHR) and submit it for assessment to the regulator. The MHR appears to be modelled on the United Kingdom's Safety Case, but would extend this significantly by requiring environmental issues to be included. Insofar as the MHR were introduced in the form envisaged, this would involve the United Kingdom in a very significant upheaval while Safety Cases were rewritten to comply with MHR requirements. Were it to be the case, on the other hand, that the MHR's objectives could be met in other ways, then the Safety Case in conjunction with, for example, Environmental Impact Assessments and Oil Pollution Emergency Plans, could be combined to achieve the desired effect.

The MHR would be based on risk assessment, though minimum standards are set out in Annexes to the proposed Regulation. Insofar as the Commission's approach draws on best practice from Member States such as the United Kingdom, it is explicitly described as risk-based. The Commission has, however, set out minimum standards in Annexes and would have the power under Articles 34 and 35 to amend those annexes unilaterally, potentially introducing a measure of prescription that would conflict with the overall risk-based approach. Insofar as a measure of prescription persists in the United Kingdom (as explicitly envisaged by Lord Cullen), this need not be a problem. The Commission's relative inexperience in relation to offshore safety does, however, raise questions in this regard.

Safety critical elements in the MHR would require independent verification. This is already a feature of the United Kingdom's approach.

Licensing authorities would assess the safety and environmental performance of applicants. This might be controversial in the United Kingdom insofar as it could be read as reintroducing elements of safety regulation to the licensing authority. Insofar, however, as the Secretary of State simply awarded points to applicants on safety and environmental criteria, in the same way that he or she does currently, for example, on criteria related to exploration plans, and insofar as health and safety remained the preserve of the dedicated specialist regulator, no problems need arise.

[21] http://www.europolitics.info/sectoral-policies/offshore-platforms-oettinger-willing-to-bend-art337207-14.html

Member States would establish a National Competent Authority for supervision of safety, environmental protection and emergency preparedness. The United Kingdom has all of these in place. The only question would be whether the Commission envisages these functions being carried out by one regulator in each Member State or whether the United Kingdom's current arrangements involving the Department of Energy and Climate Change, the Health and Safety Executive and the Marine and Coastguard Agency would be regarded as sufficient.

Member States would also be required to make information about offshore activities publicly available. While a great deal of information is available in the United Kingdom, not least in order to encourage investment from new entrants, commercial confidentiality remains a concern. There need not be any problems here insofar as Safety Reports under the onshore Control of Major Accident Hazard Regulations are publicly available subject to considerations of national security and personal and commercial confidentiality.

Oil and Gas UK criticised the proposed regulation on the grounds that while the Commission acknowledged the quality of the UK system and indeed "cherry-picked" elements of it, it was done in a clumsy way that would pose significant problems of bureaucracy, delay and cost on the UK industry and regulator.[22] Oil and Gas UK also commissioned an independent report from GL Noble Denton which exposed flaws in the justification for the proposed regulation including the assumptions the Commission made about the polluting potential of North Sea wells and about the likely reduction in incidents and costs achievable with the new regulation.[23] The Norwegian government criticised the proposed regulation, noting that despite the apparent enthusiasm of the Commission for risk-based regulation, the proposed approach was in fact rather prescriptive, and that deviated from the approach adopted in Norway (and indeed in the United Kingdom) of placing ultimate responsibility on the operator rather than the state.

A main argument was that the proposed regulation left little room for adaptation to existing legal frameworks, and that the new EU measures on offshore safety should be in the form of a directive. [24] After a period of interventions and negotiations among core stakeholders from the industry and unions the European Parliament, the Commission and the Council reached an agreement on a new directive early in 2013.[25]

[22] For details, see http://www.oilandgasuk.co.uk/ProposedEURegulation.cfm.

[23] For details see http://www.oilandgasuk.co.uk/feature1-wireline-feb2012.cfm.

[24] For details, see Norway: Comments to the European Commission proposal for a Regulation of the European Parliament and of the Council on safety for offshore oil and gas prospection, exploration and production activities, December 2011.)

[25] For the draft of the directive current at the time of writing, see http://www.europarl.europa.eu/document/activities/cont/201303/20130314ATT63222/20130314ATT63222EN.pdf

6.5. CONCLUSION

The regulation of health and safety in the offshore oil and gas industry is under more scrutiny at present than it has been for many years. Jurisdictions such as the United Kingdom where it might have been felt that the regime in place represented the state of the art find themselves having to explain and justify risk-based approaches in the face of a European initiative which, for all the purported enthusiasm for risk-based regulation, contains a strong and potentially strengthening flavour of prescription. But even if the internal incoherence of the European approach may ultimately ensure that this threat recedes, the awkward truth remains for the United Kingdom that its own initiatives both before and after the Macondo disaster have revealed apparent problems with the safety case approach. KP3, the Select Committee on Energy and Climate Change, OSPRAG – all, while strongly defending the status quo, have pointed to a gap between the way in which the Safety Case is supposed to operate (as a living document, in proactive mode) and the way it is actually operating in many instances (as a static document, in reactive mode). There is clearly a need for further research into the operation of the Safety Case, but it is surely possible to conclude in common with the UK and Norwegian industry and authorities that a shift to prescription would be a retrograde step. Whatever the weaknesses of the Safety Case, it is surely predictable that moving from a position where there is a responsibility on the operator to identify the hazards facing an individual installation, to quantify the risks and then to demonstrate how those risks will be avoided or mitigated to a position where the regulator assumes responsibility for prescriptive regulation of general application would involve the state in the assumption of risks that it is ill-equipped to deal with. The European Commission in this regard needs to consider whether the possible degree of prescription that it does in the directive (by way of the powers contained in Articles 34 and 35) it can safely say that it knows the answers to all the questions that this form of regulation assumes. Indeed, when one considers the degree of innovation that has characterised the industry over the past few decades and the degree of innovation that will be required to cope with emerging challenges, the Commission needs to consider whether it even knows what all the questions are. Of course, just such a realisation may raise questions of a more fundamental nature about the willingness of society as a whole to tolerate the risks associated with offshore hydrocarbon operations. In the context of the EU's overarching concern with energy security, however, it is difficult to see such questions being answered in the negative.

That, of course, leaves the matter of the necessary reform of the Safety Case. If it remains the best option from among available regulatory orientations, there nevertheless needs to be clearer acknowledgement of the shortcomings that have been thrown up, explicitly or implicitly, by KP3, the Energy and Climate Change Select

Committee and OSPRAG. While detailed answers must await further research, it is nevertheless possible to speculate that insofar as the Safety Case has an explicitly proactive stance, problems in that regard must relate to the implementation of the approach rather than to the approach itself. As reported earlier in the discussions surrounding the repeal and replacement of the Safety Case regulations, the HSE's guidance on risk assessment, and the findings of Key Programme 3, there have been problems with the way in which risk is assessed throughout the period in which the Safety Case has been current. These have ranged from the difficulties encountered in dealing with QRA in-house to the difficulties in quantifying risk when financial considerations come to dominate.

There appear, therefore, to be good reasons for suggesting that a fruitful direction for future research would be into the way in which risk is quantified and the way in which the results are factored into operational decisions. Furthermore, given the multiple dimensions evidently in play, there also appear to be good reasons for the research to be multidisciplinary: the Safety Case may be a legal instrument, but it is one that, almost uniquely, unites engineering, management, operational and financial inputs on an ongoing basis. There is good reason to suggest that our understanding of the functioning of this particular regulatory tool, notwithstanding that it is entrusted with the control of the most significant technological hazards, is really only in its infancy.

REFERENCES

Bergin, T. (2011) *Spills and Spin: The Inside Story of BP*. Random House Business: London.

Burgoyne, J.H. (1980) *Offshore Safety: Report of the Committee*. (Cmnd 7866), HMSO: London.

Carson, W.G. (1981) *The Other Price of Britain's Oil: Safety and Control in the North Sea*. Martin Robson: Oxford.

CRINE (1994) *CRINE: Cost Reduction Initiative in the New Era*. Institute of Petroleum on behalf of UKOOA: London.

Cullen, Lord (1990) *The Public Inquiry into the Piper Alpha Disaster*. (Cm. 1310), HMSO: London.

Daintith, T., Willoughby, G. and Hill, A. (1984) *Manual of United Kingdom Oil and Gas Law*. Sweet and Maxwell: London.

Gordon, G. (2011) "Petroleum Licensing", in G. Gordon, J. Paterson and E. Usenmez (eds.): *Oil and Gas Law: Current Practice and Emerging Trends*. Second Edition. Dundee University Press: Dundee, pp. 65–109.

Havemann, L. (2011) "Environmental Law and Regulation on the UKCS", in G. Gordon, J. Paterson and E. Usenmez (eds): *Oil and Gas Law: Current Practice and Emerging Trends*. Second Edition. Dundee: Dundee University Press, pp. 231–284.

House of Commons Energy and Climate Change Committee (2011) *UK Deepwater Drilling – Implications of the Gulf of Mexico Oil Spill, Second Report of Session 2010–11*, HC 450–1. London.

HSC (2004a) *Proposals to Replace the Offshore Installations (Safety Case) Regulations 1992*, Health and Safety Commission: London.

HSC (2004b) *A Strategy for Workplace Health and Safety in Great Britain to 2010 and Beyond*, Health and Safety Commission: London.

HSE (2006a), *Guidance on Risk Assessment for Offshore Installations*. Offshore Information Sheet No. 3/2006, Health and Safety Executive: London.

HSE (2006b) *A Guide to the Offshore Installation (Safety Case) Regulations 2005*. HSE Books: London.

HSE (2007) *Key Programme 3: Asset Integrity Programme: A Report of the Offshore Division of the HSE's Hazardous Installations Directorate*. Health and Safety Executive: London.

HSE (2009) *KP3 – Asset Integrity: A Review of the Industry's Progress*. Health and Safety Executive: London.

McNeish, J.-A. and O. Logan (2012) (eds.): *Flammable Societies: Studies on the Socio-Economics of Oil and Gas*. Pluto Press: London.

Ministry of Power (1967) *Report of the Inquiry into the Causes of the Accident to the Drilling Rig Sea Gem* (Cmnd. 3409). HMSO: London:

Oil and Gas UK (2011) *Demonstrating the UK's Oil Spill Response Capability: Report on Exercise Sula Tier 2/3 Deployment Demonstration and the Emergency Equipment Response Deployment Exercise*. Oil and Gas UK: London.

OSPRAG (2011) *Strengthening UK Prevention and Response*. Final Report. UK Oil Spill Prevention and Response Advisory Group: London.

Robens, Lord (1972), *Safety and Health at Work; Report of the Robens Committee*, (Cmnd. 5034). HMSO: London.

Steinzor, R. (2011) "Lessons from the north sea: Should 'safety cases' come to America?", *Boston College Environmental Affairs Law Review*, 38 (2): 417–444.

Tromans, S. (2010) *Nuclear Law: The Law Applying to Nuclear Installations and Radioactive Substances in its Historic Context*. Hart: Oxford.

Woolfson, C., J. Foster and M. Beck (1996) *Paying for the Piper: Capital and Labour in Britain's Offshore Oil Industry*. Mansell: London.

7

The U.S. Regulatory Regime for Preventing Major Accidents in Offshore Operations

Michael Baram

7.1. INTRODUCTION

Among the developed countries that regulate the safety of offshore drilling for oil and gas resources, the United States stands out in its dedication to prescriptive regulation and its reliance on hard law enforcement to ensure compliance by industry. Complementing this approach is another distinctive aspect of the U.S. context, namely tort liability and other liability law that are assumed to have the effect of deterring companies from acting negligently and causing harm in their conduct of offshore operations.

Over several decades, drilling activities on the U.S. outer continental shelf (OCS) and subject to U.S. jurisdiction had not incurred a major accident, during which time major accidents occurred under the jurisdiction of other countries such as the United Kingdom and Norway. However, operations regulated by the United States incurred many smaller accidents and an unacceptable level of harm to workers, which in retrospect can be seen as precursors of major accidents to come. Finally, in April 2010, a major accident occurred at the Macondo drilling site in the U.S. sector of the Gulf of Mexico (GOM), killing eleven workers and causing the worst offshore oil spill ever recorded.

Since the Macondo accident, investigations to determine how and why it occurred have produced many recommendations for improving the U.S. regulatory regime. Numerous lawsuits have also been brought by public officials to impose sanctions and by private parties to gain compensatory and punitive damages. In addition, other countries have examined their offshore regimes to determine whether they are capable of preventing a similar accident. These developments have added fuel to the ongoing global discourse on the merits of prescriptive regulation and whether other types of regulation, such as the performance-based, soft law approach taken by Norway and the United Kingdom's safety case model, offer superior approaches to preventing major accidents. Although the United States has recently made several incremental changes to its regulatory regime, it has not diminished its traditional reliance on prescriptive rules and hard law enforcement.

This chapter discusses and evaluates the U.S. regime for preventing major accidents in offshore oil and gas operations before and after Macondo. It begins by reviewing the socioeconomic, technological, political, and other contextual factors that have promoted offshore exploration and production operations and their advance into deepwater regions of the U.S. OCS. This is followed by review of what was documented about the risks of deepwater drilling before Macondo, and a critique of the regime's failure to learn more.

The legal framework for federal governance of offshore operations is then reviewed in a discussion which encompasses several federal laws and regulatory programs that have shaped the safety management systems and activities of companies working offshore. The roles assigned to several regulatory agencies, are delineated, including their rule-making, inspection, and enforcement functions and their reliance on, and adoption of, industry standards and practices. This leads to a focus on four key issues: use of prescriptive regulation, inspection to ensure compliance, adoption of industry standards, and relevant advances in safety science.

The modified regulatory regime that has evolved since Macondo is then described and examined, including the creation of new agencies and the enactment of important new rules requiring the performance of safety management functions by offshore operators, and creating worker rights regarding the safety of operator activities. Commentary follows on the design and implementation of these reforms and whether this attempt to develop a more effective regime for preventing major accidents will prove to be sufficient.

Occasional reference is made to the contrasting Norwegian regulatory regime that fosters company self-regulation. Norway, a major source of offshore oil and natural gas, has continuously sought to improve operational safety, and comparison to its differing approach is instructive.

7.2. CONTEXT

Over several decades, federal programs have promoted exploitation of the oil and gas resources of the U.S. OCS to meet growing energy needs. Numerous exploratory drilling and production operations have been conducted by companies pursuant to leasing and permitting programs devised by Congress and the Department of the Interior (DOI), regulations enacted by the DOI's Minerals Management Service (MMS), additional regulations set by the Coast Guard (CG) and Environmental Protection Agency (EPA), and numerous private standards and recommended practices developed by the American Petroleum Institute (API) and other organizations representing the offshore oil and gas industry.

Most of the seabed areas leased for offshore operations have been confined to sectors of the GOM and regions adjacent to the southern coastline of Alaska. Much of the remainder of the vast U.S. OCS has been closed to leasing and drilling activities by moratoria enacted by Congress and supported by presidential directives.[1] Generally, public opinion has supported the moratoria because of the environmental harms caused by the 1969 blowout and oil spill at the Amoco Cadiz drilling platform off the California coast and the 1988 Exxon Valdez tanker accident and spill in Alaska's Prince William Sound.

In offshore regions not closed by moratoria, federal leasing programs and permits have enabled extensive exploitation activities, especially in the GOM. There, according to MMS, more than 50,000 wells were drilled between 1947 and 2010, and by 2010, 7,000 active leases and operations at 3,600 facilities in the GOM provided 97 percent of all U.S. offshore oil and gas production.[2]

Since the mid-1990s, exploratory and production operations have advanced into "deepwater" regions of the GOM. These deepwater operations (often defined as projects conducted in seawater depths exceeding 500 feet) started to produce more oil than shallow-water operations in 1999. By 2009, nearly 4,000 wells had been drilled in seawater depths exceeding 1,000 feet, including 700 wells in depths exceeding 5,000 feet. In that year, deepwater wells in the GOM accounted for 80 percent of U.S. oil production and 45 percent of gas production offshore.

Several factors motivated deepwater ventures. One is that decades of drilling had depleted oil and gas resources in shallow regions of the GOM, and studies showed far more abundant resources farther offshore in deepwater GOM regions. For example, in 2006, DOI announced that 15 billion barrels of deepwater oil and 60 trillion cubic feet of deepwater gas had been discovered and were available for production, with high potential for discovering and exploiting an additional 86 billion barrels of deepwater oil and 420 trillion cubic feet of deepwater natural gas. It estimated that at current rates of consumption, these resources would be sufficient to replace all oil imports for almost twenty-five years and provide for all U.S. natural gas consumption for more than twenty years.[3]

Another factor is that technological advances enabled deepwater ventures on the U.S. OCS, as they had elsewhere in the North Sea off the U.K. and Norwegian coasts, the vast southwest Pacific, the south Atlantic off African and Brazilian coasts, and the Persian Gulf. New algorithms and methods of seismic imaging were being applied to locate oil and gas buried deeply in the seabed and to evaluate the suitability of

[1] The moratoria prevent drilling activities off the eastern (Atlantic) coast and the western (Pacific) coast of the continental United States.

[2] MMS Rpt., "Increased Safety Measures for Energy Development on the Outer Continental Shelf," U.S. Dept. Interior (May 27, 2010).

[3] "National Assessment Study," U.S. Dept. of the Interior (2006).

their geologic settings for drilling. Huge semi-submersible drilling rigs and new drill ships kept in position by computer-controlled equipment allowed drilling in 10,000 feet or more of seawater, depths far greater than what can be plumbed by traditional "jackup" rigs that need to be fixed to the seabed. Upon reaching the seabed, horizontal drilling made it possible to bore some 15,000 feet or more through various types of geological formations to reach oil and gas resources. The Macondo accident occurred during a well-drilling operation that exceeded 18,000 feet from rig to resource, and a 40,000-foot rig-to-resource operation off Qatar has been reported.

In addition, a mix of social, economic, and political factors in the United States motivated deepwater ventures. These included ever-increasing societal demand for fuel, the expectation that more exploitation of domestic sources would reverse the trend to higher fuel prices, concerns about the reliability of fuel imports from Middle Eastern and other troubled foreign sources, and the desire for "energy independence." In addition, lobbying by the offshore industry and states bordering the GOM emphasized that deepwater drilling would provide billions in fees and royalties annually to federal and state governments, as well as tens of thousands of jobs.

As a result, in the decade preceding the Macondo accident, Congress enacted economic incentives for deepwater projects, including royalty relief for companies with deepwater leases and suspension of company obligations to pay royalties on production.[4] Also, after many years without a major accident, opposition to offshore drilling had waned with a 2008 poll showing that 74 percent of the public supported more offshore drilling, That year, President George W. Bush proposed an end to moratoria and initiation of leasing programs for deepwater regions along the Atlantic coast and the GOM.[5]

Indications that deepwater activities posed new technical problems were ignored, such as the difficulties experienced by Chevron at its "Tahiti" site in 4,000 feet of GOM seawater and the Montara accident and spill off the Australian coast. Many assurances were given that the risks were minimal: for example, that drilling within 200 miles of the U.S. coast "had a 99% safety record," that "only .001% of the oil produced had been spilled," and that natural seeping, shipping, and runoff from land caused more contamination of the oceans than oil spill incidents.[6] Complacency about deepwater risks had set in, and on March 31, 2010, President Obama announced his plan to open up, lease, and exploit the closed regions of the GOM, the Atlantic OCS, and the Chukchi and Beaufort Seas off the northern Alaska coast.[7]

4 "Overview of U.S. Legislation and Regulations Affecting Offshore Natural Gas and Oil Activity," U.S. Dept. of Energy (2005).

5 Bush declaration, at http://www.whitehouse.gov (July 14, 2008).

6 A. Cline, "Environmentalists Say Yes to Offshore Drilling," *Wall Street Journal* (July 12, 2008); D. Lynch, "Deepwater Oil Fields are a Final Frontier," *USA Today* (June 23, 2008).

7 "Obama Unveils New Offshore Drilling Plan," *Environment Reporter*, 41 ER 724 (April 2, 2010).

Three weeks later, on April 20, 2010, the devastating accident occurred on the BP-leased Deepwater Horizon rig during deepwater drilling at the Macondo site in the GOM. The blowout, explosion, and fire killed eleven workers and injured sixteen others, and destroyed the semi-submersible mobile drilling rig owned by Transocean, which had operated the rig under BP supervision. The event ruptured the exploratory well casing at several points, which led to a high-volume release of oil and gas and several failed efforts to stop and contain the release. The uncontrollable discharge continued at a rate exceeding 40,000 barrels of oil each day for 87 days before a temporary cap on the main release point on the seabed proved to be successful. Two relief wells were then drilled and other measures taken for permanent control.[8]

The spill contaminated a large region of the GOM and the coastal areas of several states, with major impacts on wildlife, tourism, commercial fishing, and other social and economic activities. The response to the emergency involved efforts by federal and state agencies and BP to stop and contain the spill and disputes about whether CG or BP had the leadership role. Use of chemical dispersants and various methods for skimming and containing the spill proved to be of limited value, and exposed the temporary workers involved to chemicals and other risks. Environmental and occupational health agencies brought into the response effort were unprepared.

Temporary moratoria were immediately ordered by DOI on existing and new deepwater operations in the GOM, and presidential plans to repeal moratoria on other OCS regions were dropped. Investigations by several federal agencies, a presidential commission, Congressional committees, BP, and industrial and technical associations were conducted and produced various inconclusive findings about the technical, human, and managerial factors which contributed to the accident. Some investigations continue and are leading to more conclusive findings, and criminal prosecution.

Numerous lawsuits have followed by persons and organizations that claim personal injury, property damage, and economic loss, and a special federal board was created to hear, evaluate, and compensate worthy claims with funds provided by BP.[9] Federal and state governments have also brought lawsuits for damages and cleanup costs, and are imposing monetary penalties and other sanctions on BP and other firms involved in the Macondo enterprise.

According to financial analysts, BP's monetary loss, apart from the destruction of its own property and potential profits and any claim by Transocean regarding destruction

[8] C. Cleveland, "Deepwater Horizon Oil Spill," in *Encyclopedia of Earth*, at http://www.eoearth.org

[9] See, for example, Deepwater Horizon Court-Supervised Settlement Program at http://www.deepwaterhorizonsettlements.com

of the Deepwater Horizon rig it leased to BP, are likely to exceed $30 billion.[10] This estimate is based on BP's funding of the settlements being made by the specially created federal board handling thousands of claims for damages, other settlements being negotiated between BP and its business partners and contractors, and the potential outcome of a major trial in a federal district court. The trial involves damage claims by numerous non-settling parties and penalties sought by the federal government and five states for violations of several laws. In that trial, BP is joined as a defendant with several partners in its venture, Andarko, MOEX, and Mitsubishi, by rig owner and operator Transocean, and by BP's major contractors including Halliburton and Cameron. It is expected that these defendants will reach out-of-court settlements regarding their respective shares of the sums to be paid to all other parties.[11]

Among other consequences of the accident, DOI immediately terminated MMS and transferred its leasing and regulatory functions to newly created successor agencies, the Bureau of Ocean Energy Management (BOEM) and the Bureau of Safety and Environmental Enforcement (BSEE).[12] DOI and these agencies have enacted new guidelines and regulations intended to improve safety, which are discussed later in this chapter. Congress also considered whether the monetary penalties to be imposed for future spills should be increased. In addition, the findings and recommendations made by the Presidential Commission[13] and other organizations such as the Deepwater Horizon Study Group (DHSG)[14] and the National Academy of Engineering (NAE)[15] may prompt further regulatory actions.

Despite these developments, support for vigorous resumption of offshore operations soon reappeared. Offshore companies brought lawsuits seeking to enjoin and invalidate DOI's temporary moratoria and new regulations. Governors of states bordering the GOM and Alaska pressed for renewal of offshore drilling, and earlier socioeconomic and geopolitical arguments for drilling were revived because of turmoil in the Middle East and continued increase in the price of fuels. In addition, a consortium of major oil companies developed spill containment equipment for deepwater operations.

10 "The Deepwater Horizon Disaster: Insurance Market Impacts," Insurance Information Institute (June 2, 2010).

11 "Federal Trial Set to Decide Responsibility, Other Issues in Deepwater Horizon Case," 43 ER 496 (March 2, 2012); "BP Oil Spill Trial Delayed for Settlement Talks," *Reuters* (February 27, 2012).

12 Secretarial Order n. 3302, U.S. Dept. of Interior (June 18, 2010).

13 Final Report, National Commission on the Deepwater Horizon Oil Spill and Offshore Drilling (January 11, 2011) (http://www.oilspillcommission.gov/), and staff working papers. (often referred to as the "Presidential Commission").

14 Final Report on the Investigation of the Macondo Well Blowout, Deepwater Horizon Study Group, University of California (March 1, 2011), http://ccrm.berkeley.edu/deepwaterhorizonstudygroup/dhsg_reportsandtestimony.shtml

15 *Evaluating the Effectiveness of Offshore Safety and Environmental Management Systems*, National Academy of Engineering Transportation Research Board (June 2012), http://www.nap.edu/catalog.php?record_id=13434

Enticing claims of deepwater treasure were also made to persuade Congress not to overreact to the Macondo accident by limiting deepwater operations. For example, huge estimates of natural gas hydrates, which are expected to be commercially exploitable in the near future, were presented. According to a former federal official, close to 320,000 trillion cubic feet of gas hydrates can be captured from deepwater regions of the federal OCS, including more than 7,000 trillion cubic feet from deepwater sediment in the GOM. His testimony before a Senate committee further claimed that production of merely 1 percent of this resource would generate payments to the federal government of approximately $7.5 trillion, which, when added to the $4.5 trillion in prospective payments from exploiting DOI's estimates of deepwater oil and gas, would produce a total amount of federal revenue that "almost completely pays off the current national debt without raising taxes."[16]

These were among the factors that caused the federal government to reverse course. In 2011, DOI authorized resumption of offshore operations that it had suspended immediately after the BP accident, and again began to issue permits for new drilling operations on previously leased portions of the GOM seabed. Congress signaled its approval by refraining from enacting new restrictive laws and penalties. President Obama expressed confidence that offshore operations could resume safely and promised his support for rescinding moratoria and accelerated development of offshore oil and gas in new deepwater regions of the GOM and other regions off the Atlantic and north Alaskan coasts.[17]

Thus, by early 2012, deepwater drilling had resumed pursuant to the requirements of the newly modified regulatory regime. Among major developments, Shell Oil paid $4 billion for leases off the Alaskan coast and secured several of the permits it needed for exploratory operations, and BP and other firms were again drilling at numerous sites in the GOM. Similar developments also occurred under the aegis of other countries (Canada, United Kingdom, Norway, and Australia), which also claimed that lessons learned from the Macondo accident and review of their regulatory regimes indicated that deepwater operations would be safely conducted.

Nevertheless, there is continuing concern about the safety of deepwater and forthcoming Arctic operations and their potential for incurring major accidents. Investigations of the Macondo accident and other OCS incidents identified deficiencies in the U.S. regime and company safety management practices, systemic risks arising from production pressures, difficulties in management of multi-company drilling ventures, and the uncertainties and extreme physical conditions of operations in deepwater and Arctic regions.[18]

[16] W. J.Coleman testimony, Hearings on The Risky Business of Big Oil, Senate Committee on Judiciary, U.S. Senate (June 8, 2010).

[17] State of the Union address (January 24, 2012).

[18] See discussion in reports at notes 13,14 and 15 supra.

7.3. RISKS

OCS operations pose risks to workers, the environment, and a broad range of public and private interests. Of most concern are major accidents on drilling rigs, which cause deaths and injuries, and spills which contaminate ecosystems and property. But the occurrence of lesser accidents and near-miss incidents is also important to consider because these events are often precursors of major accidents and may be indicative of systemic risks. Thus, analyses of these events and other performance indicators are essential for the learning process that regulators and industry need to improve safety, as discussed further in several chapters of this volume.

Unfortunately, available information is fragmented and incomplete because the U.S. regime had not developed a comprehensive national compilation of data and studies. This has been the case at least since 1985 when a federal study found that "the lack of data makes it difficult to evaluate the level of safety achieved by oil and gas operators, safety-related equipment, and Federal regulation."[19]

The reason for this deficiency is that the two regulatory agencies responsible for ensuring offshore safety over several decades, MMS and the CG, had failed to systematically collect and evaluate such information. Doing so would have enabled them to continuously engage with industry and individual operators in learning processes to promote continuous and timely improvements of operational safety, as done in Norway and discussed in other chapters. This failure was attributable, in part, to the regulatory disarray that has characterized the relationship between the two agencies for many years, their tolerance of unreliable and incomplete reporting of safety-related events by operators, and their excessive delegation of responsibility for improving safety to the American Petroleum Institute (API), which is hostile to manadatory federal regulation, as subsequently discussed.

A month after the Macondo accident, a time of intense focus on MMS, the agency disclosed fragmentary information on blowouts and spills to support defensive assertions about the safety of OCS operations it had regulated: for example, that during the period of expansion of deepwater drilling since 1996, the blowout rate per well had not increased; that spills in the GOM resulting from blowouts from 1971 to 2010 were "not major"; and that only 30.3 barrels were spilled per million barrels produced. However, it acknowledged that the rate of spills had increased since the mid-1990's as deepwater activities increased, indicating "significant challenges" for preventing deepwater spills.[20]

The agency, however, had additional information that indicated a more problematic situation and had revealed this information in 2009 when it proposed a rule on

[19] *Oil and Gas Technologies for the Arctic and Deepwater*, Office of Technology Assessment, U.S. Congress (1985), 104.

[20] Note 1 supra.

"safety and environmental management systems" (SEMS) for OCS operators. It had justified that proposal with findings derived from its investigations of thirty-three OCS accidents between 2000 and 2007, which had caused fourteen fatalities and seven injuries, and had attributed those accidents to inadequate performance by operators of four basic safety management functions and other "contributing causes."[21]

The four functions were among those that had been initially designated in 1993 by API's Recommended Practice 75 (RP-75) for voluntary implementation by its member companies. The failed functions cited by MMS involved operator responsibilities to do project-specific hazard analysis, determine the management implications of operational change, develop and implement written operating procedures, and ensure mechanical integrity. Among the contributing causes cited by MMS were inadequate communication between an operator and its contractors, the absence of job hazard analyses and written safe work procedures, supervisor failure to enforce procedures, failure to carry out recommended maintenance, and failure to identify or correct workplace hazards.

To further support its proposed SEMS rule, MMS had also provided data on 1,443 incidents that occurred over the same years, which involved 41 fatalities, 302 injuries, 10 losses of well control, 11 collisions, 476 fires, 356 "pollution events," and 224 crane and hoist mishaps. It concluded that the majority of these incidents were related to operational and maintenance procedures or human error, and that operating procedures and mechanical integrity accounted for the greatest number of spills. It also found there had been no discernible trend of industrial improvement over the seven-year study period despite its inspection program's issuance of corrective orders for some 150 findings of operator noncompliance per year.

To MMS credit, its SEMS proposal in 2009 was an attempt to adopt the voluntary API Recommended Practice 75 (RP 75), which many operators were failing to implement, as its own rule and to thereby make it a mandatory and enforceable requirement for all operators. This had been on its agenda for several years but the enactment process had been impeded by API opposition and delayed by the President's Office of Management and Budget (OMB) whose role is to prevent an agency's enactment of a new rule if it does not pass a stringent cost-benefit test or if there is an equivalent industrial standard.[22]

The 2009 initiative by MMS stalled for these reasons. But immediately after the Macondo accident, DOI quickly re-proposed[23] and enacted the rule in October 2010,[24] because neither API nor OMB had the heart to further resist while public outrage

21 74 Fed. Reg. 28639 (June 17, 2009).
22 OMB circulars A-94 and A-119 at http://www.whitehouse.gov/omb/circulars
23 Less than a month later on May 17, 2010.
24 SEMS Final Rule, 30 CFR 250 (October 15, 2010) and supplementary NTL 2011-N09 (October 21, 2010).

over the accident was still intense. Because virtually all other offshore safety rules are prescriptive, DOI's SEMS rule is notable as the first mandatory performance-based federal rule directly applicable to the safety management functions of OCS operators, and is further discussed later in this chapter.

Additional fragments of safety performance data can also be found in studies sponsored by MMS over the years, such as a 1998 study on "The Environmental and Safety Risks of an Expanding Role for Independents in the Gulf of Mexico."[25] This study addressed concerns that an expanded role for "smaller independents" (companies with assets less than $500 million) in the GOM would pose greater risks to worker safety and the marine environment because it was assumed they lacked the technical or regulatory compliance skills of "the majors" (the largest oil and gas companies such as BP) and the "large independents" (assets of $500 million or more).

Using documentation of accidents and platform inspections, and regression models to examine the association between accidents and operational and regulatory variables, the study surprisingly found that independents outperformed majors,[26] and concluded that large and small independents were less likely to have a workplace accident or spill during exploration and production operations than the majors. There is no evidence that MMS and CG used the study to improve safety management among the majors.

Searching for additional safety information has been a frustrating experience. After the Macondo accident, Congressional research staff sought information on harms to workers arising from OCS incidents, but found that the death and injury reporting systems of MMS and other agencies had not distinguished between onshore and offshore oil and gas operations. Their only offshore finding was that MMS data showed GOM operations in 2009 caused 4 deaths, 290 injury-causing incidents, and 145 fires and explosions that "may or may not have caused fatalities or injuries."[27] Nor had additional risk information been produced by other agencies who would be expected to be knowledgeable about injurious and spill-causing incidents, such as the Occupational Safety and Health Administration (OSHA) and Environmental Protection Agency (EPA) because their legal mandates do not provide them with authority to impose such reporting requirements on offshore operators. Similarly

[25] "The Environmental and Safety Risks of an Expanding Role for Independents in the Gulf of Mexico," Publication 98–0021, Mineral Management Service (April 1998).

[26] The nominal accident rate per million platform hours was determined to be 3.34 for majors, 3.01 for large independents, and 2.08 for small independents. Similarly, the weighted accident rate that distinguished between accidents according to their severity was 8.00 for majors, 5.35 for large independents, and 3.85 for small independents. Differences between majors and independents measured in spills were found to be "similar but more extreme," with rates of 255 barrels spilled per million platform hours for majors and 24 barrels for independents.

[27] Deepwater Horizon Oil Spill: Selected Issues for Congress, Congressional Research Service (May 27, 2010).

silent were private organizations such as API, the leading association of offshore operators, and major insurers who provide coverage for operations and loss-causing accidents.

Another problem has been the lack of any authoritative studies comparing U.S. OCS safety performance data with OCS data from other countries, especially Norway and the United Kingdom, which govern highly exploited sectors of the North Sea. Major firms operate in both the North Sea and the GOM, and comparing their performance under three regulatory regimes would illuminate the influence of the differing regimes on the safety performance of these firms. Although geological, meteorological, and other variables such as water depth of operations would have to be taken into account, many other factors are relatively similar in that the same companies, types of rigs and equipment, and industrial standards are often involved.

In addition, information is available from several reputable sources, including the World Offshore Data bank (WOAD) of Det Norske Veritas (DNV), which contains data and analyses on more than 6,000 accidents from 1970, the SINTEF Offshore Blowout Data Base on major blowouts and spills since 1955, and periodic studies done for Oil and Gas UK on accidents dating back to 1990. Such a comparative analysis would be useful to the new DOI regime in which BOEM and BSEE are expected to develop regulatory reforms and to their counterparts in Norway and the United Kingdom as well. Performance indicators and the reporting and evaluation of safety-relevant data are discussed further by Blakstad in Chapter 9.

Two articles by investigative reporters at the *Wall Street Journal* (WSJ) indicate the potential value of a more thorough approach to comparative analysis of performance indicators. The first, published shortly after the Macondo accident, found that for each 100 million hours worked offshore during the years between 2004 and 2009, U.S. operations incurred 4.84 worker fatalities, more than 4 times the European rate of 1.07 fatalities, and experienced 5 major losses of well control in 2007–2008, whereas 5 other major countries (Norway, United Kingdom, Australia, Canada, and the Netherlands), with about half as much drilling activities, reported no such incidents.[28]

In explaining the disparities, WSJ pointed to overreliance by U.S. agencies on industrial organizations to develop standards and best practices, the slowness of the industrial response, and the failure of MMS to follow up on industry. One example is provided: in 2000, MMS asked industry to advise on cementing for deepwater well control and spill prevention, but ten years later, in 2010, API, the leading industrial standards organization in the United States, acknowledged that it had not yet provided the advice. The epilogue to this story is that a deepwater cementing failure was one of the main factors that contributed to the Macondo accident.[29]

[28] R. Gold and S. Power, "Oil Regulator Ceded Oversight to Drillers," *Wall Street Journal* (May 7, 2010).

[29] Note 14 supra at 85.

The second article reported in late 2010 that safety performance by operators had progressively deteriorated in recent years under U.S., Norwegian, and UK regulatory programs. The reporters found that in 2009, and taking into account hours worked on offshore rigs, the United States experienced twenty-eight major oil and gas releases resulting from the loss of well control, a 56 percent increase over 2007; that Norway had thirty-seven oil and gas releases and "well incidents," a 48 percent increase over 2008 and its highest since 2003; and that the United Kingdom, over the twelve months ending March 31, 2010, had eighty-five serious oil and gas releases, a 39 percent jump over the previous twelve months. To dispel any notion that the BP accident would bring an immediate halt to the trend, the WSJ article noted the serious near miss that occurred a month after the Macondo accident at Norway's GullfaksC site, and the David Pritchard study showing that BP's Deepwater Horizon operation was one of 43 similar ventures and raised the issue of whether the risk of another Macondo accident was 1 in 43 rather than MMS's reassuring claim that it was merely 1 in 50,000 operations conducted over many years in the GOM.[30]

The reporters concluded that the recent trend of deteriorating safety was attributable to several systemic or industrywide factors. Profit was overriding safety to a greater extent, fewer experienced workers and less competent personnel and contractors were involved in operations, and companies were pushing the limits of technology and experience into deeper waters and more complex geological contexts. It also suggested that some regulators had become "too cozy" with operators, and that safety management lapses seemed to be occurring more frequently because of relaxed regulation and inadequate agency oversight. Similar conclusions were also reached by the organizations commissioned to investigate the Macondo accident.

The value of having a coherent set of safety performance indicators is that it enables regulators and industry to detect problematic operations and industrywide or systemic risks and proactively address these vulnerabilities instead of waiting for the next big accident. A good example is provided by the Norwegian Petroleum Safety Authority (PSA), which continuously collects, reviews, and reports on performance of operations in its sector of the North Sea and works with industry and labor on improvements, as discussed in several other chapters.[31]

Some conclusions can be drawn from this discussion. MMS, other federal regulators, and the offshore oil and gas industry failed to develop a coherent set of safety performance indicators and thereby failed to systematically collect and use information that could have fueled a continuous learning process, which would in turn have

[30] R. Gold and B. Casselman, "Far Offshore, A Rash of Close Calls," *Wall Street Journal* (December 8, 2010).

[31] For example, recent PSA reports for its sector of the North Sea show a continuing decline in personal injuries, acute crude oil spills, and near-miss frequencies, but an increase in gas leaks in 2008 and 2009, an increase in well control incidents in 2009, and that four of the largest discharges occurred

enabled a proactive strategy for improving safety performance.[32] Thus, it can be said that the default policy of the U.S. regime is that improvements are to be driven by major accidents.

7.4. LEGAL FRAMEWORK

The legal framework governing offshore operations on the U.S. OCS is comprised of an uncoordinated collection of numerous laws enacted by Congress over more than 200 years. There are laws and proclamations that define and establish federal jurisdiction over the OCS and ownership of its mineral resources, divide authority between the states and federal government over coastal waters and submerged lands, and govern harbors, navigation, vessels, pipelines, and fishing. Additional laws incorporate principles of maritime and admiralty law and treaties. And many more provide for national security interests, protect endangered species and their habitats, restrict discharges of air and water pollutants and disposal of toxic waste, require environmental impact studies before major actions are taken, and authorize compensation and liability for personal injury, property damage, and harm to natural resources.[33]

At the core of this framework is the law that authorizes the conduct of OCS oil and gas operations, the Outer Continental Shelf Lands Act (OCSLA).[34] This law, amended several times since its enactment in 1953, asserts federal authority over

during the last five years. It also finds that deepwater wells in more than 600 meters of water incur a "clearly higher frequency of incidents." It concludes that such information requires it to take "new measures … to turn developments around." (See PSA reports at http://www.ptil.no).

[32] Note 15 supra at p. 8 addresses this issue.

[33] Some of these laws require agency analyses and public reports prior to making leasing and permitting decisions. For example, the National Environmental Policy Act (NEPA) requires any agency intending to take an action that may significantly affect environmental quality to prepare and publish an analysis of the potential environmental impacts, respond to comments, and adjust its intended action to prevent or minimize irreversible impacts (42 USC 4321). NEPA is applicable to several types of actions that lead to OCS operations such as the formulation of a leasing plan, the selling of leases, and the approval of an operator's exploration and production plans. An agency may claim that its intended action is exempt because it would not have significant impacts, or claim that certain types of its actions are categorically excluded for various reasons. These types of agency determinations may be challenged in court and frequently are. But MMS determinations that the lease sale of the Macondo site to BP would have no significant impacts, and that BP's exploration plan for the site was categorically excluded, were not challenged. As a result, the MMS authorized BP's Macondo operation without a site-specific environmental assessment.

The Oil Pollution Act (OPA) and Clean Water Act (CWA) similarly require agency approval of an operator's facility response plan that must be capable of dealing with a worst-case spill for a maximum of thirty days to the maximum extent practicable, and also mandate that the president and the CG respond to spills. In addition, OPA and CWA establish a complex scheme of administrative, civil, and criminal penalties for regulatory violations, and strict liability for the costs of oil removal and restoration of natural resources that have been incurred by the federal government. Roles are also assigned to state and local officials.

[34] 43 USC 1333.

the seabed and subsoil of the OCS and empowers DOI to conduct OCS leasing programs, issue permits to companies for exploration and production, and carry out a regulatory program to ensure that these activities are conducted safely. It also delegates regulation of workplace safety on the OCS to the CG.

Because Congress has not fully integrated OCSLA with the other laws, the legal framework is neither coherent nor harmonized, and companies authorized to work on the OCS are subject to regulations enacted by MMS and CG pursuant to OCSLA as well as the multitude of rules and procedures created by the other laws. As a result, operators on the OCS must comply with an extremely complex web of prescriptive rules, standards, guidelines, and procedures.

More complexity is created by judicial decisions. Under OCSLA and other laws, final regulatory decisions such as issuance of a permit or enactment of a rule can be appealed to a federal court by individuals, companies, and other private parties who claim that their interests are impacted by such agency action.[35] The subsequent judicial review may affirm the regulatory action in question or find it invalid on grounds that it is "arbitrary," or lacks a sufficient factual basis, violates procedural requirements, exceeds the agency's mandate, or conflicts with a constitutional doctrine. Because agency actions are frequently appealed, fear of judicial review and the delays and costs such cases incur, has the effect of suppressing the timely enactment of improved rules.

OCSLA stipulates that DOI and its units, such as MMS and now BOEM and BSEE, are to develop and enforce regulations for OCS operations and that leases and permits are conditioned upon company compliance with these federal regulations and with state laws that are "not inconsistent" with federal law. It further authorizes DOI to suspend an operator's activities when they threaten health, wildlife, or the environment, and to cancel any lease or permit when such threat is more likely, more serious, and outweighs the advantages of continuation. However, it also provides that under certain circumstances, DOI cancellation "shall entitle the lessee to compensation" (OCSLA s.1334).[36]

Penalties are set at $20,000 per day for an operator's noncompliance with leases, permits, and regulations if corrective action is not taken within a designated time, and increased to $100,000 for each day of a willful and knowing violation, fraud, or falsification (s.1350). In addition, OCSLA authorizes several types of lawsuits by private parties. Under specified circumstances, they may bring private enforcement actions in federal court against a company or DOI to compel their compliance with OCSLA or its regulations and may also seek compensation for damages caused by

35 5 USC 701.

36 See Mobil Oil vs. U.S., 530 U.S. 604 (2000) for a Supreme Court decision ordering government restitution of a lessee whose applications for permits were frustrated by new legal and administrative requirements.

failure of an operator to comply with a regulation or permit as well as litigation costs. A person or other private party may also may seek judicial review and revocation of DOI rules and decisions on leases and permits, as previously noted (s.1349).

When regulating the safety of operations, DOI and CG are directed to require use of the best available and safest technologies (BAST), which are determined to be economically feasible, wherever failure of equipment would have a significant effect on safety, health, or the environment, except "where it determines that the incremental benefits are clearly insufficient to justify the incremental costs" (s.1347b). [37] Thus, the agency must make several findings of fact and also pass the cost-benefit test when it seeks to require a safer technology, whether by enactment of a rule, standard, or permit requirement. If it fails to comply, it is likely that affected operators will seek judicial review and that the reviewing court will reject the agency action because it did not meet the statutory requirement, irrespective of the consequences for safety.

The OCSLA cost-benefit test is reinforced by Presidential directive. Since the Reagan administration, every president, including Obama, has issued or reaffirmed an executive order that requires federal departments and agencies, when proposing a new regulation, to present an analysis of the anticipated benefits (e.g., reductions in deaths and other harms) and costs (e.g., costs of compliance to the regulated industry, costs to consumers, impacts on the economy and jobs, etc.) to the OMB.[38] OMB's role is to then conduct a detailed review and reject the proposed regulation if it finds that the estimated benefits would not exceed the costs, even if rejection would prevent a feasible means of improving safety from becoming an industrial responsibility.[39]

For OCS workplace safety, OCSLA proclaims that any company holding an offshore lease or permit is required to protect the health and safety of their workers and contractor employees by complying with occupational safety and health standards, the "general duty" to maintain workplaces free from "recognized hazards," relevant health and safety regulations of other agencies, and additional safeguards that may be required as a condition for approval of the company's work plan (s.1348).

To meet this broad goal, OCSLA assigns responsibility to enact and modify regulations for hazardous working conditions on the OCS to the CG, a semi-military agency with many disparate policing functions (s.1347). It also directs the DOI and the CG to enforce such regulations, conduct scheduled and unscheduled on-site

[37] Note 14 supra at 114 provides insights about BAST.

[38] Note 22 supra and Executive Order 12866 as amended, at http://www.reginfo.gov/public/jsp/Utilities/EO_Redirect.jsp

[39] Monetizing the deaths and other harms to be prevented as the intended health and safety benefits and arraying these numbers against the usually inflated costs of compliance claimed by industry, and discounting long-term benefits and costs to present-day values, constitutes a very arbitrary process. Cost-

inspections of OCS facilities, and investigate and report on incidents causing death, serious injury, fires, and "major" oil spills (those exceeding 200 barrels in a 30-day period) (s.1348). Companies are also required to allow agency inspectors access to worksites and relevant records.

This mandate has had the effect of excluding the Occupational Safety and Health Administration (OSHA), the agency that regulates workplace safety in virtually all industrial sectors onshore, from playing any role on the OCS. This exclusion arises because the law that empowers OSHA, the Occupational Safety and Health Act (OSHAct),[40] provides that OSHA shall not regulate "working conditions of employees with respect to which other Federal agencies exercise statutory authority to prescribe or enforce standards or regulations affecting occupational safety or health."[41]

OSHA has therefore refrained from regulating offshore workplaces for several decades because doing so would lead to lawsuits claiming it lacked statutory authority, and has officially confirmed its abstinence from offshore regulation in several interagency memoranda of agreement with the MMS and the CG. Even after the Macondo accident and the deaths of eleven workers, when many asked "where is OSHA," the agency's response in congressional hearings was to once again state its refusal to get involved offshore.[42] OSHA's exclusion extends to other offshore workplaces subject to OCSLA, such as job sites for offshore wind power facilities. Thus, workers on OCS facilities have been deprived of the protective attention given to onshore workers by OSHA, the agency that has the most expertise about workplace safety and has developed important regulations such as its "process safety" rule.[43]

OCSLA does not create any special role for labor unions or workers in its offshore safety regime, providing only that "the Coast Guard ... may review any allegation from any person of ... a violation of a safety regulation" (s.1346). Thus, for several decades, the absence of labor unions in offshore enterprises and the lack of union and worker participation in the implementation of safety rules by operators

benefit analysis is consistently opposed by environmental and safety advocates because it restrains agencies from regulatory initiatives. See F. Ackerman, L. Heinzerling, *Priceless: On Knowing the Price of Everything and the Value of Nothing* (2005) New Press; and M. Baram, "Cost-Benefit Analysis: An Inadequate Basis for Health, Safety and Environmental Regulatory Decision-Making," *Ecology Law Quarterly*, v. 8, n. 3. (1980).

40 29 USC 651.

41 Id at s.4b1.

42 D. Michaels testimony, Hearings on Worker Health and Safety Standards Related to the Oil Industry, Oil Rigs and Drilling, Committee on Education and Labor, U.S. House of Representatives (June 13, 2010).

43 29 CFR s.1910.119. Also see M. Baran, "Process Safety Management and the Implications of Organizational Change," in *Safety Management: The Challenge of Change*, A. Hale, and M. Baram, eds. (1998) Pergamon.

has prevailed, in complete contrast to the vested role of labor in the Norwegian regulatory regime, as discussed in chapters by Bang, Engen, and Rosness. However, the Macondo accident illustrated the vulnerability of offshore workers and the inadequacy of industrial safety management systems for their protection. This has led to enactment of the SEMS rule and subsequently a supplemental rule (SEMS II) to fill this void, as discussed later in this chapter.

The U.S. offshore regime mandated by OCSLA therefore follows a detailed legalistic approach that directs agencies to develop a full menu of safety-related rules and standards and to inspect and apply sanctions to ensure that operators comply with these regulations. As a result, the MMS and the CG have created a set of federal rules that prescribe in technical detail what operators must do, and carried out policing functions. The MMS regime and its modification and implementation by BOEM, BSEE, and the CG are discussed later in this chapter.

It is also important to note that OCSLA is imbedded in a larger legal framework that expressly provides for regulatory transparency, public participation in rule making, and agency accountability for decisions that lack a sufficient factual basis. Freedom of Information law provides rights of access to agency-held records and documentation, with exceptions for proprietary business information and national security information.[44] Also, the Administrative Procedure Act (APA) enables any aggrieved person to file a lawsuit in a federal court to challenge the factual or legal validity of a rule, a permit decision, or its specific application.[45] U.S. agencies therefore operate in an open environment and know they need to have full legal and factual support for their regulatory decisions.[46]

Finally, OCSLA is one of many laws enacted by Congress with the expectation that "command and control" regulation will eliminate unreasonable risks. But history indicates the mixed consequences of this approach. It provides companies with certainty in the form of numerous prescriptive rules that must be followed by all, and disincentives for noncompliance in the form of sanctions, but does not make companies willing and capable to do more and take a proactive approach to safety. Experience also indicates that the prescriptive and inflexible rules do not address the differing aspects of industrial activities such as drilling operations and fail to keep pace with technological progress and emerging risks such as those posed by the rapid advance into deepwater locations, and new arrangements between operators and their contractors. These and other issues are discussed later in this chapter.

[44] 5 USC 553.

[45] 5 USC 551.

[46] The legal framework also includes common law tort liability, as noted earlier, which threatens companies with compensatory and punitive damages for harms caused by their negligent actions, and in many of the courts where tort claims are brought, evidence that the harm arose from an activity that did not meet an applicable regulation is regarded as conclusive evidence of negligence.

Other countries have legal frameworks that provide different approaches for governing offshore safety, as in Norway, the United Kingdom, and Australia. For example, Norwegian law addresses the same risk issues as OCSLA, but instead of requiring agencies to make rules and enforce company compliance, it requires company self-regulation based on self-determination of how to fulfill essential safety functions, and provides for agency oversight.[47] It also calls for company self-evaluation and cooperation with regulators and labor to continuously reduce residual risks and create "a sound health, environment and safety culture." The Norwegian and other regimes have been examined by the new post-Macondo U.S. regime, and several other chapters of this book discuss their strengths and weaknesses.

7.5. REGULATION

OCSLA directs the DOI and the CG to regulate the safety of offshore operations. Despite the Macondo accident and recommendations for change, Congress has not amended the law to change its mandate for DOI and CG regulation. Thus, when the DOI terminated the MMS and created two successor agencies, BOEM and BSEE, it merely transferred the MMS leasing role and regulations to BOEM and the MMS safety role and regulations to BSEE.[48] As a result, the new agencies now carry out what had been the MMS functions.[49]

According to OCSLA, BSEE's inherited responsibilities involve:

1. developing and enacting, or incorporating by reference the rules, standards, and practices needed to ensure the safety and efficiency of exploration and production operations;
2. reviewing proposed operations and facilities to determine their compliance with regulations, and stipulate additional requirements if necessary, before issuing permits;
3. inspecting permitted operations to determine if compliance is being maintained;
4. enforcing compliance and imposing sanctions for noncompliance; and
5. collaborating with the CG on workplace safety and response to accidents and spills.

[47] Framework HSE: Regulations relating to Health, Environment and Safety in Petroleum Activities (2002) as amended; and the Common Regulations on Management, Information and Materials, Facilities, and Activities, Norway (at PSA Web site, http://www.ptil.no).

[48] Order 3299, Secretary of Interior, U.S. Dept. of the Interior (May 19, 2010); and Reorganization of the Minerals Management Service, U.S. Dept. of the Interior (July 14, 2010).

[49] BSEE regulations at 30 CFR chapter II, BOEM regulations at 30 CFR chapter V, CG safety regulations at 33 CFR 142.

BSEE faces many challenges because numerous criticisms were made about the prior MMS program.[50] Although MMS maladministration has been highlighted in the media, there are more fundamental issues that BSEE needs to address if its approach to offshore safety is to be more effective than that of its predecessor: *reliance on prescriptive regulation, inspection by checklist for noncompliance, dependence on industry expertise and standards,* and *apparent disregard for new developments in safety science theory and empirical research.*

7.5.1. *Prescription*

The global discourse on modes of safety regulation, discussed by Baram and Lindøe in Chapter 2, has been highly critical of prescription in the form of detailed technical rules imposed in a command-and-control framework. This is the approach taken by MMS and transferred to BSEE. Among the charges made is that such prescription is inevitably suboptimal because it seeks conformity across industry with "one size fits all" rules, and thereby fails to recognize the unique aspects of each offshore operation and each operator's potential for developing a more cost-effective approach that would provide equivalent or superior safety performance.

Another convincing argument is that the prescriptive approach can not keep pace with rapid changes in technology, modes of operation, and advances in the state of the art for risk reduction. To modify a regime of prescriptive rules or enact a new set of rules to cope with such change is a costly undertaking that is always overtaken by more change. Failure to keep pace was seen at MMS as drilling advanced into deeper waters with new technologies, uncertainties, and multi-company enterprises.

Perhaps most convincing is the view that prescription coupled with the demand for strict compliance sends a clear message that operators are not expected to do more than demonstrate compliance – that is, are not expected to be proactive and assume responsibility for developing and applying their own expertise to minimize the risks posed by their activities. From an industry perspective, there are also pragmatic reasons for operators to refrain from doing other than what prescriptive rules require. One is that it often would require special approval or exemption by the regulator, which may involve delay and incur financial loss because rig rentals and contractor services are costly.

Another reason is that compliance with prescriptive rules provides a "safe harbor" from potential liability for the operator. For example, if the operator has fully complied but incurs a harmful incident that leads to tort liability lawsuits which allege operator negligence and seek compensatory and punitive damages, many

[50] See A. Hopkins, *Disastrous Decisions: the Human and Organizational Causes of the Gulf of Mexico Blowout Regulation* (2012) CCH Sydney, 137–154.

courts readily accept evidence of regulatory compliance by the operator as a near-conclusive defense against the negligence claim. Compliance also enables the operator to more readily obtain insurance coverage and, when an accident occurs, to have its claim of loss more readily accepted by the insurer. Given these considerations, and the greater burden of convincing a court or insurer of the soundness of a unique proactive approach to safety, the option of adhering to routine compliance may be seen as posing less financial risk.

Finally, it is simply unrealistic to expect that regulators are capable of fully prescribing the detailed design and work plan for safe and efficient conduct of complex offshore operations, each of which has unique safety-related features. Thus, a National Academy of Engineering report warns that "no regulatory system will, by itself, ensure safe drilling operations ... every company involved must develop, promote and operate in a system safety culture embraced by top management and implemented in every phase of drilling operations."[51] It seems clear that BSEE should review its prescriptive rules, examine the use of performance-based rules used in Norway and the United Kingdom, and determine where to substitute performance-based rules or at least provide the option for such rules.[52] This will be a major undertaking with implications for its inspection and enforcement functions.

7.5.2. *Inspection*

According to the MMS, its inspection responsibilities, including those it carried out for or with the CG, were to ensure and enforce compliance with regulations as directed by OCSLA. To fulfill this role, the agencies conducted announced and unannounced on-site inspection of permitted operations, reviewed operator compliance documentation, and provided training in regulatory compliance to rig managers. In claiming that its inspection program has been vigorous, MMS said that its oversight of GOM operations in 2009, for example, involved 561 drilling site inspections, 3,678 production site inspections, 3,342 "personal safety" inspections, and many other inspections.[53]

Exploration and production site inspections done by the MMS and now being carried out by BSEE involve the use of the "national checklist" of "Potential Incidents of Non-Compliance" (PINCs), a compilation of yes/no questions regarding operator compliance with all applicable regulations. For a drilling rig, MMS inspectors looked for 160 PINCs and other "verifications" to determine if detailed technical requirements

[51] *Macondo Well Deepwater Horizon Blowout: Lessons for Improving Offshore Drilling Safety*, Marine Board, National Academy of Engineering (2012), 115.

[52] Similar recommendations have been made in several major reports, including those at notes 14 and 15 supra.

[53] MMS Report, note 1 supra.

and procedures were being met. BSEE's current PINCs list for drilling operations is a fifty-page document.[54] Another type of inspection done for workplace safety by MMS for the CG involved a walk-through looking for hazards related to slips/trips/falls/railings/open gratings. Upon finding violations, corrective action notices are issued, and if uncorrected and "severe," various sanctions follow as OCSLA requires.[55]

Checklist inspection is the inevitable consequence of a regulatory approach that demands compliance with prescriptive rules and little more. It is a prime example of "proceduralization,"[56] which in the prior MMS and current BSEE program means reducing the difficult task of evaluating the safety of a complex offshore operation to a simplistic standardized routine capable of being carried out by untrained personnel. Thus, the inspection function has been seen to be a law-and-order police function focused on technical details. This has kept inspection from becoming a more holistic evaluation of safety, an evaluation that would encompass the behavioral and organizational aspects of an operation and enable BSEE to determine whether an effective safety culture has been created.[57]

Simply put, reliance on PINCs means the agency misses the forest of safety management for the trees of detailed compliance. This may be one of the reasons for the MMS finding that there had been no discernible improvement in the reduction of accidents, fatalities, injuries, loss of well control, fires, and spills over a studied seven-year period in which it issued some 150 findings of noncompliance each year in GOM operations, as noted earlier in this chapter.

Investigations and hearings before and after the BP accident also revealed that both MMS and the CG had problems in hiring and retaining inspectors with technical expertise for evaluating sophisticated operations, and had not been provided with sufficient funding to correct this problem. At Congressional hearings in 2007,[58] the CG was depicted by several witnesses as having a semi-military, command-and-control culture that had failed to build in-house expertise because of its policy of rotating the junior-level officers who do inspections throughout the broad range of its other missions (which involve many offshore semi-military and security duties) and was also accused of being insensitive to the circumstances of employees in

[54] http://www.bsee.gov/Inspection-and-Enforcement/Inspection-Programs/Potential-Incident-of-Noncompliance--PINC.aspx

[55] 43 USC s. 1349, 1350.

[56] See C. Bieder and M. Bourrier, *Trapping Safety into Rules* (forthcoming 2013) Ashgate Publ.

[57] These deficiencies and the need for holistic evaluation to determine whether a safety culture has been achieved are emphasized in major reports at notes 14 and 15 supra in particular. But the term safety culture is ambiguous and requires definition and criteria to be useful, as discussed in M. Baram and M. Schoebel, "Safety Culture and Behavioral Change at the Workplace," *Safety Science*, v. 45, n. 6 (July 2007).

[58] Hearings on Challenges Facing the Coast Guard's Marine Safety Program, Committee on Transportation and Infrastructure, U.S. House of Representatives (August 2, 2007).

business organizations. The Gulf Coast Mariners Association, created by four labor unions, testified that the CG "marginalized" workers by relying exclusively on managers for information, had too little experience with civilian marine activities, failed to compel reporting of injuries, and had not regulated offshore workplace safety in a manner comparable to OSHA regulation of onshore workplaces. The Association recommended transferring CG's workplace safety functions to a civilian agency.[59] Nevertheless, Congress was unresponsive.

Following the Macondo accident, the Presidential Commission found that the MMS had been an inconsistent and compromised performer of inspections, and that some members of its inspectorate had "inappropriate relationships" with industry personnel. As a result, such deviations had become norms in its inspection program, further contributing to its inadequacy.[60] In addition, the CG itself acknowledged that its oversight of rigs should have been more rigorous, that "the pace of technology has outrun the current regulations," and that they had inspected the Deepwater Horizon rig used at the Macondo site nine times before the accident without finding any "major issues."

Finally, the MMS-CG collaboration on workplace safety inspection had failed to become a coordinated and effective program because of jurisdictional confusion. As previously discussed, OCSLA assigned workplace safety to the CG despite its lack of qualification for this role, caused withdrawal of OSHA, which is more highly qualified, and assigned overall rig safety to MMS. Sorting out responsibilities between the agencies has been difficult and led to a series of interagency memoranda.[61] More difficulty was created by maritime law and issues such as whether a semi-submersible rig is a "vessel" and thereby requires CG safety certification and a subordinate role for the MMS.

As a result, presentations by CG, DOI and OSHA officials at congressional hearings have been ambiguous, uncertain, and conflicting. The chairman of a Congressional committee called this regulatory disarray a "jurisdictional mishmash" and pointed to the absence of OSHA's process safety management rule offshore as a critical shortcoming.[62] Nevertheless, neither Congress nor the President have taken

[59] Id. Also see Oil and Gas Management: Key Elements to Consider for Providing Assurance of Effective Independent Oversight, GAO-10–852T, U.S. General Accountability Office (June 17, 2010); and Report on Coast Guard Personnel Problems, GAO-10–268R, General Accountability Office (January 29, 2010).

[60] Note 13 supra.

[61] For example, Memorandum of Understanding Concerning Occupational Safety and Health on Artificial Islands, Installations and Other Devices on the OCS, U.S. Coast Guard, Dept. of Transportation, and Occupational Safety and Health Administration (July 17, 2009).

[62] Occupational Safety and Health Reporter, 40 OSHR 537 (June 24, 2010). Also see D. Slitor testimony, Hearings on Health and Safety Standards Relating to Oil Industry, Oil Rigs, and Drilling, Committee on Education and Labor, U.S. House of Representatives (June 23, 2010).

steps to eliminate the confusion, which obviously impairs the inspection function. The new BSEE regime that has inherited the MMS inspection function has recognized some of these problems, promised trained inspectors to supplement its use of PINC's, and now has the opportunity to bring about a more coordinated and holistic approach, as discussed later in this chapter.

7.5.3. *Dependence on Industry*

Virtually all the numerous rules and standards the MMS enacted or adopted by reference from private sources are prescriptive and technically detailed, and require company compliance in the design and conduct of a proposed operation. Most were originally developed by API as voluntary standards for discretionary use by its member companies, and when later adopted or formally incorporated by reference by MMS, became mandatory and enforceable as MMS rules. This reliance on API enabled the MMS to capitalize on API's technical expertise and ability to gain industry consensus. But it also created a situation where the regulated industry can be said to have determined the pace of risk reduction in offshore operations and which, in the case of deepwater operations, lagged behind the risks being encountered.

An example was provided earlier in this chapter. The MMS had accumulated data over many years linking most offshore accidents to inadequate company performance of several safety management functions set forth by API in its Recommended Practice 75, API's voluntary guidance on safety management since 1993.[63] Even though MMS had made such findings and discerned no trend toward accident reduction, it continued to rely on voluntary compliance until the Macondo accident, at which point it hurriedly enacted its own rule to make the API guidance mandatory and enforceable.[64] This case and others illustrate the MMS's deference to industry on safety problems, as well as the effectiveness of industry lobbying and the Presidentially mandated, industry-friendly cost-benefit test in discouraging the MMS from taking the steps it knew were needed to stop proven causes of accidents.

Similarly, the CG for many years has refrained from developing its own standards for many of its activities, and instead adopted hundreds of industrial standards. It also transferred some of its offshore workplace inspection responsibilities to the MMS. A month before the Macondo accident in which eleven workers died, the recently retired chief of CG's Office of Standards Evaluation and Development stated that "our efforts today are guided by OMB," which "directs agencies to use voluntary consensus standards in lieu of government-unique standards, except where inconsistent

[63] RP-75: Recommended Practices for Development of a Safety and Environmental Management Program for OCS Operations and Facilities, American Petroleum Institute (1993).

[64] Safety and Environmental Management Systems, 75 FR63610 (October 15, 2010).

with law or otherwise impractical." This was beneficial, he continued, because by reducing government-unique standards "to a minimum" it eliminates costs to the government, encourages economic growth, and promotes economic competition. The result, he noted, was that the CG had adopted some 450 industrial standards, which "saves potentially thousands of pages of federal regulations" and "saves the Coast Guard over $1.5 million annually."[65] Although few of these standards deal with offshore workplace safety issues, BSEE and BOEM now work with the CG and should determine if its approach has compromised workplace safety, and if so, use their authority to take corrective action.

An agency's reliance on industry standards and practices must be carefully gauged to avoid deterioration of its own technical competence and prevent industry takeover of its program to the extent that the agency does no more than accommodate "business as usual." But more than supervision is needed. It is also necessary to ensure the integrity and objectivity of the industrial and technical organizations that agencies look to for technical support and other expertise. This need is apparent when one considers the conflicting roles played by API, the leading association of offshore operators. It has developed some 500 standards and practices, many of which were incorporated by MMS in its regulatory regime – a practice now continued by BSEE as discussed subsequently. But it also spends millions of dollars annually to aggressively lobby and coordinate campaigns and public demonstrations against new laws and regulatory initiatives for improving safety because its members oppose governmental intervention and additional compliance costs.[66]

Another problem arising from dependence on API and other private organizations is that their development of voluntary standards and practices occurs in private proceedings that exclude the presence and participation of unions, workers, and other stakeholders who have personal concerns and who may also have intimate knowledge about safety problems and gaps in safety management that deserve consideration.[67] Without their presence and prompting, the performance of the private organizations is more likely to be delayed or inadequate in addressing safety-critical aspects of the operations their members are pursuing, especially those aspects that need more than incremental engineering solutions and would implicate management and higher levels within companies.

[65] H. Hine, "The Value of Voluntary Consensus Standards," *Coast Guard Proceedings*, v. 67, n. 1 (Spring 2010).

[66] "API Plans Citizen Rallies in Opposition to Energy, Drilling Reforms," *Environment Reporter*, 41 ER 1900 (August 2, 2010); also see Marcus Baram, "Big Oil Fought Off New Safety Rules Before Rig Disaster," Huffingtonpost.com (April 26, 2010); and *Environment Reporter*, 41 ER 1899 (August 20, 2010).

[67] "Worker Involvement Lowers Risk of Chemical Disasters," PETROMAKS at http://www.coe.no (July 18, 2010).

Each of these issues needs attention. In response to post-Macondo criticism, API has created a new Center for Offshore Safety dedicated to continuous improvement in the safety of deepwater operations.[68] But BSEE has yet to address its dependence on API and other industrial organizations for safety expertise and determine how to become the senior partner in public-private collaborations on safety issues. In this regard, the chapters in this book on the Norwegian regime are instructive. Finally, the private groups and BSEE need to be reminded that their offshore domain exists because of public trust in their exploitation of publicly owned resources, and that they are therefore obliged to ensure transparency and stakeholder participation in their activities. Indications that some of these issues have been recognized and may be addressed by BSEE and presidential advisors are subsequently discussed.

7.5.4. *Learning from Safety Science*

Safety science denotes the collective wisdom of experts and analysts who have sought to reduce the risks posed by many types of hazardous industrial operations. The field of safety science is interdisciplinary and populated by practitioners in industry and government, consultants, and academics who engage in research and consulting in many countries. Their activities and publications provide knowledge distilled from empirical research and experience, and backgrounds in the social, behavioral, and organizational sciences, engineering, law, and public policy. Their work and pragmatic recommendations have been recognized and adopted in many regulatory and industrial safety programs. But there is no evidence that MMS and the CG recognized or sought to gain insights from this resource or find ways to join the field and collaborate on improving offshore safety.

One important insight is the concept of "drift," developed by Jens Rasmussen, which provides that the efficacy of a regulatory program or safety management system deteriorates as operations move outside the envelope of conditions and circumstances it was originally designed to deal with.[69] Drift is a common problem in many hazardous industrial sectors and is often not recognized until a major accident occurs. In the case of the MMS, the agency continued to apply the regulatory program it had developed for shallow water operations to operations progressively moving into deeper waters, as previously noted. Thus it failed to recognize that BP

[68] Because the President's National Commission (see References) expressed concerns about API's conflicted roles, API created the new center and installed an advisory board that includes several members of the regulatory and academic communities; http://www.centerforoffshoresafety.org/

[69] J. Rasmussen and I. Svedung, *Proactive Risk Management in a Dynamic Society*, Swedish National Rescue Services (2000); and J. Rasmussen, "Risk Management in a Dynamic Society: A Modeling Problem," *Safety Science*, v. 27, n. 2/3 (1997), 183–213.

and other companies were not prepared for the full range of risks and uncertainties they would encounter at deepwater sites.

Another insight that seems particularly relevant to MMS and to the Macondo accident is the concept of legitimization or normalization of deviance articulated by Diane Vaughan in her study of the space shuttle *Challenger* disaster experienced by the federal space exploration program.[70] This concept provides that many small behavioral and technical deviations commonly occur over time and become imbedded within a regulatory program or safety management system because of convenience, cost reduction, or other reasons, without being seen as safety-significant, and eventually become norms that weaken the program or system to the point where it is incapable of preventing the accidents it was designed to accomplish.

In the case of the MMS, deviations regarding their responsibilities to thoroughly evaluate the environmental context for operations in new sectors of the GOM and assess the potential impacts of such operations deprived the agency of having information that would have enabled it to see the need for and stipulate special safeguards in the permits it issued for such operations. The evaluation responsibility is imposed on agencies by the National Environmental Policy Act (NEPA), and if met by MMS, would have involved site-specific assessment of the impacts of routine operations as well as impacts arising from foreseeable accident scenarios and highlighted the need for MMS stipulation of special equipment and methods for well control and spill containment, for example. MMS evaded this responsibility by accepting highly generalized, non-site-specific company studies and making "categorical exclusions" that exempted BP's Macondo venture and other deepwater operations in the GOM from full evaluation.

Operators have also made real time deviations from rules, norms, and recommendations for accident prevention and well control. As detailed in testimony and investigations of the Macondo accident,[71] BP did not follow recommendations by its consultants regarding the use of additional stabilizers to prevent severe gas flow, avoided doing a critical test of the cement bond for the well, did not reconcile the conflicting results of several negative pressure tests to determine if the well was properly sealed, did not check a report of damage to the blowout preventer nor have this equipment properly certified, and deviated in its use of drilling mud, for example. Rig owner Transocean made additional deviations such as disabling warning systems to avoid false alarms and failing to ensure that control panels would not cause sparks that could trigger an explosion.

[70] D. Vaughan, *The Challenger Launch Decision* (1996) University of Chicago Press.

[71] See discussion and tabulation of "decisions made during the Macondo well-drilling and completion that increased risks" in the report at Note 14 supra at 85 and similar discussion in the report at Note 13 supra.

Additional learning from safety science could have been gained and enabled MMS and the CG to work cooperatively with companies on improving their safety performance and developing the attributes of a safety culture. For example, empirical research has produced a body of knowledge that illuminates behavioral aspects of safety management, means of promoting organizational learning, and strategic and tactical aspects of adaptive management, as discussed in chapters by Mearns, Hayes, and Engen. More theoretical work of potential value to safety management has dealt with concepts of situational awareness and organizational sensemaking. The absence of agency-fostered studies and discourse on these issues in the United States stands in sharp contrast to the sustained research activities, and regulatory and safety forums for discourse and collaborations, which are fostered in the Norwegian context for improving the performance of regulators and operators as discussed in the Bang and Thuestad chapter. According to a former director of safety at a major international offshore oil company, "we have to dumb down when we operate in the US."

7.6. NEW REGIME

The Macondo accident caused the offshore industry and regulators in the United States and other countries to immediately evaluate their safety programs. A review of the changes made by DOI, BOEM, and BSEE follows to determine the extent to which they have addressed the four main issues raised by earlier appraisal of the MMS regime: its reliance on prescriptive regulation, use of checklist inspection, dependency on industrial standards and practices, and disregard for advances in safety science theory and practice.

Immediately after the accident, DOI and others sought to determine the factors that contributed to the loss of well control, explosion, fire, and spill at the Macondo site, and imposed temporary suspension of deepwater drilling. Under political and industry pressures to allow a resumption of drilling, DOI also began to consider changes to the MMS regime that would convince the public that future operations would be safely conducted.

DOI's first major change was to terminate the MMS and transfer its leasing program to BOEM and its safety program to BSEE, two newly created agencies. According to DOI, its division of MMS responsibilities and transfer to separate agencies would ensure that the economic and other benefits that motivate leasing of offshore resources would no longer compromise the implementation of an effective safety program. [72]

However, this promise may be difficult to keep in practice because both BOEM and BSEE are agencies within and accountable to their parent DOI, which has

[72] Order 3299, Secretary of Interior, U.S. Dept. of the Interior (May 19, 2010); and Reorganization of the Minerals Management Service, U.S. Dept. of the Interior (July 14, 2010).

consistently promoted leasing and benefited from its revenues over many years. A similar division of functions was made in Norway when the safety program of the Norwegian Petroleum Directorate (NPD) was transferred to a new Petroleum Safety Authority (PSA), as discussed in the chapters by Kaasen and by Bang and Thuestad. But an important difference is that this action, which kept responsibility for leasing with NPD, made PSA an independent agency not accountable to NPD, and thereby reduced the bureaucratic potential for compromising safety.

The transfer of MMS leasing regulations to BOEM and MMS safety regulations to BSEE did not include any new responsibilities. As a result, BOEM and BSEE inherited authority to implement and enforce all MMS regulations in effect at that time, including the multitude of industrial standards that MMS had adopted.[73] However, DOI soon announced that two new rules would be enacted to enable BSEE to address some of the factors that contributed to the Macondo accident: a Drilling Safety rule[74] and the Safety and Environmental Management Systems rule (SEMS),[75] previously discussed. It also sent several "notices to lessee and operators" (NTLs) advising them of additional safety requirements, and announced an internal policy restricting use of "categorical exclusions" by the new agencies. Finally, it proposed an amendment to SEMS (referred to as SEMS II),[76] which would empower workers with regard to ensuring safety offshore.

In the new regime, BOEM's role has been defined by DOI as management of the country's offshore resources in an environmentally and economically responsible way. This involves implementation of DOI-approved leasing plans and programs, performance of related environmental, technical, and economic studies, compliance with NEPA impact assessment procedures and coordination with other agencies, performance of shared responsibilities with BSEE, and other tasks such as development of an offshore renewable energy program. In addition, BOEM is expected to actively engage in advising companies on their exploration and review plans, provide them with implementation checklists, and conduct studies for development of Arctic resources that encompass infrastructure planning.[77]

BSEE administers the new regime's safety program. Its role defined by DOI is "to process permits to drill as safely as possible," enforce safety and environmental regulations, "provide sustained regulatory oversight that is focused on compliance by operators with all applicable environmental regulations," train inspectors, perform various shared responsibilities with BOEM and the CG, and implement

[73] BSEE regulations at 30 CFR chapter II, BOEM regulations at 30 CFR chapter V.
[74] 30 CFR 250.
[75] 30 CFR 250.
[76] The proposed SEMS II rule is discussed at http://apps.americanbar.org/litigation/committees/energy/articles/summer2012–0912-boemres-sems-ii-key-compliance-points-new-rules.html
[77] http://www.boem.gov/

certain aspects of oil spill response. It therefore has responsibility for ensuring compliance with the prescriptive safety regulations that were previously enacted and implemented by MMS, the new Drilling Safety and SEMS rules[78] and the numerous industrial standards and recommended practices that it and MMS incorporated by reference or otherwise adopted.[79]

The Drilling Safety Rule, formally enacted in 2012, sets new technical requirements for preventing loss of well control and spills. It adopts and makes mandatory and enforceable the voluntary practices previously defined by API's RP 65, part 2, which API prepared shortly after the Macondo accident, and has additional features. Among the matters dealt with are well bore casing, cementing, and drilling fluids, subsea blowout prevention, deepwater well control training, integrity and function testing, decommissioning, and documentation and independent professional certifications of casing, cementing, and other aspects of drilling.

Opposition to the rule from industry has been focused on the cost-benefit analysis done by the agency to justify its enactment and the legality of the enactment itself, not on its intrinsic value for preventing major accidents. In addressing comments by industry that this rule would increase costs and impede future offshore activities far beyond what was shown by BSEE in its cost-benefit analysis, BSEE stated that for the GOM, it considered the drilling of 160 deepwater wells annually in order to estimate the costs of the rule, but that this estimate may be more than what will actually be drilled because of many other factors that may influence deepwater activity. Among other comments were those challenging BSEE's estimated baseline risk of a catastrophic blowout as once every thirty-four years, and attacking its failure to consider the confusion and significant costs likely to be imposed by now making 14,000 discretionary provisions in 80 API voluntary standards mandatory requirements.[80]

The SEMS rule administered by BSEE differs from all other rules enacted under OCSLA because it directly addresses the safety management responsibilities of operators in a performance-based rule that requires operator conduct of twelve broadly defined safety management functions. It is featured and extensively discussed in all major post-Macondo reports that address the need for improving the accident prevention regime.[81] SEMS adopts and makes mandatory the voluntary management practices of API's Recommended Practice 75, as discussed earlier. As a result, the rule also incorporates by reference and makes mandatory and enforceable the

[78] http://www.bsee.gov/

[79] See Electronic Code of Federal Regulations, Title 30, Part 250 for the current list of industrial standards that have become BSEE requirements, http://www.ecfr.gov

[80] Comments submitted to DOI by U.S. Oil and Gas Association, API and other industry groups on "Increased Safety Measures for Energy Development on the OCS, 1010-AD68" (December 13, 2010).

[81] See the reports done in the United States that are indicated in the References section of this chapter.

multitude of other API voluntary standards that API had advised operators to follow at their discretion in implementing RP 75. Thus it can be said that SEMS encompasses these ancillary API standards and in doing so replaced API use of words like "should" and "may" with the BSEE word "must."

Because of this heavy reliance on prescribing many detailed industry-developed standards of a technical and engineering nature, the SEMS rule is a hybrid approach to safety management because fulfilling its performance functions requires use of many prescriptive requirements. This is somewhat similar to the Norwegian approach, previously discussed here and also in other chapters of this book. The Norwegian framework regulation sets forth broadly stated performance-based requirements but also provides technical guidances and suggests that operators consider using such guidances as well as industry standards in determining how to fulfill their performance obligations. The difference is that the Norwegian approach is permissive regarding reliance on the guidances and industry standards and provides for PSA oversight and assistance for compliance, whereas the SEMS approach is mandatory, and rigidly requires compliance and BSEE enforcement.

The SEMS rule requires that an operator develop a safety and environmental management system that fulfills twelve basic functions dealing with environmental requirements and impacts, hazard analysis, job safety, management of change, written operational procedures, selection of contractors, training, mechanical integrity and testing, pre-startup review, emergency response, investigation of incidents, safety, and environmental performance audits and documentation. The rule also informs operators that their "SEMS program must meet or exceed the standards of safety and environmental protection of API RP 75."

DOI and BSEE have developed additional safety requirements in the form of Notices to Lessees and Operators or NTLs, which, for example, require corporate-level affirmation of operator compliance (NTL-10) and operator demonstration that it is prepared to deal with a blowout and worst-case discharge (NTL-06). A DOI official has also promised that BSEE's regulatory reach will be extended to include application of SEMS to contractors without detracting from the principle of holding operators fully responsible for contractor performance. Thus, step by step, DOI and BSEE are trying to address the root causes of the Macondo accident, remedy deficiencies in the prior MMS approach, and thereby improve key aspects of industrial safety performance.

Because the SEMS rule is mandatory and requires compliance, its effect will depend to a great extent on the vigor and quality of the inspection program deployed by BSEE to evaluate each operator's implementation of SEMS and the adequacy of the underlying industrial standards that are being used to perform the functions. BSEE continues the MMS and CG inspection practice of using PINC checklists to determine noncompliance with its rules.[82] But obviously this approach is of limited value when inspecting for compliance with SEMS because each operator's activity, drilling site, and safety issues will differ in certain respects, causing variations between each SEMS program that BSEE inspects despite its requirement that each operator comply with the same set of standards.

Thus, a checklist or cookbook approach is inadequate, and inspectors will need expertise to holistically evaluate the activity-specific quality of each SEMS program, unless the statement in the rule that "Your SEMS program must meet or exceed the standards of safety and environmental protection of API RP 75" is narrowly construed by BSEE as a conclusive means of achieving compliance, beyond which further efforts will not be required. If that happens, it would establish limits for compliance and safety management, that have been defined by API's RP 75 and thereby prevent SEMS from motivating a more expansive and progressive approach to safety by operators. In contrast, the Norwegian performance-based approach emphasizes that an operator is expected to make continuous progress in reducing risk in accordance with the "As Low as Possible" (ALAP) principle which has recently replaced its earlier "As Low as Reasonably Practicable" (ALARP) principle.

Also troubling is that the SEMS rule is at odds with principles of public policy. It is a federal rule whose substantive requirements are standards originally developed by industrial groups whose deliberative procedures are not fully documented, have been closed to the public, and exclude important stakeholders such as members of the workforce and unions. As a result, BSEE's substantive requirements are created by the regulated industry without transparency, public access, and stakeholder participation.

To address this issue and ensure that this form of outsourcing regulatory work to industry is publicly accountable, BSEE should be obliged to provide that, at a minimum, its incorporation by reference of any industrial standard be carried out in proceedings that fully meet the norms of the federal rule making process. These norms include giving the public prior notice of the incorporation proceedings, the opportunity to comment and participate, have access to relevant documentation provided by the industry, and have their comments considered and addressed in the final stage of the rule-making process. In Norway, although lacking some of these procedural norms, PSA has developed a Safety Forum and a Regulation Forum that bring industry, unions, and other stakeholders together to periodically discuss existing regulations and the need for and development of new rules and other means of advancing safety, as discussed in several other chapters of this book.

Finally, a supplement to the SEMS rule has been enacted by BSEE. Referred to as SEMS II or the Worker Safety Rule,[83] it expands several of the twelve functional

[82] Potential Incidents of Noncompliance, BSEE at http://www.bsee.gov/Inspection-and-Enforcement/Enforcement-Programs/Potential-Incident-of-Noncompliance--PINC.aspx

[83] 78 Fed. Reg. 20,423 (April 5, 2013).

responsibilities by adding significant new features that empower workers. The first new feature requires operators to have procedures that enable any on-site worker to issue a stop work order against a specific task or activity when the worker perceives that it presents a threat of imminent risk or danger to a person, the public, or the environment.[84]

Other features require that the operator enable employee participation in mitigating or eliminating hazards, a feature missing from API's RP 75, and ensure that all employees have the right to report safety or environmental violations or unsafe work conditions to BSEE and request BSEE inspection. SEMS II also requires that operators specify one person on each facility who has "Ultimate Work Authority" for operational safety. Finally it requires enhanced training programs and audits of SEMS programs conducted by independent third parties who meet specified qualification criteria.

Despite SEMS and SEMS II, it seems clear that the four issues raised earlier with regard to the MMS program have not been fully addressed in the evolution of the new regime. The BSEE regulatory program remains prescriptive and its inclusion of the new SEMS rule does not change matters because the SEMS rule itself requires compliance with prescriptive standards originally set as voluntary standards by industry. The BSEE inspection program and BOEM advisory functions remain in the checklist mode, although promises have been made for advanced training of inspectors. In addition, the relationship and allocation of inspection responsibilities between BSEE, BOEM, and the CG remains uncertain, and exclusion of OSHA continues.

Nor have changes been made with regard to BSEE and CG reliance on and adoption of API standards, as indicated by DOI's transfer without modification of MMS regulations to BSEE and BOEM, and BSEE's enactment of the new Drilling Safety and SEMS rules whose substantive features for implementation are provided by a multitude of API standards and Recommended Practices. The prospect of making API and other industrial standard-setting activities transparent and open to public participation depends on Presidential follow-through on a statement of intent to address these matters.[85] Finally, there has been no visible move by DOI, BSEE, and BOEM to develop a sustained approach to learning from systematic use of safety performance indicators, and from work being done in the safety science domain on organizational learning, the procedural and substantive attributes of an organizational safety culture, and behavioral aspects of safety management.

On a more positive note, the new rules require independent third-party audits and certifications, new self-evaluation and reporting requirements, more stringent well

[84] Imminent risk or danger is defined in SEMS II as "a condition, activity or practice in the workplace that could reasonably be expected to cause' death, or serious physical or environmental harm."

[85] Principles for Federal Engagement in Standards Activities to Address National Priorities, Memorandum for the Heads of Executive Departments and Agencies, Executive Office of the President Office of Management and Budget (January 17, 2012).

control measures, and mandatory performance by managers of essential safety functions that had previously been discretionary for industry. And SEMS II empowers workers in important safety-relevant ways. These changes, coupled with the liability litigation and economic losses incurred by BP following its Macondo accident, may deter negligent conduct and improve the safety of future operations.

Accompanying these developments is the rapid revival of OCS leasing by DOI and BOEM with the intention to make "available more than 75% of estimated undiscovered oil and gas resources on the US Outer Continental Shelf" by lease sales for the GOM and the Beaufort and Chukchi Seas in the Arctic.[86] Thus, it is left to BSEE to meet this challenge in a manner that fulfills the Presidential promise that future offshore drilling will be safely managed.

7.7. CONCLUSION

Following the Macondo accident, the United States, Norway, Britain, Canada, Australia, other major offshore oil and gas producing countries, and the European Union examined their regulatory regimes. Each decided that offshore ventures subject to its authority will be continued because its safety regulations will be sufficient to prevent major accidents. The four countries discussed in this book have differing regulatory regimes, with the United States alone continuing to fully rely on a "law and order" approach with prescriptive rules and sanctions, and Norway alone in its reliance on self-governance by industry and an infrastructure of collaborative programs for improving safety performance. Such differences arise among regimes because each is shaped by the unique interaction of the traditions, values, institutions, needs, and political, economic, and social forces at play in its national context.

Nevertheless, each regime must deal with the same advancing technologies and often the same few companies that have the expertise and resources needed for offshore operations. As a result, the substantive foundation of expertise for safety regulation in each regime is a common state of the art as represented by the global industry's management practices, standards, and other norms. Thus, an important criterion for estimating the strength or robustness of a safety regime is its effectiveness in evaluating the state of the art to ensure it is keeping pace with technological change and the continuous stream of new knowledge about safety, and to then promote timely improvements and their application by industry, whether by government prescription or industrial self-regulation, or most likely by some combination of each. The chapters in this book on the differing regimes provide a qualitative basis for judging their robustness by this and other criteria.

[86] Press release, Secretary Salazar Announces 2012–2017 Offshore Oil and Gas Development Program, U.S. DOI (November 8, 2011).

REFERENCES

Analysis of Causes of Deepwater Horizon Explosion, Fire, and Oil Spill to Identify Measures to Prevent Similar Accidents in the Future, National Academy of Engineering (December 14, 2011), http://www.nap.edu/catalog.php?record_id=13273

The Deepwater Horizon Accident: Assessments and Recommendations for the Norwegian Petroleum Industry, Petroleum Safety Authority, Norway (June 2011), http://www.ptil.no/press-centre/the-deepwater-horizon-macondo-incident-article6888–172.html

Evaluating the Effectiveness of Offshore Safety and Environmental Management Systems, National Academy of Engineering Transportation Research Board (June 2012), http://www.nap.edu/catalog.php?record_id=13434

Final Report, National Commission on the Deepwater Horizon Oil Spill and Offshore Drilling (January 11, 2011) (http://www.oilspillcommission.gov/), and staff working papers. (often referred to as the "Presidential Commission"}.

Final Report on the Investigation of the Macondo Well Blowout, Deepwater Horizon Study Group, University of California (March 1, 2011), http://ccrm.berkeley.edu/deepwaterhorizonstudygroup/dhsg_reportsandtestimony.shtml

Macondo Well Deepwater Horizon Blowout: Lessons for Improving Offshore Drilling Safety, Marine Board, National Academy of Engineering (2012).

Offshore Oil and Gas in the UK: An Independent Review of the Regulatory Regime (December 14, 2011), http://www.decc.gov.uk/en/content/cms/meeting_energy/oil_gas/incident_mgt/deepwater_rpt/deepwater_rpt.aspx

8

A New Policy Direction in Australian Offshore Safety Regulation

Jan Hayes

8.1. INTRODUCTION

As described elsewhere in this volume, a series of disasters in the offshore oil and gas industry in the 1980s led to a major change in regulatory policy regarding offshore safety. Prior to that time, regulations detailing prescriptive requirements focussed on control of activities in the field. These were replaced by a risk-based form of self regulation in which operating companies are required to make a case for safety that demonstrates risk to workers is as low as reasonably practicable.

In this chapter, it is argued that regulation continues, however, to ignore important lessons about accident causation derived from the expanding body of social science research into industrial disasters. The chapter begins by outlining what has been learnt about accident causation from a specific case, a blowout during development drilling at the Montara oilfield off the north-west coast of Australia in 2009, and the regulatory failures that allowed this accident to occur. The chapter then addresses regulatory policy developments that have led to widespread adoption of the safety case concept and efforts to demonstrate the effectiveness of this form of safety regulation. Sociological explanations for industrial accidents and the contrast between these theories and the approaches to risk assessment and safety built into regulatory frameworks are then explored.

It is one thing, of course, to note the disjuncture between sociological research and regulatory policy. It is quite another to draw the two together. The chapter also, therefore, attempts to offer practical suggestions as to how social and techno-scientific perspectives on risk can be brought together in regulation to reduce accidents and their associated social and environmental consequences. There is a developing consensus that safety culture should be included in offshore safety regulation, but this is yet to extend to a framework as to how culture might be addressed or even a common definition of the meaning of the term "safety culture". This discussion proposes a different approach where issues related to people and organisations can be incrementally included in safety regulation based on factors known to impact safety performance.

8.2. THE MONTARA BLOWOUT

8.2.1. *The Incident*

The continental shelf off the north-west coast of Australia holds rich reserves of oil and gas, with hydrocarbon exploration and production activities in the area stretching back over decades. On 21 August 2009, those in charge of drilling the Montara H1 oil and gas well in this area lost control of the well, resulting in uncontrolled flow of hydrocarbons to the environment. This type of event is known in the oil industry as a blowout. The incident has been the subject of a statutory inquiry (Borthwick 2010).

The Montara field is located 250 kilometres off the coast. At the time of the blowout, the facilities in the area were the *West Atlas* drilling rig with a crew of 69 people operating over the small unmanned *Montara Wellhead Platform* and *Java Constructor*, a construction barge with 174 people on board. As a hydrocarbon cloud engulfed the facilities, the vessel rapidly moved away from the drilling rig and the drilling rig personnel evacuated. Initially, the leaking fluids did not ignite. Oil and gas flowed for more than ten weeks before a relief well was successfully put in place. The drilling of the relief well coincided with the ignition of the release. The resulting fire continued for another two days before the flowing fluids were brought under control on 3 November 2009 and the fire extinguished because of lack of fuel. A month later, the well was finally declared safe.

No one was injured or killed as a result of this incident. It has to be said that this is more good luck than good management and that, if the blowout had ignited immediately, the result could have been similar to the Deepwater Horizon incident, which resulted in eleven fatalities and many injuries (Andrew Hopkins 2012). At the time of the blowout, the construction vessel *Java Constructor* was located close to the facility. As photographs taken at the time show vividly, if the gas cloud had ignited immediately, it is possible that the crew of the vessel (as well as those on the drilling rig itself) could have been adversely impacted. Apart from the loss of the drilling rig and the platform as a result of the fire, the physical consequences of the Montara event include environmental pollution, and even in this area luck has played a part. Given the light nature of the escaping fluids and the remote location of the well, by far the majority of the hydrocarbon has simply weathered away and relatively little has impacted the Australian coast or marine life (Borthwick 2010, 26).[1]

[1] Whilst the environmental impact of the Montara blowout is limited compared to the Deepwater

Blowouts are a well-known hazard in the offshore oil and gas industry. There were thirty-nine such incidents on the U.S. Outer Continental Shelf in the period from 1992 to 2006 (Izon, Danenberger and Mayes 2007). The Australian offshore industry had also experienced six blowouts prior to Montara (with the most recent one in 1984) (Borthwick 2010). Given the potential for disaster, and because of the large sums of money involved, drilling and well construction activities in this global industry are tightly controlled, both within operating companies and by regulation. Despite this, all well control systems failed at Montara.

8.2.2. *Barriers to Prevent Flow from the Well*

At the time of the incident, the H1 well was in the development drilling phase. The well had been drilled and two sections of concentric pipe known as *casing strings* had been put into place in the well. The outer 13 3/8-inch casing ran from the surface to a depth of 1,640 metres and the inner 9 5/8-inch casing ran inside that to a depth of approximately 3,800 metres. This is the depth at which hydrocarbons were expected to be present. The detailed sequence of events that led to the blowout starts with the cementing of what is known as a casing shoe at the bottom of the 9 5/8-inch casing. This device is simply a one-way valve designed to prevent flow from the hydrocarbon zone or reservoir up the casing and to the surface.

The operating company's Well Construction Standards required two proven barriers to uncontrolled flow from the reservoir to the surface to be in place when the well was suspended. The primary pressure-containing barrier should have been the cemented casing shoe at the end of the 9 5/8-inch casing. Based on the pressure and flow profiles seen during the cementing operation, it is apparent that the integrity of the cement was never proven and that the outcome was what is known as a *wet shoe*, with the cement contaminated by drilling and/or reservoir fluids. Such contamination meant that the device could not be relied upon to prevent flow from the reservoir to the surface. Various organisational arrangements made this situation possible, if not likely. Company management failed to adequately supervise field personnel to ensure that work was carried out in accordance with documented requirements. Also, the organisational structure in place at the time of the incident did not include separate management and technical integrity functions, which meant ultimately that various key technical decisions made offshore received no technical input or review.

Secondary barriers that should have been in place (according to the final well design) were two pressure containing corrosion caps (PCCCs). The well design

Horizon incident, there have been impacts felt in West Timor, and the Commission of Inquiry has highlighted that the lack of baseline data and the slow response in putting a monitoring plan in place mean that the full extent of the impact of the Montara spill will never be known.

TABLE 8.1. *Summary of well control barriers*

Barrier	Status
Cement Shoe	Errors during installation resulted in a "wet shoe" with impaired functionality. Test results that demonstrated this were ignored.
9 5/8-inch Pressure Containing Corrosion Cap (PCCC)	Not designed for blowout prevention but included in well design partly for that purpose. Removed for operational reasons just before the blowout.
13 3/8-inch Pressure Containing Corrosion Cap (PCCC)	Not designed for blowout prevention but included in well design partly for that purpose. Never installed.
Fluids in the well bore	Could have provided a barrier, but weight not sufficient for that purpose and level not monitored.

called for these to be installed on the top of both the 9 5/8-inch string and the outer 13 3/8-inch string. Information from the manufacturer of the caps indicates that these were not designed as well control barriers, and yet the operating company chose to use them for this purpose, again illustrating the weak technical function within the drilling department. In fact, despite the fact that it was reported from offshore that both caps were installed, only one of the caps was put in place (on the 9 5/8-inch string). The 9 5/8-inch PCCC was later removed for operational reasons and then not reinstalled. The blowout occurred approximately fifteen hours later.

The well control barriers that were included in the well design and their status at the time of the blowout are shown in Table 8.1. Ultimately, when the reservoir pressure was sufficient to overcome the weight of the column of fluids in the well bore, hydrocarbons were able to flow to the surface as a result of the failure of the 9 5/8-inch cemented casing shoe. This was the only physical barrier to flow that was present on the well at the time the blowout occurred. The choices made in the months leading up to the blowout introduced a series of latent failures into the system of multiple barriers, hence negating the entire design philosophy and leaving the system in a very vulnerable state. Under these circumstances, uncontrolled flow of hydrocarbons to the surface was inevitable. It is tempting to put responsibility for these events at the feet of the individuals on the facility at the time and to see these events as issues of the technical competence of those people. This explanation is ultimately unsatisfactory. These individuals worked in an organisational context and many others within the organisation knew (or should have known) what was occurring offshore. Their actions can be directly linked to a series of flawed decisions

(Hayes in press), which can, in turn, be seen as the result of failings in organisational norms and practices. The next section explores a broader organisational explanation for the failures that led to the blowout.

8.3. ORGANISATIONAL FACTORS

Complex systems require multiple barriers in place to ensure operations remain safe. The ongoing effectiveness of such systems depends on those with responsibility for decision making to maintain their collective ability to recognise developing problems before they become critical. This section proposes three ways in which company management could have provided more effective leadership on these issues and thus have made an accident such as the blowout of the H1 well a much more remote possibility.

8.3.1. *Providing Active Supervision*

One of the most startling failures in the events leading up to the blowout is the failure to install the 13 3/8-inch PCCC followed by the reporting to onshore management that such a cap had indeed been installed. It is difficult to escape the conclusion that someone (indeed probably several people) knew that the cap had not been installed, and yet reported to onshore management that it was in place.

This event can perhaps best be considered as an indication of the relationship between offshore and onshore personnel within the organisation. The onshore management team appear to have taken a very hands off approach regarding work done offshore. In his statement to the Commission of Inquiry (CoI) the drilling superintendent stated that "if there was an issue with a forward plan that could not be resolved offshore [the Senior Drilling Supervisor] would call me to discuss the issue. … The plans were not normally sent to the [onshore] office for review unless there was an issue that could not be resolved offshore … if the Senior Drilling Supervisor … needed additional expertise from onshore staff, he would telephone me" (Wilson 2009, 33). To emphasise this point, the attitude of onshore management is to assume that offshore personnel were competent and operating in accordance with approved standards without ever conducting any checks as to whether this was actually the case.

The drilling superintendent also stated that, with regard to checking the reports on cementing operations, he had no reason to check the reports in detail and he "reviewed the [daily drilling report] to see if there was any obvious errors or issues. There were none" (Wilson 2009, 185). As the CoI report points out (Borthwick 2010, 3.123), the role of the drilling superintendent was the day-to-day supervision of activities offshore, and this involves much more than simply looking for obvious errors in written summaries of work done, especially work involving a critical safety function such as the cementing of the casing shoe. The drilling superintendent indicates again his overall attitude to supervision when he says "as there were no indications or reasons after 21 April 2009 to think that the wells were not suspended per the [drilling plan] and subsequent change control, there was no reason to conduct any form of audit to check that all work that was thought to be performed had in fact been completed" (Wilson 2009, 266(c)). In fact, there was evidence available, particularly in relation to the state of the cemented casing shoe, if only he had looked.

The fact that critical activities were apparently conducted with no supervision is a key failing on the part of the organisation as a whole. It should be emphasised that this is not a matter of considering that employees may be dishonest. The principle of active supervision is that employees take their cue as to what is important from what their superiors pay attention to, and further that people respond to positive reinforcement of appropriate behaviours. If no one ever asks about well control barriers being in place or checks that integrity tests have been done in accordance with written requirements, then the message given is that these issues are not cues that need to be considered in deciding how work should proceed.

Looking in the statements provided by management to see what was regarded as important reveals the following communication. Five days before the blowout, the drilling superintendent sent an email to a wide range of people (including on rig personnel, the overseas parent company managers and government representatives) which stated: "Whilst we have been busy drilling some our guys have been working offline to suspend Montara H3ST-1 and Montara H1. Both wells will be fully suspended by the end of the day. This has saved us about 12–18hrs of rig time by being able to do this activity offline – a job well done" (PTT 2009). In these circumstances, it is not difficult to imagine a scenario in which the 13 3/8-inch cap was found at the last minute to be unserviceable (as the organisation initially reported to the CoI) and offshore personnel decided to proceed and suspend the well without the additional barrier. It is perhaps significant that it is the Daily Drilling Report dated 17 April (the day after the previously cited general note) from the offshore team to onshore management that first reports the fictional installation of the 13 3/8-inch PCCC. By sending such a report, the offshore personnel were simply confirming what onshore management wanted to hear.

8.3.2. *Separation of Engineering Integrity and Operations Functions*

Another unsatisfactory organisational feature revealed by the quotes from the drilling superintendent in the previous section is the level of technical discretion given to offshore personnel, in particular the lack of any engineering integrity function within the organisation that was independent of line operations activity. To emphasise this point, this means that operational personnel were in control of whether or not engineering input was required, meaning that there was no separation between engineering integrity and operations functions. As is clear from the details of the cementing activity (Hayes 2011), the personnel responsible for conducting these activities offshore apparently had serious gaps in their technical ability to understand what was happening and to know at what point they needed to seek specialist advice.

This situation appears to have its roots in the roles previously held by the various individuals involved. The drilling superintendent had previously (prior to the operational drilling phase) held the post of senior drilling engineer (Wilson 2009, 12). In that role, he had been responsible for (amongst other things) the design of the wells. Once the drilling program moved into the operational phase, he took on the role of drilling superintendent with all the offshore drilling crew reporting to him, including both senior drilling supervisors and all the drilling supervisors (Wilson 2009, 16).

Importantly, this is a significant change in role from technical expert in well design to manager. Sensemaking theory (Weick 2001; Weick, Sutcliffe and Obstfeld 2005) reminds us that the cues we see depend critically on our perception of our role (or organisational identity). In this case, there was perhaps an intention on the part of the organisation that the drilling superintendent maintain some technical oversight, but he seems to have a different understanding of the new role. He seems to have adopted an attitude to supervising the personnel that was based on answering specific questions when asked, rather than retaining overall technical control and proactively providing input to the work that was being done. He goes on to say, "Although there was appropriate communication between [the Senior Drilling Supervisor] and me on 7 March 2009 there was information that I consider, with the benefit of hindsight, could have been given to me so that I would be better able to make decisions about what needed to be done in the face of the apparent failure of the float valve". Further, "my post incident analysis indicates that the 9 5/8" casing shoe most probably did not form an adequate primary tested barrier however on the day with the information supplied to me, I had no reason to suspect that it was not an adequate barrier" (Wilson 2009, 259).

This significant confusion over the role of senior management personnel with an apparently high level of technical knowledge was also seen in the decision to not reinstall the 9 5/8-inch PCCC after the cleaning of the threads on the 13 3/8-inch casing. In his statement, the well construction manager (who was on the facility at the time) said that, based on the other barriers in place, "there was no compelling reason to re-install the 9 5/8" corrosion cap" (Duncan 2009, 251). Contrary to this, he told the Inquiry during the hearing that he had expected that the cap would be reinstalled once the cleaning work was complete and that when he discovered that it had not been installed, he did not insist on the basis that "he did not want to give the impression to personnel on the rig that he was trying to teach them how to do their jobs" (Borthwick 2010, 3.187). In fact, it was the well construction manager's role both to ensure that work was done in accordance with plans (and he apparently planned that the cap would be reinstated) and to ensure the technical robustness of activities undertaken. To leave a well barrier uninstalled for such a reason is clearly an abrogation of both the managerial and technical responsibilities of this role. The blowout occurred approximately fifteen hours later.

Another source of an independent check on activities might have been the corporate Health, Safety and Environment (HSE) specialists. The well construction manager had overall responsibility for the development drilling activity. He reported to the Montara project manager, who reported to the chief executive officer (CEO) (Jacobs 2009, 4, 17–23). Corporate HSE functions report via a separate line to the CEO and appear to have played no role in integrity assurance in drilling activity, despite the statement from the chief operating officer (COO) that the well construction department works under the Corporate Safety Management System (Jacobs 2009, 4, 26).

Similar issues related to organisational design (that is, structure, roles, reporting lines related to technical specialists) have been highlighted in analyses of other accidents. Analysis of the circumstances surrounding the Texas City refinery accident (Andrew Hopkins 2008) showed that those with technical responsibility for process safety issues were marginalised by the organisational structure. This left them unable to raise their concerns with senior management in any way that was effective in initiating action. Similarly, the report into the *Columbia* space shuttle disaster discussed attributes of organisational design related to the power and authority of technical specialists that could be expected to prevent such incidents from occurring again. In particular, the report called for (amongst other things) "a robust and independent program technical authority that has complete control over specifications and requirements, and waivers to them" (CAIB 2003).

The situation here seems to have been even more unsatisfactory as there was no effective engineering input into well operations and no integrity assurance function operating in the organisation with regard to well activities at the time of the blowout.

8.3.3. *Effective Change Management – Rule Compliance versus Risk Assessment*

The lack of specific direction from onshore management then apparently leaves work on the rig being done in accordance with company-written standards and other documentation. As the COO said, "the [company] system relied upon the personnel involved in well construction following the requirements of the Well Construction Management System … it also relied on the expertise of the [rig] operator's supervisory personnel and the [company] drilling supervisors to monitor and check that the

[rig] personnel complied with the drilling programs" (Jacobs 2009, 77). This seems to indicate that what senior company management expects from offshore personnel is compliance with written rules and standards.

In his statement, the well construction manager explained the role of one of the key documents, the Well Construction Standard (WCS), as follows:

> The purpose of the WCS is to provide standards for all aspects of well design, construction, testing, abandonment and intervention that involve a risk to safety, quality or integrity. The WCS are applicable to all aspects of well design, well construction, well servicing and well abandonment. We generated and prepared the WCS through a series of reviews and workshops with the well construction team. However, the WCS was not a prescriptive set of rules to cover every possible scenario but includes processes to risk assess and manage scenarios not considered between document revisions. (Duncan 2009, 43)

He is taking a rather different view of the role of the technical requirements (on such critical safety issues as well control barriers) contained in the WCS. This material has effectively been downgraded from something that requires mandatory compliance to playing the role of a guideline that can be varied based on a risk assessment. One example of how this was working in practice is the documentation regarding the change to the well design from cement plug to PCCCs. Part way through the drilling of the H1 well, the written drilling plan was changed to include PCCCs, despite the fact that this device is not listed as a well control barrier in the WCS. Arguments used by various company personnel in favour of the change include:

- PCCCs are better than cement plugs.
- PCCCs have the same functionality as other devices that are listed in the WCS so they are, in effect, approved.
- The WCS allows for two suspension options:
  - temporary suspension "where the [rig] remains on location" and
  - long-term suspension "when the [rig] leaves the site. Wells must be suspended so that they can be abandoned with rig-less intervention to meet the standards below."

In the case of Montara, the drilling rig was leaving the site, but there was no plan to abandon the wells, so the company argued that it was reasonable to use the standards applicable to temporary suspension, even for the period when the drilling rig was elsewhere (Duncan 2009, 159; Wilson 2009, 152). Their point seems to be that the WCS does not allow for batch drilling, that is the case where wells are suspended and the rig departs, knowing that it will return for more planned work. This is indeed a case where the rules may not be applicable and a risk assessment of the batch drilling case could have been used to develop a new range of acceptable well control barriers based on the risks involved.

Putting that aside, other claims about the functionality and effectiveness of the PCCCs are not supported by the information supplied from the manufacturer. The drilling superintendent says he sent the manufacturer's instructions offshore so that "the caps could be installed as per the manufacturer's instructions. I assumed that those instructions would call for an in situ pressure test after installation and I did not note prior to sending out the manufacturer's instructions that they themselves did not call for the PCCCs to be pressure tested once installed" (Wilson 2009, 198). In fact, the Operating and Service Manual for the PCCCs (pressure-containing corrosion caps) explicitly states that they are not designed to operate as barriers against blowout and are only meant to be used on a well that has been plugged and secured. It seems likely that, in preparing the risk assessment on the proposed change, the drilling superintendent did not read any of the material provided by the manufacturer for this critical safety device (Borthwick 2010, 3.215).

Hopkins (2010) has highlighted the interrelationship between risk assessment and rule compliance. He points out that rules are often based on risk assessments and that, as far as possible, risk assessments should be formulated into rules to assist end point decision makers such as those involved in operational drilling activities. He highlights other accidents where the temptation to risk assess one's way out of specific safety requirements has contributed to accidents, as appears to have been the case with Montara.

This section has proposed three ways in which organisational systems could have been improved to the point that the blowout may have been avoided. The common factor is maintaining an organisation-wide focus on well integrity issues driven by management constantly paying attention to them. Decision making research in high-hazard organisations (Hayes 2009b, 2012b) has highlighted the value of shared stories about past failures as a way of keeping alive the level of respect that is necessary in order to make effective decisions in these circumstances. In complex systems, management decision making is not only about finding the right answer. It is also critically about asking the right question. This is a key function of managers in their role as *sense givers* for the organisation (Weick 1995, 10). It seems likely that a fly on the wall offshore or in the office would not have heard stories about past failures and the need for integrity assurance, but rather talk of operational priorities and cost savings.

Technical decisions were left to those offshore who adopted an attitude of trial-and-error learning. This is highly problematic and such attitudes can be contrasted with those that researchers of high-reliability organizations tell us are necessary to avoid accidents in the long term, such as preoccupation with failure (Weick and Sutcliffe 2001). Instead, these people as a group seem to have moved to an attitude that assumes everything is fine until proven otherwise. Use of trial-and-error learning in a high-hazard environment such as offshore drilling makes the occurrence of a serious incident only a matter of time.

8.3.4. *Regulatory Failure*

For historical reasons that are not directly relevant here, well integrity is regulated outside the main Australian offshore safety case regime. Despite this, the two sets of regulations have a common approach to safety assurance. The fundamental requirement of the type of goal-setting regulation used throughout the offshore oil and gas industry in Australia is that operating companies must set their own standards based on the hazards and risks posed by their activities, and then do what they say they will do (Ayres and Braithwaite 1992; Hood, Rothstein and Baldwin 2004). In the case of Montara, this system failed. PTT E&P, which had leased the drilling rig from Atlas Drilling, failed to comply with its Well Construction Standards (WCS) in numerous ways including:

- failure to test the cemented casing shoe and subsequent reliance on this untested barrier;
- reliance on pressure-containing corrosion caps (PCCCs) as a well barrier when these are not approved in the WCS;
- failure to install sufficient barriers to meet the requirements for long-term suspension of the well when the *West Atlas* left the field; and
- failure to monitor completion fluid parameters to ensure overbalance and subsequent reliance on this unmonitored barrier during temporary suspension (when the *West Atlas* returned to the field).

The Montara Commission of Inquiry reported that "the regulatory regime was too trusting and that trust was not deserved" (Borthwick 2010, 18). This refers to the lack of enforcement activity to ensure that the operating company complied with their own procedures and standards. It is true that stronger enforcement may have made a difference to the outcome, but to limit the regulatory response to issues of ensuring more active enforcement fails to address the fundamental issue, which is that the current regulatory regime does not address the types of issues raised in the opening section of the chapter. In fact, the fundamental management problems described earlier require a different regulatory response – one that addresses directly the need for organisational competence and capacity in recognising and learning from problems.

8.4. CONTROLLING THE POTENTIAL FOR ACCIDENTS

This section moves from the specifics of the events surrounding Montara to a summary of how both academia and government have sought to address the potential for major accidents and the means to prevent them. This section summarizes developments in each of these spheres since the 1980s.

8.4.1. *Regulation of Offshore Safety*

Paterson has described the evolution of the UK offshore safety regime in Chapter 6 of this book, The impact of the Piper Alpha disaster in 1987 and the Cullen Report were also very influential in policy making outside the United Kingdom, and the Australian regulatory regime for offshore safety is based on the same concept and premises as the UK regime.

Given that this approach to ensuring worker and public safety has been so widely adopted, it might be expected that it has been shown to be effective in reducing injuries and fatalities. To date, it has been very difficult to make an objective determination regarding the effectiveness of safety case style regulation. If this approach is successful, then the ultimate benefit should be shown in improved safety performance, but because serious incidents are relatively rare, statistical conclusions are difficult to draw in an industry where other environmental factors are constantly changing. Several studies have attempted this exercise and all have tentatively reported positive results based primarily on qualitative, rather than quantitative, arguments (HSE 1995, 2003; Rose and Crescent 2000; Vectra 2003; Vinnem 2010a). Further, a risk-based safety regime does not eliminate the need for a competent and well-resourced regulatory agency. In fact, reviews of the effectiveness of safety case regulations typically become reviews of the effectiveness of the regulator in enforcing the regulations, rather than a critique of this approach per se.[2]

This issue has come under particularly close scrutiny following the Deepwater Horizon incident in the Gulf of Mexico. At the time of the incident, the safety regulation in place for the offshore oil and gas industry in the United States took the form of a prescriptive, standards-based regime. The report of the National Commission on the Deepwater Horizon Blowout includes a summary of the development of the safety case approach in the nuclear, chemicals, aviation, and offshore oil and gas industry (National Commission on the BP Deepwater Horizon Oil Spill and Offshore Drilling 2011b, 69) and points out that the fatality rate in the offshore oil and gas industry in the United States is at least four times the fatality rate in European jurisdictions that have operated for several decades under safety case

[2] In Australia, there is a specific requirement (see Offshore Petroleum Greenhouse Gas Storage Act section 695) for a review of the effectiveness of the National Offshore Petroleum Safety Authority every three years. In practice, this judgement is made subjectively based on a comparison of NOPSA's practices with those of other regulators internationally and the views of key stakeholders, including industry.

legislation.[3] The Commission had no hesitation in recommending that the United States move to "a proactive, risk-based performance approach specific to individual facilities, operations and environments, similar to the 'safety case' approach in the North Sea" (National Commission on the BP Deepwater Horizon Oil Spill and Offshore Drilling 2011b, Recommendation A2, p. 252). It must be noted that, given the continuing strong resistance to this approach that has been historically a feature of the U.S. offshore oil and gas industry, it appears unlikely that this recommendation will be fully adopted, as indicated in Chapter 7 (Baram) of this volume.

Whilst objective evidence is somewhat ambiguous, there is no doubt that operation under a safety case regime has increased the attention paid to safety within the industry. Hale, Goossens and van de Poel (2002, 104) describe the benefit of ten years of risk-based regulation for the petroleum industry in Holland as follows: "companies have become more risk aware, safety cases and safety management systems are positive developments and there is more understanding of how and why risk control and management measures work". A major benefit of the safety case regime comes from the ongoing conversation about safety and engagement with safety-related issues within industry, especially related to relatively rare events that would otherwise generate little discussion.[4]

8.4.2. *Continuing Disasters – Social Science View*

The links between the events surrounding the Montara blowout and the actions of those in the operating company described in Section 8.2 of this chapter are not unique. In parallel with the regulatory developments described in Section 8.3, social scientists from various fields (sociology, management, organizational psychology) have taken an increasing interest in industrial safety. The work is generally of two kinds: accident analysis and normal operations studies.

A significant body of analysis has grown up based on the work of social scientists regarding the causes of specific disasters in complex socio-technical systems – for example, the loss of the space shuttle *Challenger* (Vaughan 1996), the loss of U.S. Black Hawk helicopters (Snook 2000), the loss of the space shuttle *Columbia* (Starbuck and Farjoun 2005), the explosion and fire at Exxon's Longford gas plant (Hopkins 2000), the BP Texas City refinery fire (Hopkins 2008) and the Montara blowout (Hayes 2011). This has led to generalised models regarding the organisational

3 See also National Commission on the BP Deepwater Horizon Oil Spill and Offshore Drilling (2011a) for a detailed analysis of the available fatality and injury data.

4 There is a growing unease amongst some members of the offshore industry that the generation now reaching senior management positions giving them significant decision making powers were not working at the time of the Piper Alpha incident in 1987 and hence have no personal experience of the shock such events cause.

causes of accidents, the best known of which is James Reason's Swiss cheese model (Reason 1997). In this way of thinking about accidents, there is a range of defences in place that are functionally designed to prevent any given hazard from leading to a loss of some kind (such as an accident). In practice, these defences are imperfect (like holes in Swiss cheese). The various hardware and procedural measures in place ensure that failure of any individual measure is not catastrophic. An accident occurs when the holes in the cheese line up and provide an accident trajectory through all the defences.

In this model, the holes in the cheese have two interesting features. Firstly, they may be attributable to active failures – errors made by field personnel – or they may be latent failures. Latent failures are weaknesses in the system that do not, of themselves, initiate an accident, but they fail to prevent an accident when an active failure calls them into play on a given day. Problems arise when latent failures in the system accumulate – maintenance is not done, records are not kept, audits are not done. The consequence of a small active failure can then be catastrophic as the protective systems fail to function as expected. The second quality of the holes in the Swiss cheese is that they are a function of the organisation itself. In this model of accident causation, operator actions in the field are linked to workplace factors such as competency, rostering, control room design, task design and so forth, and these issues are linked to organisational factors such as budgets, safety priorities, leadership and the like. In this way of thinking about safety defences, the performance of all components in the system is interlinked.

In parallel with this work, there has been another complementary strand of social science research looking at successful organisations and how they achieve high levels of safety performance. Researchers who favour this *high reliability* view favour an ethnographic or sociological approach to what are known as *normal operations studies* (Allard-Poesi 2005; Bourrier 1998, 2011; Hayes 2009b; Hopkins 2009; La Porte 1996; La Porte and Consolini 1991; Roberts 1994; Rochlin 1999; Schulman 1996; Weick 1987, 1993; Weick and Sutcliffe 2001). Researchers in this field have published many organisational studies of the qualities of high-performing organisations, but the only integrated theory which draws all these strands together is Weick and Sutcliffe's (2001) High Reliability Organisations (HRO) theory. These researchers claim that high-functioning, or high-reliability, organisations have the ability to detect and respond to system variations because of their state of mindfulness about their operations. In turn, mindfulness is fostered by five qualities, which are:

- preoccupation with failure – seeking out small faults in the system and using those to improve performance;
- reluctance to simplify – valuing diversity of views and resisting the temptation to jump to quick conclusions;

- sensitivity to operations – valuing experienced operating people who have a nuanced system understanding;
- commitment to resilience – using layers of protection, valuing redundancy in equipment and people;
- deference to expertise – placing appropriate value on the advice of technical experts in decision making.

Other researchers (Hollnagel, Nemeth and Dekker, 2008; Hollnagel, Pariès, Woods and Wreathall, 2011; Hollnagel, Woods and Leveson, 2006) propose that *resilience* – the ability to anticipate and plan, and also the ability to adapt and respond – is the key to excellent safety performance.

There has been significant practical uptake of these research results primarily in the field of incident investigation. It is now common for formal investigations to include a social science perspective in addition to a review of more proximate technical issues – for example, the investigation into the storage tank fire at Buncefield (HSE Major Incident Investigation Board 2008), the Baker Panel review of BP's U.S. refining operations after the Texas City refinery fire (Baker 2007), the review of the Columbia Investigation Board (CAIB 2003) and the report of the National Commission following the Deepwater Horizon blowout (National Commission on the BP Deepwater Horizon Oil Spill and Offshore Drilling 2011b). These analyses rarely reveal new technical information regarding the causes of accidents but, rather, show that existing technical information has not been applied, usually for social, rather than technical, reasons. Despite this practical use of the large body of theoretical work, these issues have largely been ignored by oil and gas safety regulators. The exception is Norway, where regulations requiring an operating company to promote a sound safety culture have been in place since 2002 (Hopkins 2007; Høivik, Moen, Mearns and Haukelid 2009; Kringen, Chapter 9 in this volume).

In summary, a series of serious accidents in the 1980s provided the impetus for some major changes in safety regulatory policy to address high consequence accidents. Despite the fact that several serious accidents had occurred, they were still relatively rare events, and a new regulatory approach based on an assessment of facility specific hazards, risks and controls was developed. This safety case style of regulation has been in place in many high-hazard industries and many (non-U.S.) jurisdictions for more than two decades. Despite the popularity of this approach, compelling evidence that such regulation is effective has proved to be elusive. In fact, major disasters have continued to occur, and analysis of the causes of these incidents by social scientists has shown that a range of organisational factors are common causes. These factors are excluded from consideration in safety cases. The next part of this article discusses some of the reasons why this is the case.

8.5 INCORPORATING ORGANISATIONAL ISSUES INTO OFFSHORE SAFETY LEGISLATION

The previous section has described how research into organisational causes of accidents has been largely ignored by regulatory policy makers to date (with the exception of Norway). This section focuses on three key issues that impact the integration of an organisational view of safety into the current risk-based regulatory policy.

According to Power (2007, 178):

> [D]espite the motivational power of the many insights derived from post-disaster analysis, and their roots in soft organizational psychology, the managerial institutionalisation of these insights has been problematic. It is not that practitioners of risk management are resistant to these ideas – quite the opposite is true; they are completely persuaded of the importance of "risk culture" and constantly make judgements about it. But these judgements are difficult to represent within rationalized designs for risk governance and the climates of auditability within which they operate. Risk insights can only acquire formal managerial significance within the conceptual and operational space of auditing and internal control systems.

The first apparent problem is that organisational issues apparently do not fit into the risk-based model of safety cases. In fact, the narrow way in which concepts of risk are used to address offshore safety effectively hides organisational causes. Because these causes are real, risk-based thinking needs to be broadened to allow such concepts to be taken into account.

The second difficulty in incorporating organisational issues into regulation is that there is no widely accepted unified framework for organisational safety on which to base regulatory requirements. This is not a reason to ignore the current substantial body of academic knowledge in this area, and some factors on which there is broad agreement are proposed as a starting point.

The third issue addressed later in the chapter is the view that using anything other than a risk-based frame of analysis is inconsistent with goal-based regulations such as safety cases. This is true only in the narrowest of views of this self-regulatory approach and, again, the approach should be broadened to allow organisational issues to be considered if safety performance is to be effectively regulated.

8.5.1. *Incorporating Organisational Issues into Consideration of Risk*

The foundation of the safety case approach is to identify and document the specific hazards, specific risks and the specific engineering, administrative and procedural measures that address them. To meet regulatory requirements, submissions to regulatory agencies must provide concrete details. Ultimately compliance and enforcement activities focus on engineering hardware and activities in the field, prioritised on the basis of risk. In this case, risk is conceptualised by operating companies and regulators as the product of frequency and consequence and is often quantified. This in turn allows the results to be compared with numerical criteria and the justification for physical changes that reduce risk to be assessed on the basis of cost benefit analysis (Paterson 2000).

In some ways, this narrow conceptualisation of risk has served society well. There is no doubt that safer designs are produced when engineering failures that could occur during operation are explicitly considered as part of the design development process. Modelling techniques for physical effects such as the prediction of explosion overpressure and the impact of fires on structures have improved dramatically in the past two decades. On the other hand, attempts to incorporate human and organisational error into this narrow definition of risk have been largely unsuccessful (Forester, Kolaczkowski, Lois and Kelly 2006). Whilst the power of the safety case approach is in the concrete and the specific (Rasmussen and Svedung 2000), the nature of the organisational contribution to safety performance is universal and diffuse. This means that, despite evidence regarding the organisational contribution to risk, the insistence on measurement and quantification of risk has led to these issues being systematically ignored.

Based on the collective imagination of technical professionals dealing with offshore design decisions in particular, risks are discussed and managed as if they are real phenomena – a physical property of the engineered system that can be measured and manipulated. The role of human judgement and experience is often unacknowledged in technical decision making (Hayes 2009b; Power 2007). Regulators and operating companies have lost sight of the view that risk is a mental model of how actual harm can be predicted and controlled (Aven, Renn and Rosa 2011; Douglas and Wildavsky 1982; Power 2004, 2007; Renn 2008; Wildavsky 1988). In making sense of the world, technical organisations and regulators have considered technical cues in constructing and enacting their reality because these are the issues that are apparently controllable. According to Scheytt et al. (2006), "we call something a risk when we seek to bring the future into the present and act upon it, i.e. make it decidable".

The technical view of risk systematically emphasises those risks that can be calculated and quantified. Whilst this has led to improvements in engineering analysis, it has also resulted in issues such as leadership, professionalism, competence, experience and judgement being ignored despite compelling evidence that the potential for harm from any hazardous activity depends critically on these factors. A broader understanding of the concept of risk is needed to encompass identification and management of relevant organisational factors.

8.5.2. *Organisational Issues to Regulate*

Incorporating requirements in safety legislation as to how operating companies should manage organisational issues implies agreement on an appropriate model which defines what issues impact safety and how they should best be managed. Unfortunately, there is no unifying practical framework that allows organisational issues to be addressed in the same way as technical safety risk, although attempts to create such a framework are often drawn together under the general description of safety culture.

Many theorists have noted inconsistencies in the way this term is used and the implications for its utility in improving safety performance (Antonsen 2009; Haukelid 2008; Hopkins 2005; Høivik et al. 2009; Klein, Bigley and Roberts 1995; Mearns, Chapter 3 in this volume; Ocasio 2005; Pidgeon 2010; Silbey 2009), for example:

- lack of agreement over whether culture is a fundamental property of an organisation or a process that enacts organisational life (or both);
- linkage (or lack thereof) between the concept of safety culture and broader organisational culture;
- the extent to which culture is homogeneous across an organisation; and
- how / if safety culture (or safety climate) can be measured.

These issues have a direct impact on the extent to which culture is manipulable by management (and appropriate for regulatory attention) in ways with predictable outcomes for safety performance. This has certainly been the goal of much of the work in this area, but the results have been mixed. Power maintains that efforts to audit organisational culture are "essentially reductive" and that "indicators lose their proxy status and become regarded as the things they stand for". His view is that good risk management is an "organizational conversation" which does not lend itself to standardised routines and auditable processes (Power 2007, 177).

Rather than seeking to regulate organisational culture overall, a better approach may be to focus initial policy in this area on some specific factors where there is general theoretical agreement. One issue on which high-reliability theorists, resilience engineering practitioners and safety culture academics all agree is the importance of learning from incidents – both small and large. This provides a potential place to start in regulating organisational issues. This is not simply a matter of ensuring that an incident-reporting system is in place, but rather requires a broader focus on organisational learning (Hayes 2009a). Various authors (Busby 2006; Hopkins, 2008; Pidgeon 2010; Pidgeon and O'Leary 2000) have noted barriers to learning from accidents and incidents. Pidgeon (2010) describes the main reasons that organisations are blind to latent failures in their systems as lack of "safety imagination" and conflicting interests as a result of organisational power and politics, whereas Busby (2006) proposes a range of factors that are either requisite for mobilisation or can undermine change once it is underway. Hopkins's (2008) analysis of the BP Texas City refinery incident proposes a number of organisational factors that prevented BP from acting on evidence of problems, including the sidelining of technical specialists and the lack of senior management performance incentives regarding process safety performance.

There is ample research available that provides indicators of the factors that are important in ensuring that organisations notice small incidents and act on the lessons that can be learnt. This would be a useful focus for regulatory attention without requiring an agreed universal framework for systematically addressing all organisational contributors to accident causation. Whilst a broader engagement with organisational issues may ultimately be desirable, it is worth noting the general experience on the utility of the safety case for technical safety issues as described earlier in the chapter. The evidence suggests that paying attention to these issues is, in itself, important. Requiring organisations to pay more attention to learning lessons from small incidents is a valid and useful regulatory goal in its own right.

8.5.3. *Safety Cases and Enforced Self-Regulation*

The safety case concept requires that operators demonstrate that the systems and processes they have adopted ensure that their facility is sufficiently safe for all those people who are possibly impacted by their activities (workers and contractors at their facility, passengers and customers, people who live and work in surrounding areas). Most safety case regulations require the demonstration to take the general form of showing that risk is as low as reasonably practicable, or ALARP (or that risk has been reduced so far as practicable – SFAP). Clearly, safety cases are an example of the broad trend towards risk-based regulation (Hood et al. 2004).

Safety case regulations can also be characterised as a type of enforced self-regulation (Ayres and Braithwaite 1992). This type of arrangement achieves regulatory goals firstly by requiring organisations to produce their own set of standards that they must undertake to meet, and secondly by enforcing compliance with those standards. Paterson (2000, 243) describes the safety case approach to regulation as follows: "[I]nstead of translating engineering or management norms into generalised legal norms, it allows the production of context-specific social norms and holds the operator to compliance with its own standards and procedures".

Requiring organisations to submit as part of their safety case a company-specific description of how small failures are identified and what lessons are learnt would fit well within this framework. Guidelines that invite companies to address such issues as organisational structures, defined roles and responsibilities, remuneration packages and bonus arrangement, separation of technical and managerial responsibilities and other organisational factors would ensure that consideration was not focussed only on traditional management system items such as databases and forms.

8.6. PREVENTING ANOTHER MONTARA

If the operating company of the Montara development had been required to make a regulatory submission about relevant failures in industry and in their own operations – how they are identified and what lessons can be learned – would the blowout have been less likely to occur?

The organisational processes described in the opening section of this chapter seem to indicate that trial and error offshore was the primary method of organisational learning regarding drilling practices at Montara, and that very little heed was taken of past experience. On the other hand, there was plenty of information about past incidents and accidents available to the organisation that was directly relevant to H1 activities if only someone had taken the time to seek it out. This could have impacted the decisions made if only systems and processes had been in place to ensure that such information was considered in well design and operations.

Firstly, in relation to the installation of the cement shoe, cementing is generally understood to be a safety-critical activity in well construction. Cemented shoes, cement plugs and similar devices are used as primary well control barriers, and cementing problems contributed to eighteen of thirty-nine blowouts on the U.S. Outer Continental Shelf in the period from 1992 to 2006 (Izon et al. 2007). If demonstration of learning from accidents is an explicit requirement of the regulations, then it is difficult to see how cementing of the shoe would not be seen as safety critical with the integrity of the final arrangement closely controlled. In that case, the company would have to make better arrangements for technical review of the results of the cementing operations and subsequent testing.

If the justification for use of PCCCs had required a review of where these devices had been used previously for well control, this could have revealed that the devices were not designed for the function for which they were proposed to be used on the H1 well.

In addition to processes to ensure that critical technical activities were identified, social-science-based guidelines regarding learning from incidents would also address issues such as the location of technical experts within the organisation. This would require organisations to consider this issue explicitly and hence give regulators access to monitoring and enforcement actions linked to these issues.

In common with the other parts of the safety case regulations, the benefit of including social science aspects is that it requires explicit consideration of and commitment regarding safety controls by the operating organisation. If such a requirement had been in place before the Montara incident, it is possible that the operating company would have been more aware of its specific vulnerabilities in this area and taken steps to address the large gaps in the way the integrity of well operations was assured.

8.7. CONCLUSIONS

Offshore safety regulation is an example of self-regulation where internal company governance processes form the basis of public trust. Safety in the offshore oil and gas industry is primarily the responsibility of the duty holder, but accidents are also an indication of the failure of government and regulation. Many researchers have shown that accidents in complex socio-technical systems such as offshore drilling rigs and production facilities are caused by organisational failures, and yet the prevention of failures of this type is not commonly addressed by risk-based safety regulation. Processes relating to identification and control of technical risks receive a generally high level of regulatory scrutiny, but, with the exception of the Norwegian regulatory arrangements, issues such as leadership, professionalism, competence, experience and judgement remain unexamined despite their criticality in preventing deaths, environmental damage and major financial losses.

Attempts that have been made to date to incorporate these issues into consideration of risk have not been very successful, primarily because of the three issues described earlier: difficulty in linking organisational factors into the current risk management tools; lack of a universally agreed framework for organisation issues; and a narrow understanding of the concept of risk leading to a misconception about what can and cannot reasonably be addressed in a safety case.

Regulation is as much about behaviour modification as it is about setting of standards (Baldwin and Black 2008; Hood et al. 2004). As Ayres and Braithwaite (1992, 4) maintain, "[t]he very behaviour of an industry or the firms therein should channel the regulatory strategy to greater or lesser degrees of government intervention". Evidence from the Montara incident and other disasters illustrates the need for regulatory change in this area. This chapter has outlined a proposal to incorporate social science research regarding learning from incidents and small failures into the technical risk management processes that currently dominate consideration of risk within the safety case framework.

REFERENCES

Allard-Poesi, F. (2005) "The paradox of sensemaking in organizational analysis", *Organization*, **12**: 169–196.

Antonsen, S. (2009) "Safety culture and the issue of power", *Safety Science*, **47**: 1118–1128.

Aven, T., Renn, O. and Rosa, E.A. (2011) "On the ontological status of the concept of risk", *Safety Science*, **49**: 1074–1079.

Ayres, I. and Braithwaite, J. (1992) *Responsive Regulation: Transcending the Deregulation Debate*. Oxford University Press: Oxford and New York.

Baker, J. (2007) *The Report of the BP US Refineries Independent Safety Review Panel*. Baker Report. London and Washington, DC.

Baldwin, R. and Black, J. (2008) "Really responsive regulation", *The Modern Law Review*, **71**: 59–94.

Borthwick, D. (2010) *Report of the Montara Commission of Inquiry.*
Bourrier, M. (1998) "Elements for designing a self-correcting organization: Examples from nuclear power plants", in A.R. Hale and M. Baram (eds.): *Safety Management: The Challenge of Change*. Pergamon: Oxford, pp. 133–147.
(2011) "The legacy of the high reliability organization project", *Journal of Contingencies and Crisis Management*, **19**: 9–13.
Busby, J.S. 2006. "Failure to mobilize in reliability-seeking organizations: Two cases from the UK railway", *Journal of Management Studies*, **43** (6): 1375–1393.
CAIB (2003). *Report of the Columbia Accident Investigation Board*. Volume 1. Columbia Accident Investigation Board.
Douglas, M. and Wildavsky, A. (1982) *Risk and Culture: An Essay on the Selection of Technological and Environmental Dangers*. University of California Press: Berkeley.
Duncan, C.N. (2009) Statutory Declaration of CN Duncan. http://www.montarainquiry.gov.au/exhibits.html.
Forester, J., Kolaczkowski, A., Lois, E. and Kelly, D. (2006) *Evaluation of Human Reliability: Analysis Methods Against Good Practices*. U.S. Nuclear Regulatory Commission: Washington, DC.
Hale, A., Goossens, L. and van de Poel, I. (2002) "Oil and gas industry regulation: From detailed technical inspection to assessment of safety management", in B. Kirwan, A. Hale and A. Hopkins (eds.): *Changing Regulation: Controlling Risks in Society*. Pergamon: Oxford.
Haukelid, K. (2008) "Theories of (safety) culture revisited – An anthropological approach", *Safety Science*, **46**: 413–426
Hayes, J. (2009a) "Incident reporting: A nuclear industry case study", in A. Hopkins (ed.): *Learning from High Reliability Organisations*. CCH: Sydney.
(2009b) "Operational decision-making", in A. Hopkins (ed.): *Learning from High Reliability Organisations*. CCH: Sydney.
(2011) "Operator competence and capacity – Lessons from the Montara blowout", *Safety Science*, **doi**:10.1016/j.ssci.2011.10.009.
(2012a) "Operator competence and capacity – Lessons from the Montara blowout", *Safety Science*, **50**: 563–574.
(2012b) "Use of safety barriers in operational safety decision making", *Safety Science*, **50**: 424–432.
(in press) "The role of professionals in managing technological hazards: The Montara blowout", in S. Lockie, D.A. Sonnenfeld and D.R. Fisher (eds.): *Routledge International Handbook of Social and Environmental Change*. Routledge: London.
Høivik, D., Moen, B.E., Mearns, K. and Haukelid, K. (2009) "An explorative study of health, safety and environment culture in a Norwegian petroleum company", *Safety Science*, **47**: 992–1001.
Hollnagel, E., Nemeth, C.P. and Dekker, S. (eds.) (2008) *Remaining Sensitive to the Possibility of Failure*. Ashgate: Aldershot.
Hollnagel, E., Pariès, J., Woods, D.D. and Wreathall, J. (eds.) (2011) *Resilience Engineering in Practice: A Guidebook*. Ashgate: Aldershot.
Hollnagel, E., Woods, D.D. and Leveson, N. (eds.) (2006) *Resilience Engineering: Concepts and Precepts*. Ashgate: Aldershot.
Hood, C., Rothstein, H. and Baldwin, R. (2004) *The Government of Risk: Understanding Risk Regulation Regimes*. Oxford University Press: Oxford.
Hopkins, A. (2000) *Lessons from Longford: The Esso Gas Plant Explosion*. CCH: Sydney.

(2005) *Safety, Culture and Risk: The Organisational Causes of Disasters*. CCH: Sydney.

(2007) "Beyond compliance monitoring: New strategies for safety regulators", *Law & Policy*, **29** (2): 210–225.

(2008) *Failure to Learn: The BP Texas City Refinery Disaster*. CCH: Sydney.

(ed.) (2009). *Learning from High Reliability Organisations*. CCH: Sydney.

(2010) "Risk-management and rule-compliance: Decision making in hazardous industries", *Safety Science*, doi:10.1016/j.ssci.2010.07.014.

(2012) *Disastrous Decisions: The Human and Organisational Causes of the the Gulf of Mexico Blowout*. CCH: Sydney.

HSE (1995) *An Interim Evaluation of the Offshore Installation (Safety Case) Regulations 1992*. In UK Health and Safety Executive: Norwich.

HSE (2003) *Literature Review on the Perceived Benefits and Disadvantages of UK Safety Case Regimes*. UK Health and Safety Executive: Norwich.

HSE Major Incident Investigation Board (2008) *Final Report – The Buncefield Incident 11 December 2005*. Richmond, Surrey.

Izon, D., Danenberger, E.P. and Mayes, M. (2007) "Absence of fatalities in blowouts encouraging in MMS study of OCS incidents 1992–2006", *Drilling Contractor* (July–August):.

Jacobs, A.C. (2009) *Statutory Declaration of A.C. Jacobs* (revised). http://www.montarainquiry.gov.au/exhibits.html.

Klein, R.L., Bigley, G.A. and Roberts, K.H. (1995) "Organizational culture in high reliability organizations: An extension", *Human Relations*, **48**: 771–793.

La Porte, T.R. (1996) "High reliability organizations: Unlikely, demanding and at risk", *Journal of Contingencies and Crisis Management*, **4**: 60–71.

La Porte, T.R. and Consolini, P.M. (1991) "Working in practice but not in theory: Theoretical challenges of 'high-Reliability Organizations'", *Journal of Public Administration Research and Theory*, **1**: 19–48.

National Commission on the BP Deepwater Horizon Oil Spill and Offshore Drilling (2011a) *A Competent and Nimble Regulator: A New Approach to Risk Assessment and Management*. Working paper 21. Washington, DC.

National Commission on the BP Deepwater Horizon Oil Spill and Offshore Drilling (2011b) *Deep Water: The Gulf Oil Disaster and the Future of Offshore Drilling*. Report to the President. Washington, DC.

Ocasio, W. (2005) "The opacity of risk: Language and the culture of safety in NASA's space shuttle program", in W. H. Starbuck and M. Farjoun (eds.): *Organization at the Limit: Lessons from the Columbia Disaster*. Blackwell: Malden, p. 103–115.

Paterson, J. (2000) *Behind the Mask: Regulating Health and Safety in Britain's Offshore Oil and Gas Industry*. Ashgate: Aldershot.

Pidgeon, N. (2010) "Systems thinking, culture of reliability and safety", *Civil Engineering and Environmental Systems*, **27** (3): 211–217.

Pidgeon, N. and O'Leary, M. (2000) "Man-made disasters: Why technology and organizations (sometimes) fail", *Safety Science*, **34**: 15–30.

Power, M. (2004) *The Risk Management of Everything: Rethinking the Politics of Uncertainty*. Demos: London.

(2007) *Organized Uncertainty: Designing a World of Risk Management*. Oxford University Press: Oxford.

PTT. 2009. Email – Montara Platform Wells Morning Update, dated 16 April 2009. http://www.montarainquiry.gov.au/exhibits.html.

Rasmussen, J. and Svedung, I. (2000) *Proactive Risk Management in a Dynamic Society*. Swedish Rescue Services Agency: Karlstad.

Reason, J. (1997) *Managing the Risks of Organizational Accidents*. Ashgate: Aldershot.

Renn, O. (2008) *Risk Governance: Coping with Uncertainty in a Complex World*. Earthscan: London.

Roberts, K.H. (ed.) (1994) *New Challenges to Understanding Organizations*. Macmillan: New York.

Rochlin, G.I. (1999) "Safe operation as a social construct", *Ergonomics*, **42**: 1549–1560.

Rose, D. and Crescent, J. (2000) "Evaluation of the offshore safety legislative regime in the UK", *SPE International Conference on Health, Safety, and the Environment in Oil and Gas Exploration and Production*. Stavanger, Norway: Society of Petroleum Engineers.

Scheytt, T., Soin, K., Sahlin-Andersson, K. and Power, M. (2006) "Organizations, Risk and Regulation," *Journal of Management Studies*, **43**: 1333.

Schulman, P. R. (1996) "Heroes, organizations and high reliability", *Journal of Contingencies and Crisis Management*, **4**: 72–82.

Silbey, S. (2009) "Taming Prometheus: Talk about safety and culture", *Annual Review of Sociology*, **35**: 314–369.

Snook, S.A. (2000) *Friendly Fire: The Accidental Shootdown of US Black Hawks over Northern Iraq*. Princeton University Press: Princeton.

Starbuck, W.H. and Farjoun, M. (eds.) (2005) *Organization at the Limit: Lessons from the Columbia Disaster*. Blackwell Publishing: Oxford.

Vaughan, D. (1996) *The Challenger Launch Decision: Risky Technology, Culture and Deviance at NASA*. University of Chicago Press: Chicago.

Vectra (2003) *Literature Review of the Perceived Benefits and Disadvantages of the UK Safety Case Regimes*. UK Health and Safety Executive: London.

Vinnem, J.E. (2010) "Analysis of root causes of major hazard precursors in the Norwegian offshore petroleum industry", *Reliability Engineering and System Safety*, **95** (11): 1142–1153.

Weick, K.E. (1987) "Organizational culture as a source of high reliability", *California Management Review*, **29**: 112–127.

(1993) "The collapse of sensemaking in organizations: The Mann Gulch disaster", *Administrative Science Quarterly*, **38**: 628–652.

(1995) *Sensemaking in Organizations*. Sage: Thousand Oaks.

Weick, K.E. (ed.) (2001) *Making Sense of the Organization*. Blackwell Business: Oxford.

Weick, Karl E. and Sutcliffe, Kathleen M. (2001) *Managing the Unexpected: Assuring High Performance in an Age of Complexity*. Jossey-Bass: San Francisco.

Weick, K.E., Sutcliffe, K.M. and Obstfeld, D. (2005) "Organizing and the process of sensemaking", *Organization Science*, **16**: 409–421.

Wildavsky, A. (1988) *Searching for Safety*. Transaction Books: New Brunswick.

Wilson, C.A. (2009) Statutory Declaration of CA Wilson. http://www.montarainquiry.gov.au/exhibits.html.

9

Safety Indicators Used by Authorities in the Petroleum Industry of the United Kingdom, the United States and Norway

Helene Cecilie Blakstad

9.1. INTRODUCTION

Contemporary offshore safety statistics is a challenging topic for several reasons. Internationally there is a lack of scientific studies based on empirical data of regulatory approaches taken by different regulatory regimes. There is also the challenge of developing robust longitudinal data which catches the relevant factors within a specific regulatory regime. However, the major challenge involves developing comparative data across regimes with different contexts, cultures/values and history.

The International Regulators' Forum (IRF) documents differences in the regulatory regimes of countries involved in offshore petroleum activities (IRF 2011a). There are many factors in each country that influence regulation itself, what is given attention by the regime, and what is relevant to measure as indicators of the efficiency or outcomes of the activities. This implies that there might be a difference between what is relevant to measure when the purpose is to evaluate effect of regulation over time in one particular country and what is relevant to measure for comparison between different countries. Hood et al. (2001) discuss the fact that variations in risk regulation regimes are often difficult to explain.

Performance data is also a recognized challenge for regulators. For instance, the Head of HSE Offshore Division Steve Walker (2010) argues that regulators need performance data for several purposes including notification of actions, immediate and long-term resource decisions and determining success or failure. IRF has provided guidelines for performance measurement data in the petroleum industry (IRF 2011c). However, the data provided by regulatory authorities are often more about measuring the activities of the authorities than the effect of these activities.[1] Companies within the petroleum industry and their industry groups also provide data that are used as indicators by the regulators and themselves.[2]

[1] Telephone interview with Torleif Husebø, PSA, 16 May 2011.
[2] Ibid.

The intention of this chapter is to contribute a positive response to these challenges.

9.2. CHALLENGES OF PERFORMING COMPARISON BETWEEN COUNTRIES

Øien et al. (2011a) find that there has been an increasing tendency to put 'everything' under the umbrella of 'indicators'. Further, a variety of perspectives and dimensions can be applied when discussing and comparing indicators, and so the focus of attention can become very different.

9.2.1. *The Term 'Indicators'*

The term 'indicator' may be used in several ways, which means that there exist many definitions (see Øien et al. [2011a] for more thorough discussion). Also, terms which may have different meanings, such as 'safety performance measures' or ' safety outcome measures' and 'safety indicators' or 'risk indicators', are sometimes used interchangeably.

Different definitions can have different purposes. For instance, OECD (2003) defines a safety performance indicator as 'a means for measuring the changes over time in the level of safety (related to chemical accident prevention, preparedness and response), as the results of actions taken'. Skogdalen et al. (2010, p 109) similarly consider a safety performance indicator to be 'a means for measuring the changes in the level of safety (related to major accident prevention, preparedness and response), as a result of actions taken'. These definitions reflect that they are made for comparisons over time to demonstrate developments as results of actions. Accordingly, they are not made for comparison between countries and they do not give attention to the needs of special users of the indicators.

Øien et al. (2011a) discuss distinguishing between the terms 'safety indicators' and 'risk indicators'. According to their view, risk indicators are developed from a risk-based approach, whereas safety indicators may be developed from various other approaches, such as a safety-performance-based approach, an incident-based approach or a resilience-based approach. Thus a probabilistic risk assessment is only one basis for the development of indicators. In this connection, the term 'safety' can be associated with something positive, whereas the term 'risk' can be associated with something negative – in other words, the terms can also reflect a difference in the focus of attention from something to achieve versus something to avoid.

Øien et al. (2011a) also associate the term 'indicator' with the term 'risk-influencing factor' (RIF). They consider a RIF to be 'an aspect (event/condition) of a system or an activity that affects the risk level of the system or activity'. A given RIF might not be measurable but can be defined theoretically. The measuring of a RIF may be performed by an indicator or a set of indicators where the indicator is an operational variable that defines a RIF measurably. Hence, it is necessary to look behind the indicators used in different countries and take into consideration the safety or other aspects that the indicators are meant to highlight for attention.

The conclusion to be drawn from these examples is that when developing indicators and comparing indicators from different countries, it is important to consider their purpose, their underlying assumptions and rationales and their focus of attention.

9.2.2. *Perspectives and Focus of Attention*

Discussion of the use of safety indicators by regulatory regimes in the United Kingdom, the United States and Norway requires looking into the differences in their perspectives and focuses of attention.

9.2.2.1. The Predictive versus Retrospective Perspective

One much-discussed distinction is between the 'predictive' perspective versus the 'retrospective' perspective (Øien et al. 2011a). Here the distinction is made between indicators to predict the possibility of having a major accident 'tomorrow' and indicators to determine causes and contributing factors after an accident or near miss.

The terms 'leading' versus 'lagging' indicators are often associated with these perspectives as discussed by Hopkins (2009a, 2009b), Hale (2009a, 2009b), Øien et al. (2011a, 2011b), Vinnem (2010), and Kjellén (2009). Øien et al. (2011a, 2011b) explain that leading indicators can be seen as activity-based indicators or proactive indicators measuring potential contributing factors and involve active monitoring to achieve the desired safety outcome. Lagging indicators can be seen as reactive indicators for measuring contributing factors after the event and related to retrospective analyses for determining why a desired safety was not achieved. Here we also find the search for indicators of 'early warnings' or signals of a dangerous development (Øien et al. 2011a), such as if and when the Macondo blowout could have been foreseen (Skogdalen et al. 2010).

In discussing the notion of early warnings, Fleming (2010) distinguishes between status indicators and curative indicators. His curative indicators relate to leadership, data management, organisational learning and communication. His status indicators apply to rate of maintenance problems, ratio of corrective to preventative maintenance to critical systems and rate of plant changes not incorporated into design documentation before next turnaround. The first group can be seen as indicators measuring the ability of an organisation to create safe conditions and handle unexpected dangerous events, while the latter can be seen as indicators measuring weaknesses of the organisation that might result in an accident.

Fleming also makes a useful distinction between outcomes and indicators. He finds that outcomes are retrospective and objective by nature and can serve as important performance measures, whereas indicators may be subjective and may be of interest only for prediction. However, he demonstrates that the distinction between predictive versus retrospective is not simple and straightforward. For instance, the same data such as lost time injury frequency rate (LTI)[3] and Leak rates can be used both for retrospective and predictive purposes. Further discussion about how LTI and Leak rates can be understood, what their foundation might be, how they might interact and whether they can be used to tell something about other topics than injuries and leaks will be elaborated in subsection 9.2.2.3.

Thus, discussion has sought to distinguish between indicators that focus upon existing, real-time conditions that can provide the ability to prevent accidents and indicators that focus upon past experience to anticipate the future.

9.2.2.2. The Technical-Human-Organisational Perspective

Another perspective is the so-called technical-human-organisational perspective; Øien et al. (2011a) have considered the nuclear industry in this regard. This perspective is related to accident investigation which looks into technical, human and organisational factors. Here the assessment of plant safety is conducted by evaluating the physical system design and performance and the operational system design and performance. The latter is about the organisational and human contribution and covers operational safety, human performance and safety culture. Included here can be indicators that focus on different types of behaviour, (Lamvik and Ravn 2006) and different types of knowledge. In Norway, this has led to discussion about whether organisational incentive systems for leaders and employees should be adopted.[4]

Other issues associated with this perspective are whether the indicators focus on direct or indirect causes, and on the sharp versus the blunt end of operations (Vinnem 2010). Øien et al. (2011b) also directs attention to different phases in the progression of work. Finally, indicators concerning the status/levels of different barriers against accidents and their negative outcomes have been discussed. This reveals that when discussing indicators, it is important to consider how the indicators give

3 LTI-rate – *lost time injury frequency rate* is defined as the number of lost-time injuries per one million hours of work. A lost-time injury is an injury resulting from an accident at work, where the injured person does not return to work on the next shift (Kjellén, 2000)

4 Workshop, Robust Regulation 17 November 2011 in Stavanger, Norway, with participants from representatives of the Petroleum Industry, Trade Unions and the Petroleum Safety Authority Norway (PSA).

attention to technology, humans and organisations and to the interactions between these elements.

9.2.2.3. Different Aspects of HSE

Indicators might be directed at different aspects of health, safety and environment (HSE), such as major hazards, occupational hazards and illnesses, helicopter transportation hazards, physical and psychosocial working environment aspects, behaviours, attitudes and safety culture, as well as perceptions of risk (Vinnem 2010). The indicators may also differ according to types of damage, such as damage to persons, assets or environment (Øien et al. 2011b).

However, applying indicators for one aspect of safety amongst other aspects can be problematic. In particular there have been discussions about the usefulness of applying indicators for occupational hazards and personal safety as indicators either for major hazards (Vinnem 2010), major accidents or system safety (Øien et al. 2011b) or process safety (Hopkins 2007, 2009a, 2009b). For example, in hindsight of the Macondo blowout, the rather high confidence in safety, based on a positive development in personal injuries (LTI-rates) as the primary safety performance measure, has been criticized (Fleming, 2010; Ryggvik, Chapter 15 in this volume). When discussing personal safety versus process safety, Hopkins (2007) argues that the distinction between leading and lagging indicators is particularly complicated when applied to process safety.

When comparing outcome indicators related to HSE, it is also important to note differences in how they are defined and reported. For instance, what does 'reported injuries' mean? Do they only include injuries leading to absence from work, or all injuries? Are serious accidents and fatal accidents included? Does 'lost time injuries' mean absence in three days or in more? Is reporting voluntary or mandatory?

Again, this reveals that when discussing indicators, it is important to be cautious about the purpose and direction of attention that the indicators represent. The discussions also underline the importance of being cautious about definitions, generalizability and use.

9.2.3. *Data and Comparability*

Comparing the use of indicators by different countries is made complicated because of varying opportunities to develop statistics and make them public. The IRF provides some self-reported background information about the regulatory regimes of different countries (IRF 2011a). This information shows that the regimes work under different conditions and apply different principles in their legislation and follow-up governance of the offshore petroleum industry. Also, through their Performance Measurement Project, IRF provides performance statistics from different countries (IRF 2011b). When looking into these, it is evident that there are variations between the countries with respect to types of data provided about safety performance, and how well the data fit into the form that IRF applies. It is a challenge that there are differences between countries in the traditions and requirements for reporting,[5] including differences in transparency and making available data open for public access and education (Steinzor, 2011; Baram, Chapter 7 in this volume).

The quality of indicators or performance data has been an issue at stake. According to Fleming (2010), good indicators are accurate (with direct relationship to system status, and difficult to manipulate), productive (related to future system states and performance) and current (giving real-time information). Skogdalen et al. (2010, 121–122) refer to the work of the American Petroleum Institute (API) about developing Process Safety Performance Indicators for the Refining and Petrochemical Industries, and finds that the success of indicators is related to the extent to which they are:

- able to drive process safety performance improvement and learning;
- easy to implement and understand by stakeholders (e.g., workers and public);
- statistically valid at industry, company and/or site level– with validity requiring a consistent definition, a minimum data set size, a normalization factor and a relatively consistent reporting pool;
- appropriate for industry-, company-, or site-level benchmarking

Skogdalen et al. (2010) add that major accident indicators must also reflect hazard mechanisms, that is, be valid for major hazards, be sensitive to change, show trends, be robust regarding manipulation and influence from campaigns giving conflicting signals and not require complex calculations. The International Atomic Energy Agency (IAEA) has also suggested characteristics of quality for safety performance indicators (IAEA 1999).

When comparing indicators from different countries, there is concern that national characteristics can influence the data in ways that make it difficult to draw conclusions. For instance, Lamvik and Ravn (2006) discuss that there might be differences in safety performance between countries because of contextual factors such as elements of national culture. One example is differences related to continuity of traditions regarding contractual terms and arrangements, workforce, crew

[5] Telephone interview with Torleif Husebø, PSA, 16 May 2011.

organisation and the role and function of management. Another example is differences related to flexibility in government, authorities and industrial relations. There may also be differences in complexity of systems, degree of multinational workforce and responsibility delegated to the operator/contractor.

Further, different regulatory regimes might themselves create conditions that complicate comparisons, For instance, when comparing management functions, especially for 'paperwork', Lamvik and Ravn (2006) found that requirements for documentation influenced the behaviour of the management. Supervisors with experiences from both South East Asia and the North Sea reported that the extensive paperwork obligations in the North Sea made them spend less time outside the office in the North Sea than when they worked in South East Asia.

The relevant level of aggregation of indicators for comparison is also an issue. Is it of interest to develop indicators at the company level according to types of activities or at the overall industry level? Each has value: the former for improving company performance in fulfillment of regulatory requirements, the latter for identifying systemic risk, for example.

When comparing data from different countries, these discussions illustrate that it is important to pay attention to regulatory requirements for reporting and certification, the sources of information, transparency and the quality and level of aggregation of the data. Also national contexts such as culture, norms and regulatory properties are relevant.

9.3. GENERAL STATUS OF INDICATORS IN THE UNITED KINGDOM, THE UNITED STATES AND NORWAY

The regulatory regimes of the United Kingdom, the United States and Norway are functioning under different conditions, and the use and status of indicators varies between the countries, as shown by IRF reports (IRF 2011a, 2011b). These differences influence the types of indicators the authorities refer to, the climate for cooperation, on indicators, and what kind of information the authorities can require from the companies.[6] Further discussion of these differences regarding indicators is provided later in the chapter. The main focus will be on the use of measurable data, the themes/content that are covered and underlying perspectives about the data. The chapter will neither go into the results of the measurements nor deal in depth with applied methods. Discussion is followed by a collection of background data from the three countries provided by IRF (2011a). Finally, several other sources of indicators that might be of interest for further research are presented.

6 Telephone interview with Torleif Husebø, PSA, 16 May 2011.

9.3.1. *Indicators Used in the United Kingdom*

Regulations of United Kingdom require companies to report injuries, diseases and dangerous occurrences (RIDDOR 1995).[7] The health and safety executive (HSE) provides yearly[8] reports of statistics (HID Statistics) on offshore accidents, dangerous occurrences and ill health. These annual reports are based on incidents reported to HSE under the requirements of RIDDOR. For offshore operations, they cover incidents directly affecting offshore installations and workers and visitors on the installations. The requirements apply to offshore wells and most activities in connection with them, as well as offshore pipelines, pipeline works and certain activities in connection with pipeline works. Incidents on offshore wind farms are included. Also included are incidents arising from certain diving operations in connection with offshore installations, that is, offshore diving and diving support activities associated with an offshore installation only. However, the reports do not include incidents arising from marine activities that are not directly connected with offshore operations (e.g. vessels or rigs in transit) or air transport activities (including transport to, from or between installations), except incidents affecting helicopters whilst located on an installation. An overview of the main indicators used for the period from April 2009 to March 2010 is presented in Table 9.1 (HSE 2010).

In addition, HSE's Offshore Division has performed programs using selected indicators (HSE 2007). The Key Program 1 (KP1), launched in 2000 to reduce hydrocarbon releases at processing plants, measured success according to the number of major and significant hydrocarbon releases.

As discussed by Paterson (Chapter 6 in this volume), Key Program 3 Asset Integrity Inspection (KP3) was initiated out of concerns for major accidents on the United Kingdom continental shelf (UKCS) (HSE 2007). The work started in 2004 and focused on the effective management and maintenance of safety critical elements (SCEs). The program report illustrates how the relationship between major hazards, development of SCEs and maintenance management is considered.[9] KP3 involved targeted inspections of nearly 100 offshore installations representing about

[7] Reportable injuries and dangerous occurrences are defined in regulation 3 of RIDDOR. Further description of 'Major Injuries' is provided in RIDDOR under Schedule 1. Reference to over-three-day injuries in this annual report includes all other RIDDOR reportable injuries that are less severe than Major Injuries. Description of reportable Dangerous Occurrences is provided in Schedule 2.

[8] From April to March of the following year.

[9] The SCE's are the parts of an installation and its plant (including computer programs), the purpose of which is to prevent, control or mitigate major accident hazards (MAH) and the failure of which could cause or contribute substantionally to major accidents, i.e. the SCEs represent the barriers which prevent, control or mitigate the major accident scenarios. The thought was that the maintenance management strategy must be developed to provide assurance that the SCEs will be available when required, they will operate with the required reliability and be able, as necessary, to survive incidents against which they are designed to protect (HSE 2007, 9)

TABLE 9.1. *Overview of indicators applied in the United Kingdom by HSE (HSE 2007, 2010)*

Source	Indicator	Description	Comments
Offshore Safety statistics (HSE, 2010)			Incidents reported to HSE under the Reporting of Injuries, Diseases and Dangerous Occurrences Regulations 1995 (RIDDOR)
	Fatal injuries to offshore workers	Number of incidents. Injury rates calculated by using offshore population data	Injury rates are calculated using offshore population data from the industry's Vantage personnel tracking system. Up to and including 2003–2004, data from the Inland Revenue was used. For 2009–2010, the estimated offshore population (based on total number of hours worked divided by 2,000 hours per worker year) was 26,598, a 5.76% reduction compared to the previous year's figure of 28,224.
	Major injuries to offshore workers	Number of incidents. Injury rates calculated by using offshore population data	Injuries are categorized according to severity and nature. The report shows the ratio of over-three-day to major injuries and calculates three-year rolling averages to smooth out variations and give a clearer picture of overall trends. Part of body injured, kind of accident, age of injured person, work process environment (type of activity being carried out when the incident occurred), agent of accident against severity of injury are also differentiated
	Over-three-day injuries to offshore workers	Number of incidents. Injury rates calculated by using offshore population data	
	Incidents of ill heath to workers offshore	Number of incidents	
	Dangerous occurrences offshore	Number of incidents. Split by type	Well-related dangerous occurrences are split into subcategories.

	Hydrocarbon Releases (HCR) Minor, Major and Significant	Number of releases, classified according to severity	The classification of a hydrocarbon release incident as 'Significant' or 'Major' implies the *potential for that release, if ignited*, to directly cause or escalate to an event severe enough to be viewed as a 'Major Accident'. Emergency action is normally required to be taken to limit the potential consequences of ignition of a reportable leak of hydrocarbon. (Severity Classification' guidance for Major, Significant and Minor HCRs can be found on the HSE Web site at https://www.hse.gov.uk/hcr3/help/help_public.asp) Detailed supplementary data relating to HCRs reported under RIDDOR are voluntarily reported to HSE by offshore operators.
KP3 factors (HSE, 2007)			
	Consistency of approach		
	Identification of common areas of good and poor practice across the industry		
	Ability to report on an industry wide basis		
	Facilitation of engagement		
	Raising the profile of integrity management across the industry		
	Performance indicators of the industry		

40 per cent of the total. The inspection program was structured using a template containing seventeen elements covering all aspects of maintenance management and a number of SCE systems tests. An element covering 'physical state of plant' was also included, allowing the inspection team's judgement of the general state of the platform to be recorded. The performance, on inspection, of each template element was scored using a traffic light system which enabled overall installation performance to be recorded on a matrix. In addition to performance indicators from the industry, the report includes evaluations of the regulatory approach taken in the program. The criteria for evaluation of the work of the program are also shown in Table 9.1.

One of the results from KP3 was the establishment of an Installation Integrity Workgroup (HSE 2007). The workgroup was set up by the industry itself and involves thirty oil companies, contractor organisations and independent verification bodies. Amongst other activities, the workgroup has developed new Key Performance Indicators (KPIs).

HSE and Chemical Industries Association (CIA) (2006) take into account that performance measurements may be divided into reactive monitoring (identifying and reporting on incidents and learning from accidents) and active monitoring (feedback on performance before an accident or incident occurs). HSE and CIA emphasise the importance of utilising both leading and lagging indicators, and use the a 'dual assurance' approach.

9.3.2. *Indicators used in the United States*

The Bureau of Ocean Energy Management, Regulation and Enforcement (BOEMRE) provides limited information about safety indicators.[10] When searching the homepage of BOEMRE for the word 'indicators', the main result is a discussion of the socioeconomic impacts of activities on the OCS.[11] These are related to ongoing social science and economic studies administered by different regions. The limited use of safety indicators by the U.S. regime is also discussed by Baram (Chapter 7 in this volume).

The main reason for the limited information about safety indicators is that BOEMRE has limited access to information, and the BOEMRE-led U.S. regime and U.S. law do not require annual updates of the offshore petroleum industry's risk level (Skogdalen et al. 2010). Traditionally the industry has been required to report LTI and oil releases. Lately the industry has also been required to report gas emissions.[12]

[10] Telephone interview with Torleif Husebø, PSA, 16 May 2011.

[11] http://www.mms.gov:8765/query.html?rq=0&qt=+indicators&charset=iso-8859–1&col=boemre

[12] Telephone interview with Torleif Husebø, PSA, 16 May 2011.

In addition, the regime calls for industry to report performance data on voluntary basis. According to BOEMRE, it replaced the Mineral Management Service (MMS) after the Macondo accident, and the MMS had collaborated with the U.S. Coast Guard and representatives of the Outer Continental Shelf (OCS) oil and gas industry to develop a suite of consensus formulas for gauging the industry's safety and environmental performance since 1997 (BOEMRE 2011). These formulas, called the OCS Performance Measures, are used to calculate twenty annual, OCS-wide performance indices. The indices provide the public with information about performance trends, and they allow OCS lease operators to compare their performance with industry 'averages'. The performance measures are presented in Table 9.2.

BOEMRE informs that the data used in calculating the OCS Performance Measures is generated by OCS lease operators who participate in the OCS Performance Measures Program (BOEMRE 2011). During the first quarter of each year, participating operators submit to BOEMRE their performance data for the previous year. BOEMRE uses the data to calculate the annual performance indices on behalf of the OCS Performance Measures Steering Committee. The OCS Performance Measures Program depends on participation by OCS lease operators. The latest statistics are from the year 2009. A look into the statistics about the participation rate in these surveys finds that it is very low.

As a result of the U.S. Chemical and Hazard Investigation Board (CSB) investigation of the 2005 BP Texas City Incident, CSB recommended the development of an American National Standards Institute (ANSI) standard.[13] The purpose was to create 'performance indicators for process safety in the refinery and petrochemical industries. The intention was to ensure that the ANSI standard identifies leading and lagging indicators for nationwide public reporting as well as indicators for use at individual facilities. The work would include methods for the development and use of the performance indicators'. As a result, the American Petroleum Institute (API)[14] issued Process Safety Performance indicators for the Refining and Petrochemical Industries in 2010, ANSI/API RP 754. These identify leading and lagging process safety indicators that are supposed to be useful for driving performance improvement. The indicators are divided into four tiers that represent a leading and lagging continuums. Tier 1 is the most lagging and Tier 4 is the most leading. Tiers 1 and 2

[13] http://www.api.org/Standards/new/api-rp-754.cfm

[14] The American Petroleum Institute (API) is a national trade association that represents all aspects of America's oil and natural gas industry. It has more than 400 corporate members, from the largest major oil company to the smallest of independents. The members are producers, refiners, suppliers, pipeline operators and marine transporters, as well as service and supply companies that support all segments of the industry. http://www.api.org/aboutapi/ Also see Baram, chapter 7.

TABLE 9.2. *Overview of indicators applied in the United States by BOEMRE/MMS*

Source	Indicator	Description	Comments
OCS Performance Measures			OCS: Outher Continental Shelf
	Recordable injury/illness	For more thorough definitions, see http://boemre.gov/perfmeas/formulas.htm	The statistics also provides background information such as Total Wells Spudded, Participants' Wells Spudded, Total OCS Platform count
	Production DART[a] case	Lost Workday Incident Rates	OCS Participant Incident rate (PIR). For production, drilling and construction, respectively
	EPA NPDES Noncompliance		OCS Participant Incident rate (PIR). For production, drilling and construction, respectively
	Fire/explosion	Number of reported non-compliances on EPA NPDES Monitoring Reports	OCS Participant Incident rate (PIR)
	Well blowout	Numbers	OCS Industry Incident Rate (IIR)
	Oil spill number <1bbl	Blowouts	OCS Industry Incident Rate (IIR)
	Oil spill number 1 to < 10bbl	Numbers and spills	OCS Participant Incident rate (PIR)
	Oil spill number >10 bbl	Numbers and spills	OCS Industry Incident Rate (IIR)
	Oil spill volume < 1 bbl	Numbers and spills	OCS Industry Incident Rate (IIR)
	Oil spill volume 1 to < 10bbl		OCS Participant Incident rate (PIR)

Oil spill volume > 10bbl	OCS Industry Incident Rate (IIR)
Oil spill volume all sizes	OCS Industry Incident Rate (IIR)
MMSb Incidents of Noncompliance (INC)	OCS Industry Incident Rate (IIR). For drilling and production, respectively
In addition, Process Safety Performance Indicators for the Refining and Petrochemical Industries have been recently developed for use. The first edition came out in April 2010.	The American Petroleum Institute (API)

[a] Unfortunately, the meaning of the acronym DART is not found.
[b] Minerals Management Service.
Note: Fatalities are not listed in the overview statistics.
Source: BOEMRE (2011)

are suitable for nationwide public reporting and Tiers 3 and 4 are intended for internal use at individual sites. The identified performance indicators are based on the following guiding principles[15]:

- Indicators should drive process safety performance improvement and learning.
- Indicators should be relatively easy to implement and easily understood by all stakeholders (e.g., workers and the public).
- Indicators should be statistically valid at one or more of the following levels: industry, company and site.
- Indicators should be appropriate for industry-, company- or site-level benchmarking.

[15] http://www.api.org/Standards/new/api-rp-754-rp-755.cfm

9.3.3. *Indicators Used in Norway*

The Petroleum Safety Authority (PSA)[16] has been engaged in measuring the safety performance of the petroleum industry on the Norwegian OCS and the effects of PSA activities (see Chapter 10 by Bang and Thuestad in this volume). The Norwegian petroleum industry is required by law to report incidents that could lead to severe accidents to PSA. This is required for both major accidents and occupational accidents. The PSA and the industry have a tradition of close cooperation and sharing of experiences.

The approach of the PSA has been to follow trends over time and interpret developments.[17] This has been done both at the level of the corporates and at the sectorial level. The main question has been whether the industry demonstrates continuous improvement or not.

The risk level of the Norwegian offshore petroleum industry has been analysed and presented on an annual basis by PSA. The scope of this risk level work covers all aspects of health, environment and safety (HES) within the authority's jurisdiction (Vinnem 2010). The first report was presented early in 2001, based on data for the 1996–2000 period. The methods used to collect data and analyse the risks were developed through the 'Risk Level Project' (RNNP). Since then, annual updates have been performed. Also, since its beginning, RNNP as a tool has been undergoing development in cooperation with different stakeholders (Petroleum Safety Authority Norway 2009b).

RNNP uses statistical, engineering and social science methods in order to provide a broad illustration of risk levels, including risk stemming from major hazards, risk stemming from incidents that may represent challenges for emergency preparedness, as well as risk perception and cultural factors (Vinnem 2010). The statistical approach is based on recording occurrence of near misses and relevant incidents, performance of barriers and results from risk assessments. Safety culture, motivation, communication and perceived risk are covered through the use of social science methods; questionnaire surveys and interviews, audit and inspection reports, as well as accident and incident investigations.

The 'Risk Level Project' focuses on risk to personnel and covers major accidents, occupational accidents and working environment factors. The risk level is assessed based on statistical analyses and subjective evaluations of risk. Øien et al. (2011b) argue that the focus in RNNP is on the total risk of all oil and gas facilities on the Norwegian continental shelf, which may conceal negative development on one or a few installations.

[16] Telephone interview with Torleif Husebø, PSA, 16 May 2011.
[17] Telephone interview with Torleif Husebø, PSA, 16 May 2011.

The following indicators have been established (Petroleum Safety Authority Norway 2009a, 2009b):

- indicators for events related to major accident risk;
- indicators for barriers related to major accident risk;
- indicators for serious occupational accidents and diving accidents;
- indicators for working environment factors;
- indicators for other 'Defined Situations of Hazard and Accident' (DSHA).

The work also includes information about:

- experiences of risk;
- HSE climate;
- qualitative evaluations related to the issues.

An overview of the main indicators used for 2009 is presented in Table 9.3 (Petroleum Safety Authority Norway 2009b).

In 2009, PSA also applied information drawn from RNNP and combined it with Environment Web databases to monitor acute discharges to the sea on the Norwegian OCS (Petroleum Safety Authority Norway 2011). Risk in this connection is confined to frequencies and volumes of acute discharges. The actual and potential environmental consequences of these spills were not assessed.

In hindsight of the Macondo blowout, Skogdalen et al. (2010) discuss the ability of the indicators used in RNNP to provide early warnings of such an accident. Their suggestions for extending the safety indicators of RNNP relate to well integrity and the two barrier principle, well planning, schedule and cost, undesired incidents and well monitoring and intervention.

As already mentioned, there is also a discussion whether organisational incentive systems for leaders and employees on safety should be included.[18] The argumentation is that some incentives, particularly economic ones, seem to counteract safety.

9.3.4. *Background Data: the United Kingdom, the United States and Norway*

As already mentioned, IRF has collected background information from the regulatory bodies of different countries involved in the petroleum industry (IRF 2011a). To enhance comparison of the regulatory regimes of the United Kingdom, the United States and Norway, the descriptions that the authorities of these countries have provided have been edited for brevity and presented in Table 9.4.

[18] Workshop, Robust Regulation, 17 November 2011 in Stavanger, Norway, with participants from representatives of the Petroleum Industry, Trade Unions and the Petroleum Safety Authority Norway (PSA).

TABLE 9.3. *Overview of indicators applied in Norway by PSA (2009b)*

Main topic	Indicator/factor	Description	Comments
Major accidents			Indicators at the sublevel of these indicators include indicators regarding barriers and maintenance and follow up
	Not ignitable hydrocarbon leaking		Information from the industry
	Ignited hydrocarbon leaking		Information from the industry
	Well incident, loss of well control		Information from the PSA
	Fire/explosion in other areas, not hydrocarbon		Information from the industry
	Ship on collision direction		Information from the industry
	Drifting object		Information from the industry
	Collision with field related vessel/facilities/shuttle-tanker		
	Damage on facility-construction (innretningskonstruksjon)/ stability-/anchoring/ positioning error		Information from the industry and the PSA
	Leakage from subsea production facilities/ pipeline/riser/wellflowline (brønnstrømsrørledning)/ loading buoy-/loading hose		Information from the PSA
	Damage on subsea production facilities/ pipelinesystems/divinig equipment caused by fishing equipment		Information from the PSA
	Evacuation (precautious [føre-var]/emergency evacuation)		Information from the industry

	Helicopter incidents		Information from the industry
Occupa-tional accidents and diving accidents			Parts of the information are derived from a survey asking employees about their own evaluation of accidental risk and HSE climate
	Occupational accident	Number of events	Information from the PSA
	Diving accident	Number of events	Information from the PSA
Work-related illness			Parts of the information are derived from a survey asking employees about their own evaluation of the working environment, own health and welfare and HSE climate
	Work related illness		Information from the industry
Other topics			
	Man overboard		Information from the industry
	Complete power failure		Information from the industry
	H_2S discharge		Information from the industry
	Falling objects		Information from the industry and the PSA

Note: For normalising the trends, the project applies data about facilities, wells, production volumes, work hours, diving hours, hours of flying helicopter etc.

TABLE 9.4. *Overview of regulatory properties of the United Kingdom, the United States and Norway, based on information from IRF (2011a)*

	The United Kingdom	The United States	Norway
Organisation	Health and Safety Executive (HSE)	Bureau of Ocean Energy Management, Regulation and Enforcement (BOEMRE) and Bureau of Safety and Environmental Enforcement (BSEE)	The Petroleum Safety Authority Norway (PSA)
Regime scope	Health and Safety. The Health and Safety regime applies onshore and offshore with specific legislation (alongside general H & S legislation) for the upstream oil and gas industry.	Authorization includes leasing and regulation of oil, gas and sulphur exploration, development and production operations on the U.S. Outer Continental Shelf (OCS). BOEMRE is authorized to lease and regulate alternative energy projects on the Federal OCS.	Responsible for safety, emergency preparedness and the working environment in the petroleum industry. The regulatory regime applies to offshore installations and exploration, production and exploitation operations, as well as to their associated onshore processing facilities and refineries.
Administering Agency/ Arrangements	HSE is a UK government, non-departmental public body reporting to an executive board.	Federal government: BOEMRE and BSEE are units of the U.S. Department of the Interior. The U.S. Coast Guard and the Environmental Protection Agency also have regulatory roles.	The PSA is an independent, government agency accountable, and providing advice, to the Ministry of Labour.
Legislation Type	National Health and Safety at Work etc. Act 1974 and supporting regulations. A combination of legislation applicable to all industries with additional specific major hazard regulations for the offshore oil and gas industry. Non-major hazard legislation is applied offshore on an activity basis.	The OCS Lands Act (OCSLA) is the primary statute governing offshore oil and gas and marine mineral activities. BOEMRE and BSEE regulations governing OCS oil and gas activities are in Title 30 of the Code of Federal Regulations. Many are based on industry standards.	Acts, Royal Decrees and Regulations. These formulate requirements in a performance-oriented fashion. The requirements are extensively substantiated by referencing recognised national and international standards.

Extent of Government Approval	Duty holders must prepare and submit a safety case to HSE for assessment and acceptance, before an offshore installation can operate in the UK sector.	Lease issuance, exploration and development/production plan approvals, drilling permits, completion and work-over permits, production safety system permits, structural permits, pipeline rights-of-way and spill contingency and decommissioning plans. Permits for geological and geophysical operations on un-leased lands	The PSA does not approve any plans or applications as such, but requires duty holders to apply for consent to commence and carry out various petroleum industrial activities/operations (specified by requirements of the regulations).
Nature of Duties Imposed	Primarily goal-setting legislation which sets required standards and objectives to be achieved by duty holders for the continued safe operation of their installations.	Hybrid regulatory approach; prescriptive regulations including 96 industry standards; performance objectives can be achieved by alternate means with BOEMRE approval.	All companies have a general duty to ensure compliance with requirements of acts, statutory rules and regulations, as well as their own, set requirements for their operations. The supervisory activities of the PSA do not exempt the duty holders from this duty.
Physical Objects in The Regime	Offshore-specific legislation applies to Offshore Installations as defined in the regulations. These include fixed and floating production installations, plus non-production installations such as MODUs, FSUs, flotels and others, according to their exposure to major hazard risks. In total, there are approximately 290 surface installations and another approximately 25 mobile drilling rigs in the UK sector on an annual basis.	Oil and gas exploration and production in the Gulf of Mexico (3,200 platforms, ~75 MODUs, 33,000 miles of pipeline, subsea production systems, wide range of support equipment), offshore California (23 production platforms, 188 miles of pipeline, development from existing facilities, record extended reach wells) and the Beaufort Sea (production from an artificial island, second production project in construction, ongoing exploration).	Offshore installations and onshore facilities. These include exploration and production installations, including FPSOs, MODUs, flotels, subsea arrangements, wells, pipelines and off- and onshore processing plants. Also, the main functions of pipe-laying barges, lifting barges, diving or other support vessels are regulated by the regime.

(continued)

TABLE 9.4. *(continued)*

	The United Kingdom	The United States	Norway
		Planning for additional oil and gas exploration in the Chukchi Sea and Beaufort Sea is underway. BOEMRE assists the U.S. Coast Guard on the review of offshore LNG gasification ports. Seventeen applications have been received and seven have been approved. Two facilities are now operating.	
Assurance Mechanisms	Government Inspectorate. Inspections of duty holders and their installations against the control of major hazards as described in the safety case, and compliance with relevant statutory provisions.	BOEMRE inspectors and investigators are based in 7 district offices (5 in the GOM, 1 in California, 1 in Alaska) and fly offshore regularly. OCS Lands Act mandates annual inspections. Inspectors have authority to issue Incidents of non-compliance which may be a warning or a facility shut-in. Civil penalties may be issued when violations pose actual harm or threat of harm to personnel or the environment. Industry self-inspections and records are required by regulation. Drills (spill response, BOPE, H_2S, evacuation, etc.) are required and may be initiated by BOEMRE without notice. Third-party reviews (design, fabrication and installation) are required for deepwater or novel structures.	The PSA assures that the industry is adhering to statutory regulations by conducting audits and verifications and – if necessary – by employing its delegated regulatory powers.

Financial Basis	Cost recovery for upstream oil and gas industry by charging an hourly rate for certain work.	Approximately 50% of the funding for BOEMRE's offshore program is from rental fees on OCS leases and cost recovery assessments. The remainder of the funding is from annual congressional appropriations.	Government general budget allocation. Expenses related to regulatory supervision, such as staff's working hours and travel expenses, must be refunded by the duty holders in accordance with rates set by the government. The refunded costs are paid into the Treasury. Typically, these amount to about 45% of the PSA's total operational budget.
Environmental Regulation Responsibilities	The Health and Safety Executive (HSE) is responsible for assessing the integrity and safety of offshore installations in the United Kingdom via the Offshore Safety Case Regulations (OSCR). DECC (Department of Energy and Climate Change) is responsible for developing the environmental regulatory framework for the UKCS. DECC administers and ensures compliance with that environmental regime in relation to offshore oil and gas exploration and production and decommissioning. This includes the approval of Oil Pollution Emergency Plans (OPEPs).	The National Environment Policy Act (NEPA) process is to help public officials to make decisions based on an understanding of environmental consequences and take actions that protect, restore and enhance the environment. The Council on Environmental Quality (CEQ) advises agencies on environmental decision making and oversees and coordinates federal environmental policy. EPA regulations deal with discharges and emissions. BOEMRE prepares environmental impact statements and environmental assessments at various stages including leasing, exploration plans and development plans.	The PSA's role with regard to protecting the natural environment is directed primarily at the preventive side by helping to ensure that environmentally harmful incidents do not occur. In addition, the PSA has a role in connection with the operator's emergency preparedness for stopping a leak or blowout.

(continued)

TABLE 9.4. *(continued)*

	The United Kingdom	The United States	Norway
		BOEMRE also coordinates with other federal bureaus and agencies in reviewing environmental im-plications of planned OCS activities and spill contingency plans.	
Oil spill response	The implementation of any counter-pollution measures deployed to minimise the pollution incident is the responsibility of the operator, their third-party oil spill responder and the Maritime and Coastguard Agency (MCA). MCA, an Executive Agency of the Department for Transport, is responsible, if required, for deploying any counter-pollution measures to minimise pollution incident, and the Secretary of State's Representative (SoSRep) has ultimate powers of intervention.	BOEMRE is responsible for planning (i.e. reviewing preparations) for potential oil spills related to oil and gas operations on the federal Outer Continental Shelf. Operators are required to submit Oil Spill Response Plans for approval by the Bureau. These plans must address worst-case discharge and response capabilities including equipment and response time. BOEMRE conducts annual inspections of equipment and conducts drills to test industry response. BOEMRE shares responsibility with the U.S. Coast Guard (USCG) in incident response. The USCG acts as the federal on-scene coordinator and leads the operational spill response, while BOEMRE focus is on source control.	Issues related to oil spill response is the responsibility of the Norwegian Climate and Pollution Agency (Klif). The Norwegian Coastal Administration (NCA) is responsible for the operational side of such clean-ups. Both the NCA and the Norwegian Clean Seas Association for Operating Companies (Nofo) will answer questions about the equipment used to clean up spills.

Transparency	The offshore section of HSE's Web site (www.hse.gov.uk/offshore/) provides a wide range of related information, including annual offshore health and safety statistics, reports of key intervention programmes, full details of HSE's internal assessment procedures and standards, and safety alerts and information.	BOEMRE and BSEE functions and responsibilities are described at the Web site, http://www.doi.gov. The Web site addresses the organisation and responsibilities of each division and region of the bureau. All environmental studies and technical research is posted to the Web site. Statistics and investigation reports conducted by BOEMRE are posted or linked within the Web site. All Notices to Lessees which provide clarification of regulations are posted on the Web site. The site also provides a Fast Fact query function that allows the public to query many aspects of the data that BOEMRE controls, including, but not limited to: • Applications for Permit to Drill • Boreholes Drilled • Exploration and Development Plans • Facility Measurement Points • Lease Owner • Offshore Statistics by Water Depth • Pipeline Locations • Pipeline Permits • Platform Structures • Plugging and Abandonment Liability Production Data	The PSA's Web page is the most thorough source of information about its regulatory system and activities. Almost everything PSA does as a regulator is accounted for here. Most of the postings are also published in English. (This is costly but done for the benefit of media and colleagues in other countries.) The Web site accounts for PSA organisation and introduces people working with the PSA, day-to-day supervisory activities, and features articles related to regulatory and professional issues. Statistics are posted as well as annual reports, such as the annual Risk Assessment Report and the annual PSA report. Should a major accident occur in Norwegian waters or onshore, the PSA mobilizes its Emergency Response Team, not least in part to be able to inform the central government and the public about the position. The PSA Web site will also change to bring the latest news about an accident.

9.3.5. *Other Sources of Indicators*

During work on this chapter, the author has compiled some of the main sources of indicators that might be useful for those who work with regulation and indicators. The sources are listed and commented on in Table 9.5.

9.4. DISCUSSION AND CONCLUSIONS

Many issues need to be considered when discussing development and use of indicators for regulatory purposes. Amongst these issues are:

- the national context, for instance, culture and regulatory properties;
- the regime's requirements for reporting, available sources of information, traditions of transparency and reporting results to the public;
- the intended use and purpose of the indicators:
  - For whom and for what – for instance, for process safety, or working environment?
  - How instructive are the indicators?
- the approach that is applied to the development of indicators, positive (safety) or negative (risk);
- the aspects that are given attention (the safety aspect or risk-influencing factor that the indicator is trying to measure);
- whether the applied indicators are of a predictive or retrospective nature;
- how the indicators give attention to technology, humans and organisation and the interactions between these elements;
- the definitions of the indicators, the quality of the data, the level of aggregation in the reporting and the generalizability and use of the indicators.

When comparing the United Kingdom, the United States and Norway, it appears that there are regulatory differences between the three countries. In addition, there are differences in the legal requirements and traditions of cooperation with respect to reporting and transparency. Hence, access and quality of information needed for developing indicators varies.

Regarding focus of attention, there are differences between the indicators of the three countries, especially with respect to their preventive nature. Norway and the United Kingdom have done more to develop and apply predictive indicators than the United States has. Also, the indicators in the United Kingdom and Norway reflect more attention to early stages in the development of an accident than those in the United States. However, if taking the work of the API and the discussions

TABLE 9.5. *Sources of Indicators*

Other Sources and Guidelines	Type	Comments
International Regulators' Forum (IRF)	Performance measurements	Statistics reported by national authorities of different countries to IRF Performance Measurement Project. The guidelines of the IRF focus on the following themes (International Regulators' Forum 2011): • Fatalities • Hours worked • Gas releases – Number of releases • Gas releases – Amount of releases • BOE (Barrels of Oil Equivalent) gas production • Collisions, fires and losses of well control • Well-related activities
International Association of Oil and Gas Producers (OGP)	OGP Safety performance indicators	OGP's main aim is to coordinate the members' activities vis-à-vis regulatory authorities (Ryggvik 2011). The statistics are about fatalities and injuries.
Oil & Gas UK	A system of Key Performance Indicators (KPIs)	Directed at hydrocarbon releases, verification compliance, safety critical backlog (Paterson 2010a)
Center for Chemical Process Safety (CCPS)	Guidelines for Process Safety Metrics ISBN: 978-0-470-57212-2	The Global CommunityCommitted to Process Safety

(continued)

TABLE 9.5. *(continued)*

Other Sources and Guidelines	Type	Comments
Center for Chemical Process Safety (CCPS)	Process Safety Leading and Lagging Metrics	The Global CommunityCommitted to Process Safety
National Offshore Petroleum Safety Authority (NOPSA)	According to CEO Jane Cutler,b the National Offshore Petroleum Safety Authority Australia (NOPSA) has industry performance data.	National Offshore Petroleum Safety Authority (NOPSA) is a statutory agency regulating Commonwealth, state and territory coastal waters with accountability to the relevant ministers.
SINTEF Offshore Blowout Database	SINTEF Offshore Blowout Database http://www.exprosoft.com	See also Holand (2010) for overview of indicators, data sources and evaluation of quality of data from different countries that are included in the database. Holand makes references to requirements that could have prevented blowouts. In particular he compares conditions in the United States, the United Kingdom and Norway.

[a] Paterson, R. (2010) Presentation at International Regulators' Forum, 18–20 October, Vancouver.

b Presentation at International Regulators' Forum, 18–20 October, Vancouver.

in hindsight of the Macondo accident into consideration, it seems that attention towards preventive indicators is increasing in the United States. The discussions in hindsight of the accidents underline how important it is to be cautious about conclusions drawn on the background of indicators and in particular to give attention to their context and generalizability.

Regarding evaluation of regime efforts to prevent accidents, it is worth noting that regimes seem to use statistics as indicators of their own performance. However, without explanations of causal links between statistics of activities and developments in safety and/or control of risks, such statistics do not say anything about the efficiency of the authorities' efforts.

The overall conclusion is that it is very difficult to compare safety status of the petroleum industry in the United Kingdom, the United States and Norway and the efficiency of regulatory regimes by using existing safety indicators. However, comparing views and uses of indicators might be useful for learning more about different purposes of indicators and how they can be applied, how they can function and for whom they might be useful. It is always important to keep in mind that the regard for indicators, their applications and their value depends on regulatory culture and contextual factors.

REFERENCES

Bureau of Ocean Energy Management, Regulation and Enforcement (BOEMRE) (2011) *OCS Performance Measures Program*. http://www.boemre.gov/perfmeas/

Fleming, M. (2010) *Know Where You Are Going, Not Where You Have Been*. Presentation at the International Regulators' Forum, 18–20 October, Vancouver.

Hale, A. (2009a) "Editorial special issue on process safety indicators", *Safety Science*, 47: 459.

(2009b) "Why safety performance indicators?" *Safety Science*, 47: 479–480.

Health and Safety Executive (HSE) (2007) *Key Programme 3. Asset Integrity Programme*. A report by the Offshore Division of HSE's Hazardous Installations Directorate UK. http://www.hse.gov.uk/offshore/kp3.pdf

Health and Safety Executive (HSE) (2010) *Offshore Injury, Ill Health and Incident Statistics 2009/2010*. HID Statistics Report HSR 2010–1. http://www.hse.gov.uk/offshore

Health and Safety Executive (HSE) and Chemical Industries Association (CIA) (2006) *Developing Process Safety Indicators. A Step by Step Guide for Chemical and Major Hazard Industries*. Health and Safety Executive: London.

Holand, P. (2010) *Offshore Blowouts 1980–2007*. Presentation held at The Bellona ONS Seminar, August 23, Stavanger.

Hood, R., Rothstein, H. and Baldwin, R. (2001) *The Government of Risk. Understanding Risk Regulation Regimes*. Oxford University Press: Oxford.

Hopkins, A. (2007) *Thinking about Process Safety Indicators*. Working Paper 53, prepared for presentation at the Oil and Gas Industry Conference, November, Manchester.

(2009a) "Thinking about process safety indicators," *Safety Science*, 47: 460–465.

(2009b) "Reply to comments", *Safety Science*, 47: 508–510.

International Atomic Energy Agency (IAEA) (1999) *Management of Operational Safety in Nuclear Power Plant*. INSAG–13. International Nuclear Safety Advisory Group. International Atomic Energy Agency: Vienna.

International Regulators' Forum (IRF) (2011a) *IRF Performance Measurement Project*. Homepage of International Regulators' Forum. http://www.irfoffshoresafety.com/country

International Regulators' Forum (IRF) (2011b) *IRF Performance Measurement Project*. Homepage of International Regulators' Forum. http://www.irfoffshoresafety.com/country/performance

International Regulators' Forum (IRF) (2011c) *IRF Performance Measurement Project*. Homepage of International Regulators' Forum. http://www.irfoffshoresafety.com/country/performance/scope.aspx

Kjellén, U. (2000) *Prevention of Accidents Through Experience Feedback*. Taylor & Francis: London.

(2009) "The safety measurement problem revisited", *Safety Science*, **47**: 486 – 489.

Lamvik, G. and Ravn, J.E. (2006) "National performance – offshore drilling. Safety and reliability for managing risk", in Guedes Soares, C. and Zio, E. (eds.): *Safety and Reliability for Managing Risk*. Taylor & Francis: London, pp. 363–370.

Organisation of Economic Co-operation and Development (OECD) (2003) *Guidance on Safety Performance Indicators*. Guidance for industry, public authorities and communities for developing SPI programmes related to chemical accident prevention, preparedness and response. OECD Environment, Health and Safety Publications. Series on Chemical Accidents, No. 11. OECD: Paris.

Øien, K., Utne, I.B. and Herrera, I.A. (2011a) "Building safety indicators: Part 1 – Theoretical foundation", *Safety Science*, **49**: 148–161.

Øien, K., Utne, I.B., Tinmannsvik, R.K. and Massaiu, S. (2011b) "Building safety indicators: Part 2 – Application, practices and results", *Safety Science*, **49**: 162–171.

Petroleum Safety Authority Norway (2009a) *Trends in Risk Level. 2008. Norwegian Continental Shelf* (in Norwegian). http://www.ptil.no

Petroleum Safety Authority Norway (2009b) *Risk Level in the Norwegian Petroleum Industry RNNP. Main Report – Trends in 2009 – Norwegian Continental Shelf* (in Norwegian). http://www.ptil.no

Petroleum Safety Authority Norway (2011) *New RNNP Report Surveys Acute Discharge Risk on the NCS*. http://www.ptil.no/trends-in-risk-level/new-rnnp-report-surveys-acute-discharge-risk-on-the-ncs-article7407–155.html

Skogdalen, J.E., Utne, I.B. and Vinnem, J.E. (2010) "Looking back and forward: Could safety indicators have given early warnings about the Deepwater Horizon accident?" In *The Macondo Blowout. 3rd Progress Report*. Issued December, 5, 2010 by The Deepwater Horizon Study Group which was formed by members of the Center for Catastrophic Risk Management (CCRM), Appendix H.

Steinzor, R.I. (2011) "Lessons from the North Sea: Should "Safety Cases" come to America?" *University of Maryland School of Law*, 2011–3. http://ssrn.com/abstract=1735537

Vinnem, J. (2010) "Risk indicators for major hazards on offshore installations", *Safety Science*, 48 (6): 770–787.

Walker, S. (2010) *Presentation at the International Regulators' Forum*, 18–20 October, Vancouver.

PART III

Norwegian Self-Regulation

Challenges and Lessons Learned

INTRODUCTION TO PART III

Part III includes six chapters based on case studies assessing the Norwegian regime and reflecting its striving for continuous improvement and robustness. Within the scope and frame of the book these studies are unique in that other regimes have not been studied in such detail. In Chapter 10, the paradigm shift from prescriptive regulation towards a system of government-supervised self-regulation is presented with risk assessment, the delineation of safety functions, and performance-based requirements as basic elements. The development of this new regulatory approach to operational safety and the working environment also emphasizes the importance of tripartite cooperation and the responsibilities and accountability of the players. That is further developed in Chapter 11, which explores how this co-regulatory approach is continuously challenged in negotiations involving vague and floating legal standards as well as contested risk indicators. It further discusses how purpose- and principles-based regulation may evaporate, dissolve, or become irrelevant, unless industrial realities, management practices, and regulatory expectations are considered. The issue is illustrated by two cases, firstly a dispute over sleeping facilities for workers and working hours and secondly an assessment of the regulatory concept of a "good and sound HSE-culture" and the burdens taken by the regulator and the industry in the ensuing its implementation. "Boxing" and "dancing" are used in Chapter 12 as a metaphor of the shifting patterns of adversarial and co-operative modes of industrial relations and social partnerships in assessing and maintaining the tripartite collaboration on safety. It reveals how a conflict involving the level of safety level on the NCS between trade unions, employers' federations, and the political and regulatory authorities was handled and solved. Chapter 13 discusses how a technological transformation process (NORSOK) took place on the NCS and threatened the tripartite model. This technological transformation created the premises for the regime we see today, and some of its current vulnerabilities and problems. Two major incidents on the NCS are analyzed in Chapter 14. The Gullfaks C incident took place in 2010, two months after the Macondo accident and challenged the fundament of the tripartite system and visualized the presence of the risk of a major accident. The incident contributed to a "risk aversion climate" and influenced the regulatory and safety community in Norway as well as in EU. Finally, inspections, audits, and enforcement measures are discussed in Chapter 15. The qualitative aspects of several audits of companies and installations in the year prior to the Macondo accident are assessed, showing how communications and conflicts between the involved parties influenced where, when and how audits were conducted.

10

Government-Enforced Self-Regulation

The Norwegian Case

Paul Bang and Olaf Thuestad

10.1. INTRODUCTION

The offshore oil and gas activities on the Norwegian Continental Shelf (NCS) in the 1960s introduced a new and complex industry to the nation with no prior traditions and bonds to which it could directly relate. Nevertheless, a number of important factors and preconditions were already in place for responding to this new business. Norway had already established a land-based industrial tradition through the development of hydropower and power-intensive metallurgical plants in the early twentieth century, and a strong and international maritime tradition through its shipping sector. Growth backed by Marshall Aid after 1945 had also strengthened the social democratic philosophy and further developed this into a "Nordic welfare model". Collectively, these factors made an important contribution to Norway's ability to develop a foundation for the oil sector based to a great extent on Norwegian values and ways of thinking.

As early as the late 1940s, the government's creation of the Norwegian State Educational Loan Fund – which gave grants and loans to students – had allowed the higher education sector to recruit intellectual capital from all levels of society. This led to a valuable growth in such capital during the 1950s and 1960s, while increasing mobility between social classes. This in turn helped to lower class barriers and ensured a marked boost in the level of knowledge amongst the general population.

Adding to this, social democracy tradition in Norway already had a long history of collaboration between government, employers and unions, and worker protection legislation established rights and duties which naturally could apply to the new industry as well. This also laid the basis for the high level of unionisation in Norwegian workplaces, which also gradually was implemented in the petroleum sector.

All the same, the petroleum industry represented, as mentioned earlier, something new and challenging for Norway. Its operations took place offshore, and the sector gradually became technologically demanding and had a high risk potential.

The industry was also accordingly characterised by a high degree of technological development, where new and adapted solutions were continuously emerging.

This chapter presents the regulatory response to these challenges with a focus on the paradigm shift from prescriptive type of regulations towards a system of government-enforced self-regulation with risk assessments and principle-based requirements as basic elements. The development of a new regulatory approach to safety and the working environment also emphasised the importance of involving the tripartite arenas for cooperation and the responsibility put upon the players. The legal framing and some core issues of the Norwegian regulatory system is presented in Chapter 5 (Kaasen) of this volume.

10.2. THE REGULATORY RESPONSE

The government's attention in the early years was concentrated on securing national interests and ownership to resources on the NCS and territorial rights, while simultaneously developing a licensing system which safeguarded ownership and control for the Norwegian state. Through an approach to licensing based on application rounds for the award of exploration acreage, where the licensees[1] were grouped in defined production licences,[2] the regulator had by this established a systems for selecting competent companies which through synergy could bring forward the best results. In administrative terms, the sector was subject to the Ministry of Industry up to 1979. The latter was accordingly charged with taking care of both resource management and safety interests.

Initially the regulations of safety and the working environment were largely based on adapting prescriptive regulations developed for other industries and supplemented with new regulations in new technological areas such as drilling and production equipment. Where the working environment was concerned, a harmonisation and integration process during the 1970s culminated in the Norwegian Working Environment Act from 1977 (Karlsen and Lindøe 2006) This was also very significant for the oil and gas industry because it was extended – with certain modifications – to the production activities offshore.

The legal requirement for cooperation between employers and employees became important, helping to lay an important basis for the participation and collaboration which also later developed into the particular tripartite collaboration between companies, unions and government in Norwegian oil and gas industry.

[1] Petroleum act Section 1–6 definition: licensee, physical person or body corporate, or several such persons or bodies corporate, holding a licence according to this Act or previous legislation to carry out exploration, production, transportation or utilisation activities. If a licence has been granted to several such persons jointly, the term "licensee" may comprise the licences collectively as well as the individual licensee.

[2] Petroleum Act, Section 2–1 Granting of exploration licence etc.

The new working environment legislation opened the way to a more modern socio-technical way of thinking and prompted a shift towards regarding working life from a human/technology/organisation (HTO) perspective. That in turn laid the basis for new forms of collaboration and management. By treating commercial companies as dynamic entities where people, technology and organisation mutually interacted, employee participation, expertise development and collaboration between the key stakeholders provided the best basis for promoting growth and progress at the companies and prudent operations.

Where the regulators were concerned, this helped them to introduce and develop ideas of government-enforced self-regulation or internal control, also supported by evolving trends nationally and internationally. These concepts were not new, and had been applied, for instance, in Norway's nuclear power sector as early as the 1950s.[3]

The idea of self-regulation first found expression in the oil and gas sector through safety regulations adopted in the 1960s and 1970s, before the Working Environment Act was passed.[4] However, more detailed safety provisions at the time formed part of a fragmentary body of prescriptive regulations developed by many different regulators and based on different supervisory methods, which relied principally on checklist-oriented inspections and government-based approvals. A unified approach was accordingly lacking to facilitate the development of a system for management through internal control of the kind which developed later.

In this early, pioneering period, Norway also suffered several major accidents. Both the Ekofisk Bravo blowout in 1977 and particularly the *Alexander L Kielland* disaster in 1980, in which 123 people died, shook the nation and had major political and administrative consequences. The Bravo accident in 1977 also focused attention on dilemmas posed by handling overriding resource management and safety concerns within the same ministry. Responsibility for oil and gas matters was consequently placed in a new Ministry of Oil and Energy which was established in 1978, whilst also partly as a result of this incident, responsibility for following up on safety and the working environment was subsequently transferred to the Ministry of Local Government and Labour in 1979, which was already responsible for working

3 "Internal control has been practised in the Norwegian nuclear power sector (the former Institute for Atomic Energy, now the Institute for Energy Technology) from as early as 1951 with regard to both production safety and radiation hygiene, and systematic measures have also been established to safeguard this system. Government requirements to adopt such measures have first been introduced in recent years" (NOU 1981, 1).

4 "Earlier safety regulations issued in 1967 and 1975 also contained provisions which required every participant in the industry to verify compliance with the safety rules. A similar requirement was incorporated in section 4, sub-section 1 of the 1976 regulations. Pursuant to sub-section 2, an overall duty rested on the licensee to see to it that all involved parties act in compliance with the regulatory provisions. … It was first in 1976 … that official regulation would change to basing itself on the licensee's internal control system" (*Op cit.*). See also NOU 1987: 10A Internal control in an integrated strategy for working environment and safety.

environment legislation onshore. That made the division of roles and responsibilities between these two overriding concerns more clear and transparent. The NPD, which had been a subordinate agency of the Ministry of Industry (from 1972 to 1978) and later the Ministry of Oil and Energy (from 1978) responsible for following up on oil and gas operations, thereby reported along two different channels for the oil and gas sector and had to organise its supervisory activities so that both considerations could be handled in a predictable and consistent manner.[5]

The investigation into and follow-up of the *Alexander L Kielland* accident identified clear weaknesses associated with the "traditional" regulatory approach adopted by the regulator. Development and following up of prescriptive regulations in the form of inspections which reactively identified faults and deficiencies in the industry, and which in many respects could be perceived as making the government, as regulator, more responsible than the operator for ensuring that the operations were acceptable, had turned out to be unsuitable and inadequate. Unclear interfaces between the many regulatory agencies involved also contributed to complexity and the lack of a unified development of regulations and follow up of safety conditions.

A clear recognition accordingly emerged that a more integrated regulatory regime was required, which would provide better coordination of Norwegian government administration. That paved the way for the paradigm shift in 1985.

10.3. THE PARADIGM SHIFT

10.3.1. *Requirement to Prudent Operations*

Norway passed its first Petroleum Activities Act in 1985, almost two decades after oil exploration had begun on the NCS. This legislation consolidated important legal provisions which had been extended to or developed for offshore resource management and safety regulation, and contributed to a unified regulatory approach to the sector. The Act included important elements from existing safety legislation and thereby also conferred greater legitimacy on these.

A complex industrial activity will always involve many players who participate at different levels in the value chain. A hierarchy consisting of the oil companies, contractors and subcontractors was developed at an early stage in the oil and gas business nationally and internationally. In the Norwegian context, this led to the establishment of a formal hierarchy with the oil companies at the top of the "ladder" being granted licences and having the overall responsibility for ensuring prudent operations. Further down the chain came a cluster of contractors and subcontractors who developed an advanced contractual system for delivering goods and services.

[5] Annual reports from the NPD-1977–79.

It became clear to the regulator that clarity and consistency were required over the issue of who was responsible for the activity and how that responsibility was to be discharged.

Underlining the responsibility of everyone involved in the oil and gas industry accordingly became an important element in the Petroleum Activities Act[6] stating that "[o]perations pursuant to this Act must be conducted in an acceptable manner, and take account of considerations related to the safety of personnel and the environment". A commentary on the Act amplified this provision: "[It] establishes as the general safety standard that operations must be conducted in an acceptable manner. The other provisions of this section must be regarded as casuistic amplifications of the conditions which the section seeks to safeguard. In sub-section 1, the section makes special mention of showing consideration for the environment and personnel safety. Sub-section 2 emphasises the consideration which the petroleum industry must show to other activities taking place in and on the sea and in the air over the continental shelf". Also, "the principal responsibility for maintaining the level of acceptability required by the section rests with the operating party." Also, "this term covers not only the licensee, but also anyone who conducts or exercises such activities. That follows from the connection with section 58, see section 4 of the safety regulations, whereby "everyone" is required to verify that operations are conducted in compliance with the requirements specified in and issued by virtue of the legislation. Section 45 represents the Petroleum Activities Act's main provision on safety. The connection with section 1 of the Act indicates that the requirement to operate acceptably will be significant for all activities embraced by the petroleum industry" (Hagen et al. 1989, pp. 448–449).

In other words, this "safety provision" emphasised at the legislative level that "anyone who conducts operations" is responsible for seeing to it that the activity is pursued in an acceptable manner, while identifying the licensee as primarily responsible for ensuring compliance. As a result, the safety regulations expanded and emphasised that the licensee who was selected as the operator[7] for a licence simultaneously became responsible for the activity being conducted in a fully acceptable manner, and for ensuring that both it and those who were in a contractual relationship with it were also operating acceptably. The Petroleum Act places a duty[8] to verify compliance with the provisions – the licensee undertakes to see to it that anyone who conducts work, either in person, through employees or through contractors or subcontractors, complies with the provisions specified in and by virtue of the Act. (This section partially replaces the provisions in section 3 of the 1975 decree and section 4 in the 1976 decree.) (*op cit*, p. 449).

6 Section 45 of the 1985 Act on the requirement to operate acceptably.

7 Petroleum Act, section 1–6 Definition Operator: anyone executing on behalf of the licensee the day-to-day management of the petroleum activities.

8 Section 58 of the 1985 Act.

This is underlined in today's HSE regulations,[9] specifying that "[t]he operator shall see to it that everyone who carries out work on its behalf, either personally, through employees, contractors or subcontractors, complies with requirements stipulated in the health, safety and environment legislation." This duty *to see to it* represented a principle which was unique at the time. It gave the licensees in general and the operators in particular a responsibility not only to comply with the regulations themselves, but also for the licensees to see to it and facilitate compliance for the operators in the licence context and furthermore for the operators to see to it that all involved in the value chain comply with the provisions of the legislation and statutory regulations. In safety work, this duty of verification has been a driving force for the way safety and working environment are managed throughout the value chain, and the industry has responded to this by developing tools and solutions which support such management.

10.3.2. *Leading the Business from Norway – Norwegian Organisation*

From the start, the main players in Norway's oil and gas industry have been well-resourced and powerful multinational oil companies and contractors. The industry is international and the oil companies operate globally. It became clear to Norway as resource owner and regulator at an early stage that the country had to make appropriate arrangements for managing the industry in accordance with Norwegian administrative principles, and thereby also prevent "regulatory capture" by powerful companies which were allowed to set the agenda for developments.

The 1985 Petroleum Act thus made a provision to address this by specifying that the industry had to lead their activities from Norway.[10] This requirement extended and strengthened the existing practice under which major oil companies had established affiliates in Norway. From the early 1970s, these affiliates had appointed Norwegians to a steadily growing number at every organisational level. This development was paralleled by the emergence of Norwegian companies which similarly built up capacity and the expertise of national oil workers.

The fact that the operators as well as the major contractors acquired pools of Norwegian technological and legal expertise opened the way for a greater degree of involvement and participation in the way the regulator was to set the terms for prudent operations and develop predictable framework conditions. Expertise and knowledge about Norwegian regulation and regulatory principles thereby acquired increased significance, and the subject became an important area of collaboration

[9] Section 7 of the framework regulations on responsibilities pursuant to the regulations.

[10] Section 48 of the 1985 Act on management of activities, bases, etc: "The licensee must have an organisation able to lead operations on an independent basis from Norway. To achieve this, the ministry can set specific requirements for organisation and company capitalisation".

in terms both of regulating and supervising the activity as well as the important tripartite partnership which developed after the 1985 paradigm shift.

10.3.3. *Expanded Safety Concept*

Before 1985, Norway's safety and working environment regulations for the oil sector were fragmentary and bore evidence of, to some extent, still being adopted from other industries. The understanding of safety had its origin from legislation for fire and explosion prevention, and was incorporated in the regulations in the early 1970s and gradually further developed in the oil and gas industry. With the introduction of the new Petroleum Act in 1985, safety acquired an enlarged significance. Defining safety[11] to cover people, the environment, and material and operational assets secured greater acceptance and recognition amongst all players in the sector.

It also provided the basis for a new possibility for a dialogue in areas of both mutual and diverse interest between the regulators and the regulated. Recognising the joint interest of commercial business and interest of the society at large in that petroleum activities should take due account of the safety of personnel, the environment and of financial values, including also operational availability, paved the way for development of a broader understanding and recognition of safety in a wider sense. By expressing safety with all these elements included the opportunity to develop greater trust, openness and common interests between regulator and regulated. That in turn created a wider space for developing a new supervisory regime based to a greater extent on collaboration without the use of formal instruments as an alternative to the traditional control-based system to ensure compliance with legislation and statutory regulations, and to achieve a greater degree of safety. Such dialogue also contributes to developing knowledge and expertise amongst all the players, and promotes technological development and continuous improvement.

10.3.4. *Adjusting to Continuous Technological Improvement*

Additionally to the petroleum industry being a diversified and complex business with a "high risk" potential, it is also characterised by a great many different players and by integrated decisions and decision processes. The industry involves the use of technology and operational methods which are under continuous development. The management of risk and performance of safety evaluations was gradually perceived as key elements in every decision with consequences for the safe operations, necessitating

[11] Section 45 of the Petroleum Activities Act in 1985. While the petroleum industry and the fire and explosion prevention legislation included the worker's physical environment in the concept of safety, the Working Environment Act used the broader concept of "working environment" (NOU 1987: 10A, p. 10).

the adoption to a proactive attitude to learning and further development of operational safety – principles, an underlying intention in the regulations governing the industry.

Experience from the major accidents Norway suffered up to the 1980s showed that the offshore oil and gas industry involved complex organisations with a high level of risk. From the regulator's perspective, it became clear that setting specific requirements for each activity was impracticable. The regulations, which then contained extensive and detailed requirements about the methods to be used by the operating party, had proved in practice to have seriously negative aspects. Because petroleum technology was constantly advancing, detailed requirements quickly became outdated and presented an obstacle to innovative thinking on safety. The discussion in Chapter 13 on new technologies and the NORSOK-program illustrate this point.

That recommendation paved the way to a transition from prescriptive requirements to more goal-based requirements expressed in functional terms, which gave the operating party opportunities to choose which detailed solutions could be used to operate safely, and in accordance with the level of standard set in the regulations. "That the activity is to be conducted in an 'acceptable manner' expresses a legal standard which provides no direct specification of the level of safety required. The content of the standard will depend on current social perceptions, practice in the petroleum industry, and so forth. The requirement to maintain an acceptable level of safety can thereby change in line with developments" (Hagen et al., *op cit*, p. 449.)

Continuous improvement and a focus on quality and on quality improvements developed during the 1980s into a universal trend in mostly all important industries. Methodological approaches to quality were developed within different areas of technology, also embracing user requirements which were based on the systematic approach taken in engineering-based manufacturing. This trend paralleled and eventually merged with socio-technical principles of organisational theory concerning HTO processes and management (Lindøe and Hansen 2000).

The development of management systems based on internal control and the need to manage activities under a more goal-oriented regulatory regime became a distinctive feature of the Norwegian oil and gas industry during 1980s, and strengthened the dialogue-based supervisory methodology between the regulator and those being regulated.

10.4. NEW REGULATORY APPROACH TO SAFETY AND THE WORKING ENVIRONMENT

In the white paper on the *Alexander L Kielland* disaster,[12] the government expressed an overall goal of simplifying and enhancing the efficiency of supervising oil and gas

[12] Report no 67 (1981–82) to the Storting.

operations on the NCS. The accident had underscored the importance of being able to establish and maintain a high level of safety with clear regulatory boundaries.

In the old regulatory setting, the industry had to deal with many different government agencies and sets of regulations and also regarded the supervisory methods as resource intensive, in that the operating party had to relate to inspectors from, and inspections by, many different regulators checking compliance in their respective areas of authority. The responsible party from this perspective had to ensure that it had sufficient internal capacity and expertise to maintain an overview in a number of regulatory areas in order to ensure that it complied with different sets of legislation and statutory regulations. In the wake of the *Alexander L Kielland* accident, this "single purpose" regulatory approach came to be regarded as inadequate for the oil and gas industry. The supervisory regime which was established consisted of two elements. A single agency – at that time the NPD – was appointed as responsible for assessing safety and working environment and given the responsibility to draw up detailed regulations and to make overall safety and working environment assessments.

A Royal Decree[13] appointed the NPD to be the coordinating regulatory authority for following up all aspects concerning safety, health and environment by coordinating the fewer government agencies with independent regulatory authority and given responsibility for safety for all remaining areas. As pointed at by Kringen (Chapter 11) in this volume, efforts to modernize regulatory regimes and supervisory agencies has been pushed by OECD towards member states, and administration of risk management has been part of the public agenda for a long time (Hovden, 2002). Even if the need for improved coordination, harmonisation and reduction of the number of regulatory agencies has been recognized, the offshore oil and gas industry with the NPD (PSA) is the only exception of a strong coordinating agent (Braut and Lindøe 2010). Drawing up coordination instructions and entering into coordination and support agreement were an essential part of the regulatory reform, and were intended in part also to ensure that the government administration did not duplicate expertise. The instructions and agreements were also intended to ensure that interfaces with other regulations and regulators were clarified and dealt with in a way which avoided unnecessary overlapping or conflicts.

Formal agreements with the pollution control authorities for the natural environment and with the health authorities for health-related issues were established. Agreements on technical support were also entered into with a number of other regulatory agencies which no longer had any formal responsibility for the oil and gas industry. Coordination and support between many different government agencies represented a new approach to administration and differed from the common model

[13] Regulatory supervisory activities with the safety etc. in the petroleum activities on the Norwegian Continental Shelf. Royal Decree, 28.6.1985.

which implied that single agencies were responsible for their individual aspects of safety and working environment.

When the new administrative model for the Norwegian oil and gas business was established in 1985, one regulator – the NPD – was chosen as the coordinating agency. This coordination of supervision gradually identified the need for developing a joint set of regulations based on a number of overall considerations, including safety and the working environment, the natural environment and occupational health.

Since 1985, ongoing developments has thus resulted in an integrated set of safety, health and environment regulations, issued and followed up by the responsible government agencies jointly and coordinated by one agency. From the industry perspective this "multi-purpose" approach was welcomed, making regulatory compliance easier for the operating party. It secured a *single* main point of contact with the regulator and a *single* set of safety regulations.

The industry response was not only that the reform represented a clear simplification and efficiency enhancement, but the more integrated HSE-regulations also made it possible and stimulated to a greater degree of participation and collaboration in influencing the operating parameters in the regulatory development processes. It was also perceived as a simplification and efficiency enhancement by the regulator, which could now supervise the business on a more unified basis. On the other hand, the regulator's administrative job became more demanding in that it had to handle collaboration issues and possible conflict of interests across government agencies through formal procedures and agreements in order to ensure that the system functioned well in practice.

10.4.1. *Functional Regulations*

The processes of developing new regulations after 1985 focused more consistently on how the government now could regulate and follow up the industry by using several options such as specifying requirements for technology and operations, administrative aspects or for management and decision-making processes. A goal-setting and risk-based approach and risk management turned out to be the chosen options and also increased their importance in the regulatory development. As the party principally responsible, the operator was required to establish a safety goal for its overall operations. In reducing the risk, the responsible party had to focus on technical, operational or organisational solutions that, according to an individual and overall evaluation of the potential harm and present and future use, offered the best results, provided the costs was not significantly disproportionate to the risk reduction achieved. All identifiable risks therefore had to be assessed with a view to adopting measures which could reduce them. Planning, establishing and maintaining emergency preparedness for defined hazards and accidents was an integrated part of these evaluations.

Part of this work was initiated after the Petroleum Activities Act of 1985 on simplifying the regulations and ensuring that they accorded with the presumptions of the change. In that context, emphasis was to be given to:

- establishing a systematic regulatory collaboration between everyone concerned;
- undertaking a general clear-out of provisions based on a detailed audit of controls, documents, approvals and technical specifications;
- reducing the number of regulations;
- highlighting the responsibility of each participant to set their own criteria for, and to follow up, their own operations in line with the basis specified in applicable legislation;
- making an assessment of and utilising accepted standards as non-mandatory reference documents in the area as an alternative to preparing dedicated regulations where this could be justified on the basis of safety consideration.

As a result, the regulations came to be primarily risk-based. They rested on principles expressed as functional requirements, supplemented by detailed guidelines which provided more specific information on the intentions of the regulations or specified ways in which the regulatory provisions could be met. If a solution other than the ones suggested in the guidelines was chosen, the operator – as the party principally responsible for the activity – had to be able to document that it was better or equally good. The regulations could thereby provide greater scope for adopting alternative technology or approaches, while ensuring that the level of safety was clearly defined by the regulator. At the same time, the regulations provided those regulated with predictability when choosing well-tested solutions and using known technology.

By and large, the technical parts of the regulations were issued by virtue of safety legislation under the Petroleum Activities Act, where the safety concept was more extensive than it was in the context of a working environment. Although the safety legislation provided a broad enforcement basis, it nevertheless did not embrace key working environment conditions. This led, in an interim phase, to detailed regulations also being developed under the Working Environment Act. These were integrated at the end of the 1990s in the unified safety and working environment regulations which currently apply to Norway's oil and gas industry.

10.4.2. *Introduction of Internal Control*

In accordance with the traditional regulatory approach in the 1970s and early 1980s, regulation of the petroleum industry on the NCS involved a great deal of detailed management and control by the regulators. However, the major accidents just before and at the beginning of the 1980s meant that the government recognised that the oil and gas sector was:

- a complex and diversified industry, with different players as well as integrated decisions and decision-making processes;
- by its nature a user of technology which was being continuously developed;
- a resourceful industry in terms of companies and employees, with a substantial self-interest in operating prudently and safely;
- an industry whose interests coincided in many important areas with those of the government.

These conditions made it possible to introduce new instruments in a regulatory and supervisory context. One of these was to make clearer the individual participant's responsibility for conducting its own operations acceptably within applicable safety and working environment parameters. The government amendments to the regulations to make them more risk- and principle-based also meant that the responsible party itself had to formulate detailed requirements on how it might ensure compliance in a systematic and documentable manner where the official regulations were functional and gave no specific instructions on how to meet their requirements. Requirements based more on the functional principle allowed the operating party itself to choose options and solution for meeting the requirements. But this approach also required the companies to adapt to new management principles and systems in order to document and show in practice how compliance would be achieved.

The "safety provision"[14] meant in practice that the operator, on the basis of its overall responsibility, was the party which had to document the administrative and decision-making systems required to ensure compliance with the obligation to conduct this form of internal control. Introducing the latter as a guiding principle for shaping regulations and regulatory supervision was initially challenging for both regulator and regulated. The petroleum industry was initially uncertain both about the intentions underlying this new approach and over how it should relate to the requirement for internal control both organisationally and administratively. For the regulator, the change meant that it became necessary to shift its focus from detailed checks to a system-based supervision which to a greater extent set framework requirements for the industry and left far more of the detailed management to the undertakings themselves, while at the same time being more open to discussing alternative solutions.

As mentioned earlier, the internal control philosophy was also supported by key socio-technical principles which underpinned such innovations as the legal right of employee participation and co-determination enshrined in the 1977 Working Environment Act. The introduction of a legal requirement for internal control also meant that management principles derived from organisation theory became more widely recognised in organisational contexts, both on the industry level and the individual-company level. The linkage of internal control to the WEA of 1977

[14] Section 45 of the 1985 Petroleum Activities Act.

is important. The revised law was based on the action-research-based Industrial Democracy Project in the 1960s involving workers in the improvement of products, processes and working conditions. The new principles became a merger between developing work environment in the workplace and strengthening industrial democracy (Gustavsen and Hunnius 1981). Based on an assessment of prevention at the workplaces in Europe, Vogel (1998) says it is a "model that is genuinely different from those found elsewhere in Europe, although obviously not uniform". The approach is also reflected in the EU Framework directive (89/391) encouraging improvements in the safety and health of workers.

In many respects, it can be said that the complex and risky nature of the oil and gas industry made it possible to develop an enlarged context for the safety concept, which in turn could lay the basis for a new interactions between regulators and regulated. Both sides could agree on the importance of safety for people, the environment and material assets, when viewed collectively. Embracing the whole sector, this means that everyone, individually and collectively, must contribute to ensuring compliance. Furthermore, it means that technology development and continuous improvement must be viewed in relation to the development of knowledge and expertise by everyone involved in the industry. Responsibility for and the management of such activities are moreover essential and must form part of a collaboration between everyone involved – both formally, through operational parameters established by legislation and statutory regulations, and more widely by developing and making provision for such collaboration to occur at sector level between the key stakeholders, as well as in the individual workplace. In essence, the internal control principle embraces many of these elements and has developed in parallel with them in the Norwegian oil and gas sector.

10.4.3. *Cooperation with Those Involved – Tripartite Collaboration*[15]

The stipulation in the petroleum reform that "the regulator should establish a dialog between employees, employers and government on the issues relating to development of regulations should be established"[16] meant that the NPD had to develop new forms of collaboration after 1985 to fulfil this condition. Through its chapter on the legal right of employee participation and collaboration, the Working Environment Act[17] had already laid the basis for further development of collaboration between employer and employees in the individual workplace. The development of social democracy in the Scandinavian countries, too, had similarly established extensive

[15] See also Rosness and Forseth (Chapter 12) in this volume for the dynamic relationship within the tripartite collaboration.

[16] White Paper "st. meld. nr. 51 (1992–93).

[17] Working Environment Act, chapters 6, 7 and 8.

collaboration between central employer and labour organisations. Although this had largely been intended to deal with industrial safety interests as well as pay and working conditions for employees, it eventually acquired an expanded significance in the oil and gas business.

The level of unionisation in the offshore sector increased to become extensively high. That provided a basis on which the government could establish a tripartite collaboration, which involved officials of the main employer and union federations in the development of the regulations through an organised partnership with the regulator. In practical terms, an external reference group for regulations (ERR) was established[18] with representatives from all those involved. Supplementing the general rules for involvement in developing new regulations through a formal consultative process, this allowed companies and unions to become more actively involved in the actual work of framing regulatory provisions from the very start. The ERR developed into a permanent forum which met monthly at times when much work was under way on the regulations. In addition to chairing these meetings, the NPD acted as the secretariat. All relevant government agencies also had the right to attend when appropriate.

10.4.3.1. The Regulatory Forum

The main purpose of establishing the ERR was to facilitate openness and information on documents which determined the parameters for oil and gas operations in Norway, and to create a good basis for all stakeholders to be involved and committed.

Important matters discussed in the forum were:

- the development of new safety, health and environment regulatory strategies;
- follow-up of ongoing work on regulations;
- adaptation to European Union/European Economic Area and other international parameters, follow-up of national and international norms and standards;
- the exchange of views on the content of and experience with the practice of and interpretation of the principles underlying the regulations.

Although the ERR later has changed its name to the Regulatory Forum, its organisation and mode of working has remained more or less unchanged. It also became the model for the other established institution which currently characterise tripartite collaboration in Norway's oil and gas sector.

[18] Regulatory Forum Mandate, http://www.ptil.no/regulatory-forum/category168.html

10.4.3.2. The Safety Forum

The Safety Forum was established in the autumn of 2000[19] with representatives from government, companies and unions. Its main job was to promote work on safety and the working environment in the petroleum industry by:

- serving as a forum to discuss, initiate and follow up relevant safety and working environment issues;
- facilitating good collaboration between companies, unions and government;
- serving as a reference group for projects under way or planned by the companies, unions or government.

Generally speaking, the Safety Forum was intended to confine itself to discussing issues which fell within the ambit of the NPD – since 2004 the Petroleum Safety Authority Norway (PSA). That excluded conditions governed by union-management agreements or other private-law deals. In addition, the forum was to have a role as a reference body for key projects in the oil and gas sector, and thereby contribute to the creation of a shared understanding of realities among the players. This is regarded as the best basis and a precondition for establishing a collaboration to solve common challenges. The Safety Forum would also be a reference and management body for important joint projects, and a reference body for the preparation of white papers of relevance for safety, health and environment for oil and gas industry.

Establishing the Safety Forum gave tripartite collaboration in the Norwegian oil and gas business a further boost, resulting in a greater commitment among, and broader involvement by, all the key stakeholders. The creation of this body meant that the regulator has been organising and running two active fora – on regulations and safety, respectively – since 2000, where employees, employers and government meet regularly to monitor and discuss conditions which promote safety and the working environment in Norway's petroleum industry, and continue to develop an integrated set of regulations which provide predictable and transparent conditions. These bodies also support the goal of developing the greater trust, confidence and shared interests between regulator and regulated, envisaged by the 1985 reform of the regulatory regime. Collaboration without the use of formal instruments by the regulator to ensure compliance with legislation and statutory regulations, and to achieve a greater degree of safety, has also been extended to new contexts through these arenas. The development of knowledge and expertise by all players also has been reinforced.

[19] Safety Forum, http://www.ptil.no/safety-forum/category167.html

10.4.3.3. Working Together for Safety

In parallel with the establishment of the Safety Forum, the Norwegian petroleum industry launched its own Working Together for Safety (SfS)[20] project in 2001. Bringing together representatives from oil companies, suppliers, contractors, unions and employer organisations, its main purpose initially was to improve the safety of human behaviour on vessels and installations and to focus attention on conditions which influence such actions. The NPD – later the PSA – was involved as an observer, which ensured that the regulator could participate actively in the work of enhancing safety in this area as well.

The principal job of SfS today is to improve safety in the oil and gas industry on a general basis. This embraces safety on installations and vessels on the NCS and at land-based plants in Norway. Much of the work is pursued through working parties, which prepare recommendations for the industry. These can take the form of best-practice documents or represent a harmonisation of varying practice so that people working in different locations avoid constantly having to learn new routines and procedures. Contributing to experience transfer by producing safety films based on actual incidents has also been an important job for SfS.

Through its work, SfS follows up and reinforces the goal of dialogue and collaboration aimed at developing trust and credibility between those involved. At the same time, it gives a practical face to the development of knowledge and expertise in order to promote continuous improvement. To date, SfS has prepared and revised more than thirty different recommendations in every area covered by the safety and working environment regulations.

As the changes to the regulations were extensive in terms of both structure and form, an educational programme for the petroleum industry was launched to increase knowledge of the regulations and awareness of the requirements for HSE. This project was named "competence in rules and regulations for the petroleum industry" (RVK).[21]

10.4.3.4. Competence in Rules and Regulations for the Petroleum Industry (RVK)

Based on the established tripartite collaboration between government, employers and employees, the RVK project became an educational arena responsible for the educational programme and for administering the teaching packages developed. Basic training programmes were established to provide familiarity with the structure and content of the regulations and more in-depth knowledge of how to make active

[20] http://www.samarbeidforsikkerhet.no/category.aspx?CatId=140

[21] http://www.rvk.no/

use of the regulations in promoting continuous improvement of safety and working environment conditions.

At present, RVK is an active project collaboration which has adapted its teaching programmes in line with the continuous development of the regulations. An estimated 15,000 participants have so far attended an RVK course.

10.5. CHALLENGES IN A GOVERNMENT-SUPERVISED SELF-REGULATION REGIME

10.5.1. *Developing an Open and Transparent Dialogue*

In a government-supervised self-regulation regime like the Norwegian petroleum sector, a great challenge is to create and sustain an open and transparent dialogue with the companies and collaborating with all the stakeholders to achieve results. How a proactive role of the authorities has been decisive in restoring trust and dialogue among the stakeholders is presented by Rosness and Forseth in Chapter 12 of this volume.

A necessary condition for giving emphasis to dialogue and alternative instruments is that a mutual sense of trust exists between the government and the players in the petroleum industry. Trust is, in turn, closely connected to credibility. This in turn calls for a combination of competence and trustworthiness.

For the PSA as the responsible regulator to trust the companies, the latter must be perceived as credible or have the ability and competence to accept responsibility for their operations.[22] At the same time, an expressed willingness must exist to be open in the communication relationship, which forms the basis for perceptions of reliability.

Similarly, the trust of the companies in the government is conditional on the latter being regarded as credible.[23] That calls, in turn, for the PSA, as a regulator for the petroleum sector, to be competent and perceived as reliable in all types of issues, and for the industry thereby to regard the regulations and the PSA's enforcement of these as just and predictable.

[22] Framework Regulation – Section 7 Responsibilities pursuant to these regulations:

"The operator and others participating in the activities are responsible pursuant to these regulations. The responsible party shall ensure compliance with requirements stipulated in the health, safety and environment legislation.

The operator shall see to it that everyone who carries out work on its behalf, either personally, through employees, contractors or subcontractors, complies with requirements stipulated in the health, safety and environment legislation.

In addition to the duties imposed on licensees and owners of onshore facilities by individual provisions in these regulations, they are also responsible for seeing to it that the operator complies with the requirements stipulated in the health, safety and environment legislation.

According to Section 2–3 of the Working Environment Act and Section 25 of the Fire and Explosion Protection Act (in Norwegian only), the employees have a duty to contribute."

[23] See Tharaldsen (2011).

A supervisory regime based on government-enforced self-regulation embraces all the activities conducted by the PSA to take steps to ensure that the responsible parties in the petroleum industry perform prudent and safe operations in compliance with the regulations.[24]

In practice, follow-up can be pursued in a number of ways, where some "tools" fall within the traditional regulatory concept and are applied through regular control-based supervision. But other activities also contribute to checking, monitoring or influencing the players. These involve processes and professional dialogues between government experts and industry experts and can accordingly be regarded as activities which fall clearly within knowledge and expertise development. Nevertheless, all these activities are part of the total follow-up directed at the players. On that basis, the regulator's follow-up incorporates a number of expertise-enhancing activities which all, in their way, build up knowledge of health, safety and the environment in a broad perspective. Such activities could take the form of general/technical meetings, seminars or various courses/training activities. They can be pursued in the actual follow-up context between the PSA and one or more companies or within the tripartite setting, initiated by the government, the employers, the unions, individual companies or research institutions. Examples of activities which incorporate clear expertise-building aspects and which can thereby also be defined as part of health, safety and environmental training include:

- the Safety Forum's annual conference;
- technical seminars with participation by the industry collectively or by individual groups, such as contractors;
- technical meetings on all subjects, where the discipline teams in the PSA join with their counterparts in the companies;
- participation by and contributions from PSA personnel at courses and conferences organised by companies, employer associations or unions;
- collaboration with research institutions on developing HSE training within a broad range of disciplines – contributions in the form of technical papers from PSA personnel;
- PSA participation in follow-up by the licensees of their "duty to see to it", and in the organisation of systematic work on health, safety and the environment in the licences.

[24] Framework Regulation, Section 10, Prudent activities:

"The activities shall be prudent, based both on an individual and an overall assessment of all factors of relevance for planning and implementation of the activities as regards health, safety and the environment. Consideration shall also be given to the specific nature of the activities, local conditions and operational assumptions.

A high level for health, safety and the environment shall be established, maintained and further developed."

10.5.2. *Dilemmas Presented by Checks and Balances*

Dilemmas will always exist in a government-supervised self-regulation regime over balancing the need for regulatory control against opportunities to make a free choice of solutions as an incentive for the regulated. Freedom to choose solutions at the detailed level calls for knowledge of all aspects of the core business, including technology, design and operational methods, both at the regulator and in the individual company. Continuous development of knowledge and expertise among all stakeholders involved will accordingly be crucial for establishing a well-functioning regulatory system under such conditions. Formulating the regulations in such a way that they primarily specify performance requirements which make it possible for those responsible to determine solutions at a detailed level stimulates the pursuit of planned and managed knowledge development between the regulator and the regulated, including provision for exchanging experience and utilising peer reviews. This also assumes that the regulator participates actively in standardisation efforts and references the industry's own industrial standards in its guidelines.

Awards to, and composing groups of, competent oil companies of licences, requirements for integrated management of operations, a clear allocation of responsibility at decision makers, and duties for licensees in general and operators in particular to follow up those who work for them are essential regulatory instruments in the Norwegian petroleum industry. Furthermore, necessary training for all players, active worker participation in planning and decision-making processes, legal protection of whistle-blowers and safety representative arrangements at each workplace are also significant in such a context. Together these factors can be regarded as necessary checks and balances on a government-supervised self-regulation system which, together with competent regulators, forms a robust regulatory regime.

The regulatory and supervisory regime governing safety and the working environment in the Norwegian petroleum sector is based on trust, competent players and facilitation of decisions at the detailed level by those responsible. When the regulations were reshaped to make greater use of performance requirements, the industry also requested that the regulator should ensure a form of predictability in the application of the regulations – or, put another way, ensure that those responsible could be confident that solutions chosen were in accordance with the requirements without constantly having to check this out with the regulator. The need for predictability was contrasted with the question of whether the reference to prescriptive non-mandatory industry standards could stifle technological development because it would be simpler for the industry to opt for suggested solutions in the regulator's guidelines.

In connection with the regulatory reform in 1985, it was resolved to take care of the need for predictability by converting earlier detailed regulations to non-mandatory guidelines. This also ensured that experience gained from following up activities, and which had been incorporated in the earlier regulations, was not lost. The guidelines were not legally binding, facilitating the choice of solutions which could be documented as meeting the same requirements. Nevertheless, the long-term goal was that the regulator's guidelines should be replaced in the more technical areas with references to standards prepared either under the auspices of the national and international standardisation organisations or in the form of industrial standards developed by the industry itself (recognised standards), so that the post-1985 regulations were directed more at managing the activities and operational and organisational conditions.[25]

The principles for using recognised standards in the petroleum industry were enshrined in a specific provision of the regulations.[26] One consequence was that the regulator had to participate more actively and at different levels in standardisation efforts to ensure that the standards developed could also take care of conditions which concerned the government, thereby making them acceptable for application to the NCS without the regulator having to impose possible additional requirements – in the form of guidelines – through its own regulations.

10.5.3. *Follow-Up in a Government-Enforced Self-Regulation Regime*

The white paper "Report to the Storting no 12 (2005–06)" discussed key conditions which provide guidance for regulating health, safety and the environment.[27] Among other considerations, the government here emphasises that the right of those workers who could be exposed to occupational risk to participate in decision-making processes is important from an ethical perspective. The same preconditions for worker participation must also be applied when the regulator draws up risk-based regulations.

A regulatory regime with the focus on performance requirements is demanding for everyone involved, and not least for the workers, who will not always be able to have the same resources available for participation during the preparation, consultation on and interpretation of regulations as the other players. At the same time, workers possess knowledge and experience which are important to take into account. That

[25] White Paper no. 51 (1992–93).

[26] The framework regulation, Section 24, Use of recognised standards: "When the responsible party makes use of a standard recommended in the guidelines to a provision of the regulations, as a means of complying with the requirements of the regulations in the area of health, safety and the environment, the responsible party can normally assume that the regulatory requirements have been met. When other solutions than those recommended in the guidelines to a provision of the regulations are used, the responsible party shall be able to document that the chosen solution fulfils the regulatory requirements. Combinations of parts of standards shall be avoided, unless the responsible party is able to document that an equivalent level for health, safety and the environment can be achieved. Existing documentation, including maritime certificates issued by Norwegian or foreign flag state authorities, can be used as a basis to document compliance with requirements stipulated in or in pursuance of these regulations."

[27] Report no. 12 to the Storting (2005–2006) on health, safety and the environment in the petroleum activity, section 3.3.4, ethical considerations.

makes it essential for the workforce to have a genuine opportunity to participate in the work of developing regulations and building knowledge of these, so that they can also participate in internal company decision-making processes.

Provision has accordingly been made for such participation through the established Regulatory Forum, through facilitation of participation during the preparatory phase and regulatory provisions making participation in the workplace mandatory. Furthermore, provision has been made for developing regulatory expertise through the educational programme on competence in rules and regulations for the petroleum industry (RVK).[28]

In the white paper discussing the RVK,[29] the ministry identifies a clear relationship between those who have received adequate training in health, safety and the environment and those who express satisfaction with the new regulations in this area. To support this scheme and to create the best possible common arena for following up key areas where all the regulators have a regulatory responsibility, a common set of health, safety and environmental regulations has also been developed. These are prepared, adopted and enforced jointly by the health, safety and environmental regulators.

It is nevertheless the case that collaboration between government agencies with an independent regulatory responsibility can be demanding in areas where these bodies can give differing weight to issues, so that close and good dialogue between regulators and the preparation of formal collaboration agreements become essential. This was also the intent of the coordination regime established in 1985 – that such issues would be resolved jointly by the regulatory agencies.

Converting from a regulatory and supervisory system based on detailed requirements, government checks and approval procedures to a government-supervised self-regulation regime can be demanding both for the industry and for the regulator. While the regulations gave the industry scope to continue developing technology, development solutions and production methods, they also created uncertainty concerning the need for predictability and removed the opportunity for companies to use government checks as part of their own quality assurance or internal checking. This became more visible during the 1990s and led to over-specification of details and unnecessary cost increases in many projects. The issue is further developed and related to the NORSOK-program by Engen in Chapter 13 of this volume.[30]

The challenge for the regulator has been to shift its follow-up from detailed checks to system supervision, to draw up operating parameters for the industry and

[28] This is intended to contribute to increased knowledge and awareness of regulations concerning health, safety and the environment. Since the RVK programme was launched in the autumn of 2001, roughly 15,000 people have taken its courses.

[29] Report no. 12 to the Storting (2005–2006), section 4.7.5.3, project on regulatory competence.

[30] In section 5.2.4 of Report no. 7 to the Storting (2001–2002), the Ministry of Labour and Government Administration noted that a number of reports showed that the industry's knowledge of the regulations

to "dare" to leave more of the detailed follow-up to the industry itself. It also had to be open to – and competently prepared for – a discussion on the appropriateness of alternative solutions rather than pointing to deviations from a specific recommended solution. This has been particularly demanding where genuine disagreement has prevailed over what level of safety is required or should be achieved. In such circumstances, responsibility for interpreting the regulations nevertheless rests with the regulator. Meanwhile, the obligated parties have the opportunity to appeal to the ministry. They make less use of that possibility today than in the past.

A supervisory regime based on government-enforced self-regulation embraces all the activities conducted by the PSA to ensure that the responsible parties in the petroleum industry perform prudent and safe operations in compliance with the regulations.[31] Follow-up embraces three key elements: checking, monitoring and exerting influence. Checking (control) is a central part of the supervisory concept and includes sanctions in the event of non-compliance. In order to monitor how the industry is complying with requirements related to health, safety and the environment, the PSA needs to have sufficient information about the players and the industry as a whole. It is important, for instance, to know how the various players and the industry as a whole conduct their activities, and which health, safety and environmental challenges are the most important. Exerting influence is the third important element, which can be very significant for achieving improvements in health, safety and environmental results.

In practice, follow-up can be pursued in a number of ways, where some "tools" fall within the traditional regulatory concept and are applied through regular control-based supervision. But other activities also contribute to checking, monitoring or influencing the players.

10.6. THE MONITORING PROGRAM (RNNP)

Partly as a result of these diverse opinions, the regulator initiated collaboration with external specialists in 1999 on a pilot project to establish the status of and trends in the level of risk on the NCS. [32] This work concentrated initially on the 1996–2000 period. It was basically launched to develop indicators which could highlight major

was deficient in the years leading up to the presentation of this white paper. The ministry also maintained that this not only affected the level of health, safety and the environment in the industry, but also had major documented financial effects on the industry – not least as a result of inaccurate interpretation of the regulatory requirements. On that basis, the Norwegian Oil Industry Association (OLF) and the Norwegian Shipowners Association took a joint initiative to establish the extensive RVK programme to improve knowledge of the regulations throughout the industry.

[31] See note 24 to this chapter.

[32] See the Petroleum Safety Authority Web site: "Trends in risk level 2012 – The 2011 study of trends in risk level in the Norwegian petroleum activity." http://www.ptil.no

accident risk in relation to loss of life and occupational accidents (Vinnem 2010; Skogdalen, Utne and Vinnem 2011). The pilot project was intended to identify and develop key risk indicators and develop a methodology for evaluating major risk relating to fatalities and personal injuries, which could provide an overall picture of trends in the risk level on the basis of these indicators – both individually and collectively. Emphasis was given to identifying the statistical risk in terms of near misses, actual incidents and perceived threats by identifying a compromised set of indicators. Blakstad gives a detailed overview of the indicators (Chapter 9) in this volume. The different elements embedded in the program are shown in Figure 10.1.

The first report from the pilot project in 2001 initiated a series which has developed into annual studies of risk trends. Through further systematic development of methods and base data, these reports helped employers, unions and government to monitor trends in risk level in the industry through their established tripartite collaboration. The reports also made an important contribution to a joint understanding of changes in risk levels and thereby also identification of measures to improve the level of risk and analyse the effects of corrective actions taken by the responsible parties.

Over time, the RNNP (Trends in risk level in the petroleum industry) process developed into the most important tool for this purpose, and was expanded in 2007 to embrace the PSA's responsibility for land-based petroleum plants. The RNNP is organised through a collaborative network embracing research institutions, industry, employers and unions and the government as shown in Figure 10.2.

Involving all the important stakeholders ensures consensus over working methods and ownership of the conclusions reached in the annual status reports on trends in risk level in the petroleum activity. The PSA is responsible for day-to-day operation of the RNNP process, and also organises the network collaboration. Research institutions advise on developing the methodology and further development of models for processing data and information, and also make a big contribution to securing acceptance for, and understanding of, the consensus-based conclusions. The industry itself contributes data and information and participates actively with relevant expertise.

Norway's model of tripartite collaboration also represents an important factor, and collaboration between all parties is an essential part of the network. In practical terms, this has been facilitated by creating a management model with PSA having an administrative and leading role, using Safety Forum as a reference group. In this way, work on the process has helped to revitalise the tripartite collaboration after the major conflicts of interest around 2000. Through its involvement in and commitment to the RNNP network, this partnership has moved in a more constructive direction, where the various participants have, by and large, been able to develop a new shared understanding of the realities relating to the level of health, safety and the environment (HSE) in the industry. A shared understanding of that kind is

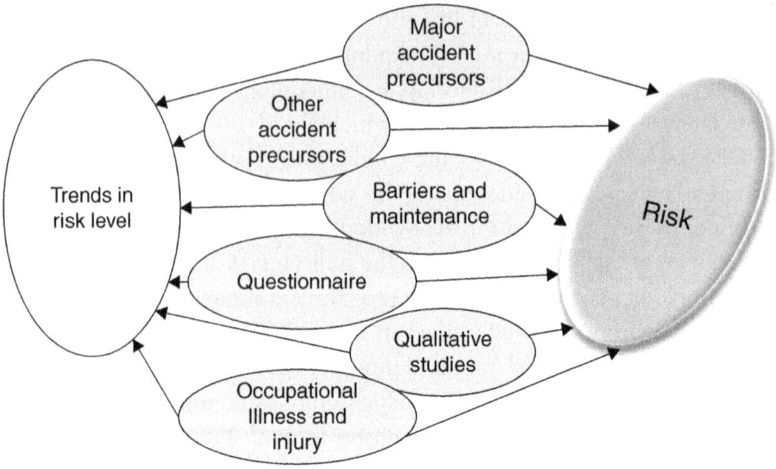

FIGURE 10.1 Elements in the "Trends in Risk Level" program.

FIGURE 10.2 Participants in and contributors to RNNP.

widely regarded as a precondition for constructive collaboration on finding solutions to HSE challenges.

The RNNP has made a substantial contribution to a position where risk level trends in the Norwegian petroleum activity are currently of concern to everyone involved in the industry, and also of general interest. Through the development of methodology and knowledge, it has been possible to establish the RNNP as an

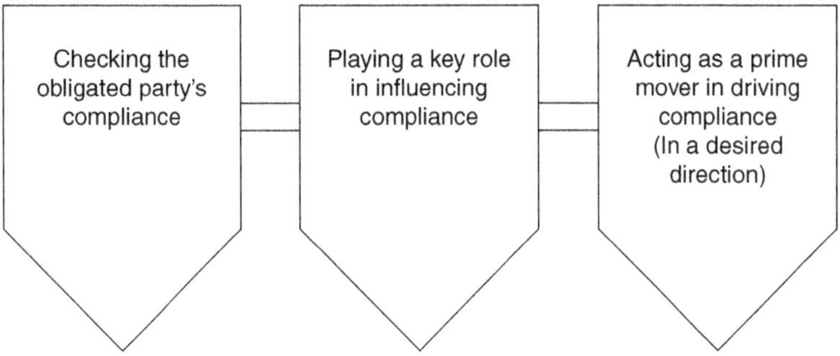

FIGURE 10.3 Modes of regulator's role in an authority-based self-regulation context.

instrument for measuring the impact of overall HSE work in the industry. This development has occurred in a tripartite collaboration which has agreed that the chosen form of the network is sensible and rational as a basis for a common perception of the level of HSE and its development in an industry perspective. The work has acquired an important position in the petroleum sector through its contribution to shaping a shared understanding of the level of risk.

This process also plays an important role for the regulator's supervisory activity. As mentioned earlier, it is important in a government-supervised self-regulation regime that the regulator develops new modes of regulatory oversight. This means that the traditional way of enforcing regulations, in the form of checking the obligated party's compliance with the regulations, is no longer sufficient. New administrative approaches and methods must be developed in order to influence the industry to think continuous improvement in order to operate acceptably, and to be a prime mover in areas where the companies, individually or collectively, lack sufficient incentives or interest to develop good solutions. The regulator's principal role will thereby be expanded to embrace three important components as shown in Figure 10.3.

This expanded role comes across in the RNNP process as an important means of ensuring compliance with the regulations. The network pursues a systematic development of knowledge about risk status and condition in the industry at the same time as this knowledge is applied in a regulatory context. Direct checking of the obligated party's compliance is replaced here by other important regulatory mechanisms. Data and information on all relevant conditions in the industry are processed in a robust model which is generally agreed to provide a credible picture of the risk level. The broad spectrum of different indicators provides the basis for assessing the condition of a number of individual areas. Should a negative trend be identified in any of these areas, it will be discussed and assessed in the various institutional contexts by the network, and will often result in the adoption of measures, either one-off or in the form of campaigns. In this manner, the RNNP process serves as an additional instrument for ensuring compliance with regulatory requirements and continuous development of knowledge, technology and methods.

Maintaining an integrated overview of data and information also allows the regulator to use this to concentrate attention on important individual areas. The industry can thereby be influenced to focus on areas where action will be important in a regulatory context. It also provides opportunities for the regulator to be a prime mover in a longer-term perspective, in part by making active use of the tripartite context in cases where the industry itself does not immediately see the need to develop solutions.

In such a context, the RNNP process serves as an important tool not only to ensure compliance with the regulations but also to go on developing the application of communication, collaboration, learning and scientific methods to ensure that the industry continuously develops and improves its safety work.

10.7. CRITERIA FOR SUCCESS

10.7.1. *Licensing Legislation and the Selection of Competent Players*

As mentioned initially in this chapter, a licensing system was developed from the very start of Norway's petroleum activity based on licensing rounds for awarding exploration acreage, where licensees were grouped in production licences. In legal terms, all oil companies were defined as licensees with the associated rights and duties. Under the licensing system, oil companies which apply for an exploration or production licence will normally be placed in a licence group comprising several licensees. One of these is appointed to serve as operator having an overall responsibility, which implies more specified duties to operate prudently and safely. All licensees have a "duty to see to it" under Norwegian law, which means that they must facilitate the operator's activity through their work on budgets, in decision-making contexts and in developing knowledge and technology. They must also see to it, in part through audits, that the operator is fulfilling its special responsibilities for both safety and resource management. These duties are additional to the responsibilities of the licensees in respect of individual decisions. Such a system had permitted the establishment of a partnership in which several oil companies could collaborate and utilize their special areas of expertise in a complementary context. Formal agreements on the way this licensee collaboration is to be organised have made it possible to develop an open, transparent community of interests, where commercial oil companies cooperate and, through this, share experience which benefits the whole industry. The government has been able to follow up this cooperation by having an opportunity to participate as an observer in licences.

In addition to the actual licensing system, a prequalification process was introduced in the second half of the 1990s. Such a prequalification did not entitle the applicants to be awarded licences, but was introduced in order to simplify the evaluation of applicants in the licensing round or acquisitions. In practical terms the result was that oil companies applying for licences on the NCS would be evaluated in key areas in advance. Prequalification involves an assessment of the company in terms of its financial capacity, core expertise in geology and geophysics (G&G) and competence for managing operations in general and the health, safety and environmental aspects of these in particular. Prequalification begins before a company has been granted a licence, and will be repeated if the company wishes to apply for operator status in a licence – which involves wider and more specific duties. The prequalification scheme is not a legally binding instrument, but can be regarded more as an arrangement which expresses an expectation by the regulator that oil companies intending to conduct business in Norway have sufficient core expertise to operate acceptably in line with the applicable operating parameters. Similarly the Ministry of Petroleum and Energy must approve all transactions concerning ownership of licences. The licensing system therefore sends clear signals to the players in an international industry that Norway requires all participants to have the capacity to conduct their business prudently and in compliance with applicable legislation and statutory regulations.

10.7.2. *A Model Deeply Rooted in the Institutional Tripartite Collaboration*

A distinctive feature of the Norwegian governance model is that employers and unions have redefined and expanded their roles as participating stakeholders. These relationships have been further strengthened by practising a sector-based performance-oriented regulatory regime in the petroleum activity, which opens opportunities for the regulated to identify alternative technical and operational solutions also in compliance with the regulations. Involving the union and employer confederations and establishing relevant institutions closely tied to the regulator's supervision of the activity have made it possible to develop collaboration between leading elected officials for employers and unions and the regulator, which provides opportunities to supplement the monitoring of safety and the working environment in the industry. The regulator organises and administers two of the tripartite arenas, where employers and unions take the opportunity to influence regulation and follow-up of the industry through active commitment and involvement. Collaboration processes in the tripartite context are open and transparent, and are also shared in the public arena through the publication of everything which takes place on PSA Norways Web site (www.psa.no).

The very high degree of unionisation among workers in the Norwegian petroleum industry also allows the regulator to make active use of employers and employees by involving them in the exercise of supervision. Union representatives are given the opportunities to participate in all supervisory activities directed at individual players. This brings forth knowledge and views from the employee side in a clear and constructive manner. In a general perspective, it could be a challenge that such involvement calls for a high level of unionisation.

The Safety Forum is also involved in the process of producing policy documents on health, safety and the environment in the petroleum industry in forms of white papers. The parties can thereby also actively influence the policy framework in a strategic perspective.

10.7.3. *Knowledge and Expertise Development*

In a government-enforced self-regulating system, development of knowledge and expertise represents important and integrated elements. Capacity building is therefore included in all aspects of the activities. An active collaboration and dialogue between government, industry and research institutions facilitated through regulations and supervisory models has made the integration of a risk-based management of the industry, and the required expertise associated with this, possible. As a result, active internal training and expertise development with safety, health and the environment are pursued in the various institutions – collaboratively through industry organisations, in a tripartite context and in direct collaboration between stakeholders.

Legislation and statutory regulations provide the basis for the way the industry pursues its operations in a systematic and transparent manner and in compliance with the regulations. At the statutory level, key provisions formulated in the Petroleum Activities Act and the Working Environment Act are determinative for expertise development with health, safety and the environment. The Petroleum Act specifies[33] that "the King may issue rules relating to the licensees' obligation to undertake training of civil servants". This has given the government the right to utilize core expertise within oil companies to develop their own expertise fully on a par with that possessed by the companies across the whole area, including health, safety and the environment.

The Working Environment Act contains several provisions which deal specifically with training both of the company's chief executive and for legally required safety and environmental work.[34] Statutory requirements on the primary level of legislation (act level) are further elaborated and clarified in the secondary level of

[33] Petroleum Act Section 10, subsection 11.

[34] Section 3, subsection 5 of the Working Environment Act on the employer's obligation to undergo training in safety, health and environmental work:

"(1) The employer shall undergo training in health, safety and environmental work."

regulations. Where the regulations are formulated as goal-setting requirements, the responsible party may choose a specific solution best suited for the purpose; this requires capacity building and provides a basis for collaboration between all the relevant institutions and fora established to share experience and further develop knowledge and best practice relating to safety, health and the environment.

The framework regulations on health, safety and the environment in the petroleum sector provide the operating parameters for prudent activities within the petroleum industry, and include a number of important provisions which govern the way the players pursue active expertise development and training in safety, health and the environment. Section 12 on requirements for organisation and competence has accordingly been important for the way companies must organise themselves in order to pursue petroleum operations in an acceptably safe manner, and the expertise they need to obtain in order to be able to do this.[35]

The framework regulations also elaborate on the special "duty to see to it", which rests on operators in the petroleum activity. This confers a duty on the operator to ensure that everyone working within its area of responsibility is adequately qualified in terms of health, safety and the environment, and complies with regulatory requirements.[36]

Section 23 of the management regulations highlights the significance of continuous improvement as an important instrument for pursuing prudent operations. Improvement efforts are directly related to continued development of knowledge and expertise, and accordingly represent a driver for such work.[37]

[35] The employer is responsible for the cost of training and other expenses related to the work of the safety representatives. Safety work which must be conducted outside normal working hours pursuant to section (10, subsection 4) qualifies for overtime pay.

[36] Framework regulations, Section 18:

Qualification and follow-up of other participants
When entering into a contract, the responsible party shall ensure that the contractors and suppliers are qualified to fulfil the regulatory requirements relating to health, safety and the environment. Furthermore, the responsible party shall follow up to ensure that the participants comply with the requirements while performing the assignment in the activities covered by these regulations. The operator shall ensure that any deficiencies in other participants' management of health, safety and the environment are corrected, and that the necessary adjustments are made with respect to its own and other participants' management systems, to ensure the necessary uniformity.

[3] Section 65, Training of public employees.

[37] Management regulations, Section 23:
"Continuous improvement
The responsible party shall continuously improve health, safety and the environment by identifying the processes, activities and products in need of improvement, and implementing necessary improvement measures. The measures shall be followed up and the effects evaluated. The individual employee shall be encouraged to actively identify weaknesses and suggest solutions, cf. Section 15 of the Framework Regulations. Applying experience from own and others' activities shall be facilitated in the improvement work."

The activities regulations have a separate chapter on competence. This provides the basis for ensuring that all personnel who participate in the industry have the necessary expertise to perform their duties in a prudent and safe manner. This is also an important instrument for the way the government in general assesses training activities in the companies and how they should acquire expertise for the various jobs and functions.[38]

The Norwegian government has expressed an ambition that the country's petroleum industry should be a world leader for health, safety and the environment. Such an ambition means that it is not sufficient to follow the example of others or to be satisfied with fulfilling minimum requirements. It can only be fulfilled by paying continuous attention to knowledge development in connection with continuous improvement, and collaboration with all the parties involved.

[38] Activities regulations, Section 21:

"Competence

The responsible party shall ensure that the personnel at all times have the competence necessary to carry out the activities in accordance with the health, safety and environment legislation. In addition, the personnel shall be able to handle hazard and accident situations, cf. Section 14 of the Management Regulations and Section 23 of these regulations.

Personnel who will carry out bell diving or surface-oriented diving, shall have a valid certificate. The Petroleum Safety Authority Norway appoints suitable enterprises to issue certificates on its behalf. Payment can be charged for issuance of certificates."

Activities regulations, Section 22:

"Safety and working environment training pursuant to the Working Environment Act

Managers and others with responsibility for decisions that affect the working environment, shall be provided the same training as members of working environment committees and safety delegates.

The individual employee and manager shall be provided with training in working environment factors of significance for conducting their work.

Managers with direct responsibility for work with radioactive sources shall have completed theoretical and practical radiation protection training.

The employees shall be provided necessary training in health and safety matters, and the training shall take place during working hours. Criteria shall be set for what constitutes necessary training.

Training as mentioned in the fourth subsection, shall be provided upon employment, transfer or change of work tasks, introduction of new work equipment or changes to the equipment and upon introduction of new technology that applies to the individual's workplace or work tasks.

The training shall be adapted to the changed or new risk in the enterprise, and repeated when necessary."

Activities regulations, Section 23:

"Training and drills

The responsible party shall ensure that necessary training and necessary exercises are conducted, so that the personnel are always able to handle operational disturbances and hazard and accident situations in an effective manner."

REFERENCES

Braut, G.S. and P. H. Lindøe (2010). "Risk Regulation in the North Sea: A Common Law Perspective on Norwegian Legislation." *Safety Science Monitor* 14(1): 1–9.

Gustavsen, B. and G. Hunnius (1981). *New patterns of work reform: The case of Norway*. Oslo, Universitetsforlaget.

Hagen, L., U. Hammer, T.G. Michelet and T. Stang (1989). *Petroleumsloven med kommentarer*. Tano, Oslo.

Hovden, J. (2002). The Development of New Safety Regulations in the Norwegian Oil and Gas Industry, in B. Kirwan, A. Hale and A. Hopkins (eds.) *Changing Regulation. Controlling Risk in Society*. Pergamon Elsevier Science, Amsterdam, pp. 57–78.

Karlsen, J.E. and P. H. Lindøe (2006). "The Nordic OHS Model at a Turning Point?" *Policy and Practice in Health and Safety* 4(1): 17–30.

Kaasen, K. (1984). *Sikkerhetsregulering i petroleumsvirksomheten*. Sjørettsfondet, Oslo.

Lindøe, P. H. and K. Hansen (2000) "Integrating Internal Control into Management Systems. A discussion based on Norwegian case studies", in K. Frick, P. L. Jensen, M. Quinland and T. Wilthagen (eds.): *Systematic Occupational Health and Safety Management*, Pergamon, Elsevier, Amsterdam, pp. 437–455.

Skogdalen, J. E., Utne, I.B. and Vinnem, J.E. (2011). "Developing safety indicators for preventing offshore oil and gas deepwater drilling blowouts." *Safety Science* 49: 1187–1199.

Tharaldsen (2011) "'In Safety We Trust': Safety, Risk and Trust in the North Sea Petroleum Industry." PhD thesis, University of Stavanger.

The Petroleum Safety Authority (2011). "Trends in risk level 2011 – The 2011 study of trends in risk level in the Norwegian petroleum activity."

Vinnem, J.E. (2010). "Risk indicators for major hazards on offshore installations." *Safety Science* 48: 770–787.

Vogel, L. (1998). *Prevention at the workplace*. Brussels, European Trade Union Technical Bureau for Health and Safety.

11

Contested Terrains in Risk Regulation

Legitimacy Challenges in Implementation Processes

Jacob Kringen

11.1. INTRODUCTION

This chapter reviews and analyses recent developments in the Norwegian petroleum safety regime, focusing on accountability and legitimacy challenges arising from regulatory content as well as from institutional context.[1] As outlined in Chapter 5 and 10 in this volume, the regulations have over a period of years established a purpose- and systems-oriented self-regulatory system, embedded within a largely trust-based tripartite institutional framework. This involves a delicate balance between the parties involved, notably the authorities, the industrial actors and the unions. Legal provisions seldom provide clear-cut thresholds for acceptable risk and regulatory compliance. Because of reputational concerns and a 'compliance-friendly' regulatory incentive structure, enforcement policies are largely accommodative and few cases involve sharp confrontations over legal standards, which are virtually unexposed to judicial review in court. Still, the regulatory system is continuously challenged through negotiations over vague and floating legal standards, as well as contested risk indicators, and may erupt into open conflict when trade-offs between safety and profitability are perceived as disproportionate. On the other hand, purpose- and/or principles-based regulation may also evaporate, dissolve or become irrelevant if not properly attached to industrial realities, management practices and regulatory expectations.

This chapter addresses how these two threats are encountered and coped with in the 'normal state' as well as in cases of 'disturbance'. Two such cases are reviewed. The first case deals with sleeping facilities and working hours, where basic features of the regulatory system have been severely tested through regulatory enforcement practices, and shows how industrial and union interests are mobilised in negotiations over economic costs, levels of protection, acceptable risk and welfare considerations.

[1] The chapter is primarily based on an extensive qualitative study of the Norwegian petroleum regime, including documentary studies, fieldwork and interviews with government officials, unions and industrial organisations (Kringen 2009).

The second case addresses a regulatory invention requiring the industry to develop a 'good and sound HSE-culture' and the substantial and interpretive burdens put on both the regulators and the industry in the ensuing processes of implementation. The chapter discusses how the regulators must handle the burden of proof in accounting for convincing and evidence-based causal links between regulatory practice, prudent safety levels and risk management strategies, and examines some of the costs and dilemmas arising from purpose-/systems-based self-regulatory regimes. The discussion deals primarily with the accountability and legitimacy problems encountered, both from an empirical and a normative perspective.

Policies for smarter, better and more flexible regulations have dominated regulatory agendas in most OECD countries during the past two decades (OECD 2010; 2011). Such policies reflect unstable and contested trade-offs between fulfilling the overall goals of regulation against the regulatory burdens and costs put on the regulated enterprises. Regulatory goals, be they welfare promoting (e.g. public service functions) or risk reducing and protective, reflect general welfare levels as well as public expectations in terms of protective demands and risk tolerance. Regulatory burdens and costs cover substantial and 'justified' as well as 'unnecessary' administrative burdens and 'disproportionate' regulatory interventions not perceived as comparatively effective in achieving the goals. Sometimes, the goals themselves may be contested, such as in the determination of 'acceptable risk'. The use of quotation marks above indicates the deep controversies that are often involved in these issues, scientific and factual as well as political and normative.

Controversies and debates may tend to cluster around the 'means and method' issues rather than around normative issues. This is understandable, because advocates of any position often ascertain their commitment to the goals, be it 'safety first' or varieties of 'zero-harm' philosophies. Politicians, at least in the Norwegian context, rarely are very precise about cost-benefit trade-offs or risk acceptance criteria. More for less, and a general call for administrative efficiency, is the general elixir for most public reform measures, often subsumed under the New Public Management agenda. Similarly, regulatory reform programs follow the 'means and method' strategy and do not challenge the overall policy or goal behind any specific regulation. Actual trade-offs and normative valuations are often left to lower-level actors within regulatory systems.

11.2. PERSPECTIVES AND CONCEPTS

The two cases reviewed raise several distinctive and yet similar problems of accountability and legitimacy well known to modern regulatory systems and regimes (Lodge 2004; May 2007; Black 2008). Within the context of this book, such problems may well be phrased in terms of regulatory robustness, given that such robustness will depend heavily on precisely the degree of acceptance the regime enjoys from key stakeholders within the relevant regulatory context.

Legitimacy issues can tentatively be clustered around two sets of problems: (1) the legitimacy of the regulatory goals in terms of how much risk the regime is prepared to tolerate, and (2) the legitimacy of the regulatory means and instruments designed for achieving these goals. Legitimacy issues related to governmental and administrative exercise of discretionary powers involve both aspects and appear critically relevant when dealing with vague and goal-oriented regulations and legal-democratic mandates. Black (2008) analyses legitimacy issues as a series of dynamic relationships, thus drawing attention to the dialectical and communicative aspects of how accountability is constructed, maintained and/or undermined within different institutional environments. Legitimacy clearly rests on normative foundations as perceived by relevant parties and communities. Subjected to empirical analysis, studies of legitimacy will seek to discover the social acceptability, desirability or appropriateness of any acts or agents. In the regulatory context, legitimacy will certainly reach beyond the purely governmental or legislative mandate of the regulators, including also the perceived credibility and appropriateness of how this role is in fact enacted. Legitimacy thus rests in the values, interests, expectations and cognitive frames within different communities, which may themselves represent divergent, conflicting and changing legitimacy claims. Legitimacy is generally put to test through different accountability claims, in this case holding the regulators accountable for choices and responsible for defending these with reference to factual as well as normative justifications. Within the regulatory context, we may distinguish four overlapping and not mutually exclusive aspects of accountability (and legitimacy): legal, bureaucratic, professional and political (May 2007). May defines legal accountability as the perceived fairness and appropriateness of the rules and standards, including the rules and standards of rulemaking itself. Bureaucratic accountability refers to the implementation and enforcement processes, stressing the need for regulators to be reasonable and fair in interpretations and assessments of rules and facts when enacting their authority. May draws special attention to the increasing importance of professional accountability, where bureaucrats and front-line law enforcers, as well as regulatees, must also meet standards of professional peers, and ultimately of the scientific basis of these. Political accountability refers to the responsiveness of elected officials to regulatory shortfalls, and the various oversight mechanisms available for detecting these.

The widespread adoption of purpose- or systems-based regulations contributes to the processes of delegation and the accompanying call for accountability structures. Broad-brush formulations of basic obligations implies a mixture of legal and non-legal (soft law) norms, referring also to (self-imposed) professional codes of conduct, industrial standards and company-specific rules and systems of implementation.

The field of regulation lacks standard terminology, and a plethora of concepts and definitions have gradually evolved (Coglianese and Mendelson 2010). The issues raised in this chapter deal with aspects of regulation where the proliferation of terms has been particularly apparent and often applied in contrast to their terminological counterparts (such as command-and-control regulation or prescriptive regulation). These terms refer to a cluster of regulatory inventions now dominating many regulatory regimes and reforms: self-regulation, enforced self-regulation, meta-regulation, purpose-based regulation, principles-based regulation, performance-based regulation, systems-based regulation, management-based regulation and so forth. Several attempts have been made to specify their meaning, to define underlying dimensions and to achieve more analytical clarity. Common denominators, underlining their family resemblance, have also been suggested, for instance, through the term 'process-oriented' regulation (Gilad 2010).

For the purpose of this chapter, it may be useful to follow a distinction provided by Black and Baldwin (2010) between regulations addressing inherent risks and management or control risks. Regulations addressing the former deal directly with the risks as such (e.g. dangerous substances or work processes), whereas the latter deal with the organisational capacities to cope with them (e.g. management systems). The distinction is not clear-cut, because regulations addressing management and control risks typically also involve the responsibility of the regulatee to deal directly with inherent risks without reliance on legally defined criteria for risk acceptance. A similar distinction is provided by Gilad (2010), distinguishing between the regulation of first-tier and second-tier operations. The first address core production processes whereas the second address the management structures and controls that the regulatees must have in place in order to effectively monitor compliance with first-tier requirements. She also introduces third-tier regulations, addressing the organisations' overall and continuous evaluation of risk and its management. Taken together, these levels of regulation correspond roughly with Parker's concept of *meta-regulation* as a dynamic and process-oriented evaluation of the efficacy of existing practices and outcomes, and subsequent experience-based learning and improvement. Meta-regulation involves the incremental improvement of performance of not only the regulatees but also of the regulators, using the experiences within the regulatory domain as lessons for improving regulatory strategies and practices (Parker 2006).

In the following we use the term 'purpose-based regulation' in addressing the first level of inherent risks – or first-tier regulations. Such regulations vary according to how precisely the purposes are defined. If purposes are clearly specified in terms of outcomes, they will often be referred to as performance-based, but the term is avoided here as such precise criteria are largely absent in the cases discussed. The second level of management-based regulations (second-tier) is referred to as systems-based. In referring to the third level of regulation, we apply Parkers notion of meta-regulation.

Taken together, such regulatory designs could be seen exclusively in instrumental terms, as simply better ways of achieving regulatory goals in dealing with complex and dynamic risk contexts. But they also have normative justifications in terms of how regulatory responsibilities are distributed and the degrees of freedom granted to regulatees. These regulatory systems imply a reallocation of responsibilities for working out the practical application of the regulatory provisions, which raises several critical questions in terms of normative legitimacy and legal accountability, as well as empirical and explanatory issues regarding distributed governance, decentered regulation and hybrid forms of normative control in the (post) regulatory state (Scott 2004).

These systems have inherent accountability and legitimacy challenges, in large part following from discretionary powers of the regulatory authorities and the associated problems of judging just which means and measures are effective and proportionate in promoting the regulatory goals. But the goals themselves may also be vaguely described, leaving the more precise determination of acceptable risk and what is 'safe enough' to be resolved in the implementation processes.

Still, there are obvious justifications for implementing purpose- and systems-based forms of regulation. First, rules should not exceed the scope of their purpose, thereby granting the regulatees the freedom to choose different solutions. Second, the regulatees are often in the best position to implement knowledge-based solutions adapted to their specific operational contexts, assuming that the regulatory purpose is adequately met. Third, purpose-based rules sustain and underscore the responsibility of the regulatee. Fourth, rules cannot be updated with sufficient pace in accordance with technological and other advancements in different areas, and must therefore have sufficient flexibility to be able to absorb change.

11.3. THE REGULATORY SYSTEM IN THE NORMAL STATE

The petroleum safety regime is fairly coherent and comprehensive because it is organised according to a sector principle. It thus covers a broad range of risk areas in the petroleum industry, with its complex array of actors, technologies, processes and natural resources. Health and safety risks cover an extensive number of areas both in terms of causes of hazards (e.g. helicopter transport, fires and explosions, well blowouts, lifting and crane operations and falling objects) and in terms of possible outcomes. The latter include fatalities as the worst case, in particular within the scenario of major accidents, but also occupational injuries, from cuts and bruises to serious and incapacitating accidents and occupational illness, often as a result of long-term exposures to various hazards (like noise, chemicals, bad ergonomics etc.). Risk indicators are now broadly categorised in terms of major accidents, occupational accidents and occupational health. The relationship between risk factors has been increasingly focused, however, drawing attention to the effects that the general working conditions have for operational safety. Physical and mental strain and stress may affect not only the health of individual workers but also operational safety through the execution of decisions and tasks. And the regulatory provisions, addressing the underlying causes of risk (management, organisation, operations etc.), do not distinguish between the various outcomes that may result. Regulations are holistic in terms of management requirements but more fragmented in terms of specifically addressing the actual risk themes and contexts (e.g. drilling operations, structural integrity of platforms, working conditions etc.).

The Petroleum Safety Authority (PSA) executes the day-to-day implementation of statutory regulations. The agency also has a much broader mandate, including the preparation of legislative proposals and the production of (secondary) regulations. Although the PSA is an 'ordinary' governmental agency in legal and constitutional terms, it enjoys considerable discretionary autonomy and must be seen as the major strategic policy maker within the governmental apparatus (this in contrast with the much more politically and hierarchically governed resource management regime). Almost all discretionary authority is delegated from the ministry. Also, there are relatively few instances of formal complaints on agency decisions, which minimises opportunities for ministerial interference. Generally (historically), the ministry has expressed great trust in the professionalism and vocation of the agency. The role of the ministry has basically been that of administrative control and overall governmental management according to administrative and budgetary rules and processes.[2] Regulatory reform proposals and amendments are coordinated and discussed with the ministry prior to the ordinary formal public hearings and reviews. Within the

[2] This trust could be more bluntly expressed by a former Director General of the Department of Working Environment and Safety (1990–1997): 'Generally, we had a deep-seated confidence in the safety department in the NPD; they were leading experts in this field and Magne Ognedal [General Director] was really an international capacity. The information asymmetry was obvious; if they said it's safe to drill a 5000 meter hole, we could not but trust them. We did not, and could not challenge them on their professional competence. We had lawyers, and economists and social scientists, whose basic expertise was that of maneuvering within the administrative and political interface'. In terms of regulatory process, the basic role of the ministry was largely described in terms of benevolent inquisition: ascertaining that due processes had been followed, relevant parties involved, expert judgements requested and so forth. As to the substantive issues of establishing a 'prudent' level of safety, the agency appeared relatively self-governed. Some critical issues could surface, however, but largely on a case-by-case basis, depending on activism from below or political attention from above (or both). Basic division of responsibilities has not changed significantly in the last decade; in a survey conducted by The Directorate of Public Management in 2002, the ministry even suggested that the delegation of powers to the agency had reached a level where additional delegations would remove their governing capacity altogether (Statskonsult 2002).

overall legislative framework and mandate, participation 'from below', codified within the tripartite system, appears equally important in shaping the actual content of the regulations. In fact, the maintenance and management of this system is a basic feature of the political mandate of the PSA. Since 2002, the regulatory system has been reviewed and politically endorsed in white papers issued every four years (see also Chapter 10 in this volume).[3]

The combination of purpose- and system-based regulation has been the major governmental strategy in Norway during the last three decades, with the petroleum safety regime being at the forefront (regularly referred to as functional regulations). Overall, the strategy has been supported by all stakeholders as a necessary condition for industrial flexibility and value creation (see also Chapters 5 and 10 in this volume for a more detailed discussion). These regulatory strategies emerged gradually from the late 1970s, aiming to mobilise regulatory resources and to promote prophylactic, prudent and self-corrective mechanisms within all managerial and operational facets of the industry. Prescriptive requirements have now largely been reformulated in industrial standards, and the statutory provisions are goal- and management-oriented. Most provisions are briefly formulated target statements, supplied with non-statutory guidelines that refer to the relevant standards. However, some important deterministic requirements remain regarding safety barriers, redundancies and 'defence-in-depth' principles. More loosely defined rules for risk management have replaced quantified acceptance criteria, in particular by specifying that the 'As low as reasonably practicable' (ALARP) principle requires the duty holders to improve and implement safety measures beyond simple risk-cost-benefit considerations, thus shifting the burden of proof in favour of safety.[4] Regulations also include requirements for 'continuous improvement', underscoring the ambitious and dynamic character of the governmental safety policy. There are now five general and purpose-oriented regulations, with guidelines supplementing each, and with reference to a large number of associated national and international standards as possible but not mandatory paths to compliance.[5]

Although the regulatory framework is statutory and state-imposed, it is also firmly set within the Norwegian (Nordic) tripartite system, with long traditions for cooperation between industrial players, the unions and the authorities as addressed in Chapters 10, 12 and 13 in this volume. This system gradually replaced the more adversarial regime of the pioneer era during the 1970s, with weak and fragmented unions, union-hostile and centralised U.S. oil-companies and industry-friendly

[3] Ministry of Labour (2002; 2006; 2010).

[4] According to Hovden (2002), these proposals met some opposition in the industry, as they no longer allowed for justifying cost cutting based solely on (controversial) risk-cost-benefit analysis.

[5] The PSA participates extensively in the preparation of standards (primarily national) through working groups, reference committees and review processes.

government policies reflecting an interest in extracting economic benefits from the petroleum resources. Strong unions developed from the late 1970s, able and willing to voice their interests and to exploit statutory rules as those were being produced. The regulatory development reflects increased risk awareness, particularly after serious accidents in 1977 and 1980.

Several bodies, collaborative arenas and projects are established for making the tripartite system a deeply institutionalised part of the regime. This includes the *Safety Forum*, comprising the management level of unions, industry organisations and the authorities, the *Regulatory Forum* with a key role in the development and amendment of regulations, and several working groups and joint project groups promoting the implementation of regulatory objectives within the industry (see Chapter 10 in this volume for a detailed description).

The present regulatory system is thus a continuation of a long regulatory development, firmly established within the tripartite system. This presupposes close cooperation with the industry and a dialogue with the involved parties largely based on at least some level of consensus and mutual trust. If union and employee representatives become more critical of the industrial practices, the system is vulnerable.[6]

The regulatory strategy has recurrently caused discussion, and criticism, of the regulatory role in standard setting. Unions and researchers have occasionally considered vague purpose-based rules to be a circumvention of responsibility by the authorities – that the latter did not take a stand by providing explicit standards. But the process of mobilising the self-regulatory capacities of the industry has been the dominant rationale. The industry had to make safety-critical judgements on its own, not just rely on government prescriptions.

Defensiveness and blame avoidance may have been elements in this strategy, but the justifications were obvious. Partly it has been considered a matter of simple necessity. As one of the agency veterans explained, with reference to the pioneer days, they received 'piles of paper from the industry' and were severely under-resourced in terms of making expert-based reviews of all the technical documentation (see Kringen 2009). But there were also more positive justifications. It would not only 'empower' the industry, but also engage them actively in the risk management processes and make them accountable for the solutions adopted.

6 Indeed, it was argued that the 'functionality' of loosely defined targets increasingly served as a smoke screen for reduced HSE standards, and that it made it difficult for safety representatives and union representatives to object to measures or to demand improvements if no standard was available as threshold or norm. An evaluation of the regulations conducted in 2004 revealed highly disparate opinions among the parties, however, including also among different unions. The industry was moderately satisfied with the target orientation and the flexibility of the regulations, but more critical in terms of adequate cost-benefit assessments and predictable enforcement practices.

At the same time, the authorities have been continually confronted with the need to know what is 'good enough'. The primary response has been to provide guidelines so that the functional requirements could be optionally met by adhering to such guidelines, but that other options would be possible if an equally satisfying effect could be documented. The next step has been to substitute guidelines with industrial standards. The maintenance of detailed guidelines generally had high costs, and the technology advanced ahead of them. By replacing much of the content in the guidelines with references to industrial standards, the role of the authorities has largely been to participate in standardisation groups on a professional basis. To the extent that these standards are referred to in the guidelines, compliance with the standard would normally equal compliance with the regulation to which it is linked. And if an incident occurs despite compliance with a recommended standard, this could have bearing on the subsequent legal assessment.[7] Still, it is of some importance that a distance is upheld between the authorities and the standards. References to standards are always made in the guidelines, not in the regulations. If standards are referred to in supervisory reports, as exemplifying conformity or non-conformity, it must be formulated so that the standard does not appear as the only path to compliance.

Still, the regulatory system does not relieve the authorities of establishing what in a given case should be considered 'good enough'. Deviation from a standard that may lead to increased risk must be compensated for by other measures, as will any change in circumstances that may make the 'standard' inappropriate. All operations must at any time be 'prudent' in terms of risk and safety as judged by the authorities, and enforcement practice cannot always, or solely, rely on prespecified norms, as will be discussed later in the chapter.

Importantly, however, this regulatory system presupposed organised involvement from the parties within the tripartite framework, as outlined by the (former) regulatory director of the PSA:

> Obviously, when you operate a regime where the involved parties use the 'freedom' of the regulations to proactively improve HSE-standards, that's an ideal situation. But if that 'freedom' is used to look for minimum solutions it creates a different and more problematic situation. There is of course some flexibility. The employee

7 In a court ruling from 2004 (HR-2004-763A – Rt-2004-698), the Norwegian Supreme Court acquitted an oil company in one case of pollution where an incident occurred despite adherence to an industrial standard. In this case a tank that was built and inspected according to API (American Petroleum Institute) standards had a leakage. The Norwegian pollution authority (and the lower-level courts) claimed stricter requirements than the standard specified, but because the standard was considered to be widely accepted, the Supreme Court sided with the company. This is still a grey zone, however, and just recently there has been a controversy between the PSA and the industry regarding life-boats. The guidelines originally had a reference to the general marine regulations for such boats, but the agency found these to be insufficient for the offshore facilities, removed the reference and reinforced the requirements.

> side may observe, in their view, a reduction in standards, which we may still judge to be in accordance with justifiable interpretations of the regulations. This regulatory policy is clearly founded on participation from the industry and all the parties. One thing is participation in the regulatory process. But the regime also presupposes participation in the 'in-house' self-regulatory process. And this is a critical and sensitive issue. In the old prescriptive regime with detailed rules, if the industry wanted to adopt a different solution or approach, they would have to apply for exemptions from the regulations. This decision could then be subject to appeals from the employees. This provided a formal access to the decision process. In the case of self-regulation, the parties must participate. We transfer decisional powers to the industry on the condition that they have a dialogue with the employees when choosing specific solutions. If the employees or their representatives then say 'we are not sufficiently included in the decision processes', we would be in conflict with an important premise for the development of the regulations. This is an important principle for us. But thus far, the parties have generally been satisfied with this regulatory policy and the access to the regulatory process, and we're really happy with that. (see Kringen 2009, p. 120)

Negotiation within the regulatory system thus involves not only arguments about specified acceptance criteria for risk but also more fluid agreements and disagreements between principal stakeholders, with the regulator in a mediating role. And as noted, within the overall legislative framework and mandate, participation from below appears as equally important in shaping the content of the regulations.

The regulatory system has obvious implications for law enforcement, which is no trivial pursuit in the petroleum sector. Regulatory intervention can generate large costs, such as those requiring temporary production stops or technological investments. Historically there have been cases of controversy over regulatory standards and the proportionality of interventions, but the basic philosophy of purpose-based regulation has still been a common interest for regulators and regulatees alike, disregarding at this point some noted objections from the unions.

The PSA enforcement policy is however generally presented as accommodative and dialogue-based, but must be interpreted against the reputational concerns of the industry in being 'well-behaving', and ultimately against the powerful role the authorities have in influencing the probability for industrial market access through the licence system. The enforcement measures are often displayed as steps on a staircase, the uses of which ascend according to the severity of the regulatory transgressions. It starts with 'dialogue', and is followed by orders, coercive fines, stopping activity, legal prosecution and finally expulsion. The enforcement strategy is displayed as a downward arrow to illustrate the preference for accommodative alternatives. The term 'dialogue' covers a broad spectrum of encounters with the industry. In terms of enforcement, dialogue subsumes a number of informal strategies, such as 'requests', informal warnings, identical letters to groups of actors about trends or general observations, copies of letters and reports to higher company management levels, distribution of safety notices, summoning meetings with company management and so forth; in short, all non-statutory initiatives taken to influence company behaviour. Even the posting of audit and investigation reports on the PSA Web site may be considered as part of an enforcement strategy.[8]

A 'Handbook of enforcement measures' provides rough guidelines for how the various instruments are to be applied in various circumstances. 'Requests' are the most common response used in the follow up of audits. Requests are used as a first response even when non-conformities can be categorised as regulatory violations if these do not justify the use of orders. They might also be the response to observations that cannot be strictly categorised as non-conformities (referred to as 'potentials for improvement'). Such responses may also be based on the more evaluative and discretionary judgments not specifically related to single observations but to overall 'impressions' that are not easily substantiated by factual 'evidence'.

An order is a formal and legally binding enforcement measure. It is considered a 'powerful preventive policy instrument' that clearly signals that the violation is of a serious kind with a considerable potential for causing harm. It may also be used when the actual violation (in isolation) is of a less serious kind but when the company in question has a bad record of responding adequately to the milder requests. Orders shall always be preceded by a notification, in itself seen as a strong signal to the company. Occasionally, the notified orders are not issued, or they may be modified or altered, because of faulty assumptions from the supervisors or corrective action already taken by the company. But as a rule, notifications result in orders, indicating that there has been a preceding process of trying to come to an agreement. Normally, requests (and orders) do not specify changes or solutions to be implemented. This follows from the regulatory philosophy and the nature of the regulations, and the regulatee is rather prompted to 'do something', that is, to analyze causes, identify alternative interventions, and to act on these. This call for self-diagnosis and self-treatment is as integral to the enforcement philosophy as self-regulation is to the regulatory philosophy. Responsibility must be firmly placed on the regulatee and auditee, not on the regulator and auditor.

The frequency of enforcement measures decreases with their strength. In fact, the strength of formal instruments may be indicated by their infrequent use. On average,

[8] This risk of sanctions was indeed phrased in conjunction with the doctrine of dialogue, making dialogue conditional to the 'cooperativeness' of the firm, as evident from the following statement: '[T]here is a continuous dialogue between the Petroleum Safety Authority Norway and the players on the Norwegian shelf. For us, this dialogue is a means of influencing decisions/actions, and it is a central element in our supervision strategy. If the dialogue is unsuccessful, we can give notice of orders, and then orders. In serious cases where safety is endangered, we can demand that an activity be temporarily stopped. The authorities can also file charges with the police and impose fines' (http://www.ptil.no/English).

only some ten to fifteen orders are issued each year; some of them are also related to the same incidents but issued to several companies involved. Orders are thus unusual events, as are the legal instruments further up the enforcement ladder.[9] Legal prosecution will normally follow only very serious incidents or accidents, and will regularly result in just a monetary fine. All formal PSA decisions (like orders) can be appealed to the ministry. This occurs very seldom, perhaps once or twice each year, and if it does, the PSA decisions are confirmed in some 50 per cent of the cases.

The enforcement handbook, supported by internal training programs and administrative systems, provide clues as to how due processes are to be achieved; however, developing a uniform practice and culture of law enforcement has been a constant and ongoing challenge, involving the comparison of quite diverse cases, across a range of complex socio-technical contexts and contested risk domains. As was explained by one senior legal advisor:

> [W]e have this stair-case, indicating degrees of seriousness. The steps are meant to specify how measures should reflect this. But how do you define what is 'very serious'? What criteria characterize a minor incident? And so on. It turned out to be very difficult. We tried to formulate criteria, but we simply had to give it up. It was too difficult. Like, you cannot link it to specific requirements, that transgressions of this or that provision were less serious or very serious. Even if some requirements are more critical ... as an administrative authority you're obliged to treat similar instances uniformly. But what is 'similar'? So it's more like ... the various disciplines need to agree within their field, as a minimum. And then there must be a correspondence across disciplines. It takes a lot of coordination.... It's not really law in that sense. It requires very good processes. You need to compare instances. Complications arise if you have a case, and you have some previous case, and then you must compare. ... And no one knows what the future will bring.... You need good processes to assure yourself that you find the right level. That people talk together. (see Kringen 2009, p. 154)

Thus, it is not altogether clear how internal work processes have contributed to proportionate, transparent and harmonised enforcement practices. Certainly, there have been rumours and some complaints from consultants and other external observers about 'hidden orders' and lack of accountability.[10] The important point, however, is the general lack of close legal scrutiny and pressure from the industry to test out the

[9] Coercive fines are extremely rare, and are used when orders are not complied with within the deadline. Stopping activity is also rare, and shall only be done when there is an immediate danger. As explained by the agency: 'Normally, the company will stop such activity themselves', referring also to the rights of the safety representative to stop dangerous work according to the WEA. Expulsion in terms of withdrawal of license or change of operator is the ultimate enforcement measure, but has never been used.

[10] Accountability in law enforcement was one of the apparently few critical issues in the evaluation report commissioned by the Ministry of Labour in 2006.

precise thresholds for intervention. Some key stakeholders described the regime as an 'almost empty legal space', referring to the fact that there was hardly any case law and that the legal standards were never tested in courts and seldom challenged by the industry. Still, lack of transparency and unclear legal warrants for agency decisions has clearly been of some concern to the industry through the years.

There is thus a considerable amount of discretionary judgement involved in deciding upon the proper use enforcement measures, requiring a corresponding reluctance as to which interpretations can be made from simply 'counting' them as indications of the 'enforcement policy' of the PSA. Several mechanisms contribute in keeping enforcement and compliance questions out of the legalistic spotlight. Legal advisors of the PSA confirmed a suspicion that companies were reluctant to voice objections to enforcement practices, let alone complain formally, reflecting a (mistaken, in their opinion) fear that signs of regulatory recalcitrance would negatively influence government goodwill and ultimately market access in terms of future licences (Kringen 2009). The regulatory 'dialogue' thus appears as a largely local 'problem-solving' encounter between peers, based on highly contextual judgements about the case-specific conditions. These also include considerations about the probable results of legal instruments as compared to informal accommodations, facilitated, of course, by the rich contents of the regulatory toolbox. These features were roughly summed up by one senior legal advisor in the PSA:

> You have engineers on both sides. Lawyers and economists in the companies tend not to be involved. So they just consider the technical aspects, if it seems all right. So they don't pick on our foundation for requiring … if they find our requirements reasonable. This may be combined with an exaggerated respect for what we require. They might have objected, but they may fear a negative effect if they do. But mistakenly so. It's quite legitimate to disagree with the agency. (see Kringen 2009, p. 155)

To some extent, it was in the bilateral encounters between agency officials and industry staff that law was made, at least in the past. Although there is a clear division of legal and non-legal measures, there is still no direct or transparent relationship between a legal transgression (non-conformity) and the use of legal measures. The supervisory report is largely an account of the observations made. Based on that, the measures are decided upon. Normally, as noted, only a request is made to the companies for taking action in terms of analyzing causes and finding remedies. The idea is to differentiate, and to use legal measures only when these appear justified by the severity of the transgressions. But these overall considerations were also tempered by additional concerns, or rather subject to a holistic appreciation of a variety of contextual factors made relevant in the specific case. Regulatory responsiveness included sensitivity to a variety of concerns, such as timing, priority of resources, company-specific compliance records, their amenability to regulations etc. The restrictive use of formal legal enforcement instruments can thus be explained in cultural as well as in instrumental terms. While the regulatory processes and frameworks are strategically crucial and generally awarded high priority, the legalistic outlook is not dominant in the actual enforcement policy. As was observed by one senior legal advisor:

> It's not without reason that they talk about 'Supreme Court engineers'.... A lot of the engineers feel they can manage this themselves. Companies and other places, they say the engineers are a 'breed', in their training or whatever, they want to solve problems, and they want to do it themselves. Engineers tend to be problem-solvers, you know. They're world champions in most matters, so they handle the legal aspect too. (see Kringen 2009, p. 155)

The PSA is dominated by engineers with a practical and pragmatic problem-solving approach. The role of the lawyers has been quality control in the decision-making process in order to guarantee some degree of legal prudence, confirming the impression that legal criteria and boundaries are not severely tested. Few lawyers have been included in the management, and even the directors of regulation and legal affairs have been engineers.

Legal measures thus appeared not primarily as instruments for getting things done, but rather as occasional 'signals' to the industry about the behavioural record. 'Dialogue', implicitly within the context of the basically asymmetrical relation, was the favoured approach. A further discussion of the enforcement regime of PSA can be found in Chapter 15 (Ryggvik) of this volume.

11.4. TESTING BOUNDARIES AND CHALLENGING ACCOUNTABILITY

The two cases presented here appear in their own specific way to challenge the robustness of the regime and may thus serve as vantage points for exploring the boundaries of its legitimacy. They address, respectively, regulatory inventions related to (1) inherent risks of first-tier regulations dominated by purpose-based rules, and (2) management risks of second-tier and system-based regulations. In the first case, agency policies regarding the proportionality and adequacy of specifying means for fulfilling purposes were challenged, instigating the entry of legalistic and judiciary evaluation into the problem-solving and collaborative context. In the second case, the regulators extended the meaning and scope of management regulation (and meta-regulation) to include the company souls – that is, their safety culture.

11.4.1. *Case I: Risk and Rest(itution)*

The exploration of interpretive 'freedoms' gave the agency much power in terms of determining what was 'good enough'. Although the regulatory process was highly participative and consensus seeking, one requirement appeared to have unexpected consequences for the industry. Section 31 in the Activity Regulation regarding the 'arrangement of work' stated that 'the work shall be planned so that as much work as possible is done daytime, and so that the employees are assured necessary restitution and rest'. The general and goal-oriented wording was, however, further detailed in the guidelines, endorsed in the 2002 white paper, implying strong restrictions on night work and so-called joint sleeping or hotbedding (that is, workers sharing cabins). These restrictions had potentially far-reaching implication with respect to night work during special operations, manning levels in critical phases, the availability of cabins and so forth.

The wording of the guideline reads as follows: 'The requirement to do as much work as possible daytime ... implies, inter alia, that night work should be limited to tasks and functions that are necessary in order to maintain prudent activities. The requirement to necessary restitution and rest ... implies, inter alia, that all personnel are allowed to sleep undisturbed and normally alone'. As for night work, the PSA expressed their policy in a joint letter to the industry in July 2002, stating that the new regulation would maintain and continue current legislation, but specifying these to the effect that night work was to be restricted for tasks necessary for keeping up ongoing production in a safe manner. So-called pre-emptive and prophylactic maintenance work was to be planned and carried out during daytime shifts.

These interpretations and the following enforcement practices (including some 'letters of interpretations' issued jointly to the industrial actors) caused much discussion and controversy, and were seen by key industrial actors as unduly disproportionate. Costs were estimated at possibly several billion NOK, and the Norwegian Oil and Gas Association later published a report which addressed costs and benefits of the new requirements (Norwegian Oil and Gas Association 2006).

This report was issued in March 2006, just prior to the finalisation of the 2006 white paper, and was submitted to the PSA and to the Ministry of Labour. It addressed a number of concerns related to the cost-benefit balance, and argued that authority interventions in several cases jeopardised the purpose orientation in the regulations by requiring specific solutions that were not sufficiently justified from a cost-benefit perspective. The interpretation and enforcement of the provision regarding working hours, rest and restitution appeared as prominent examples. Furthermore, a general argument was made for a stronger reliance on internal risk-cost-benefit analyses, based on 'realistic and operational demands', and considering also the need for a broader perspective on 'societal value creation'. Reference was also made to the governmental regulations for the public administration about assessment of economic and administrative consequences and impacts of reforms and proposals ('Instructions for Official Studies and Reports'). These are laid down by royal decree and concern impact assessment, submission and review procedures in connection with official studies, regulations, propositions and reports to the Storting (Norwegian parliament). Although quite ambitious in substance and intent, these regulations are not very strongly followed up and variably practiced (Difi 2012). No centralised control system exists for enforcing the regulations, and compliance is the responsibility of the individual ministries.

The enforcement of the provision was considered a breach with the established functional goal orientation in the regulations. These rules were more difficult to apply in cases where the relationship between means and ends was dubious or contested, as it of course was in the case of determining the impact of working hours and sleeping conditions in relation to health and operational safety. The NPD/PSA appeared firm in their responses and produced their own counter-report, contesting what they saw as highly exaggerated cost estimates and defending the importance of the work arrangements, in terms of both health and safety. The justifications related to the latter were based on the risk for fatigue and reduced alertness in the execution of safety-critical tasks. Although research and documentation were referred to (such as the claim that a higher proportion of accidents occurred on night shifts, on facilities with hotbedding etc.), the results were contested in terms of scope, strength and criticality. But in referring to the potential consequences of having 'tired workers', a strong argumentative rhetoric could still be mobilised.

On the whole, this issue caused much disturbance in terms of confidence and mutual trust; but it also revealed the powers of the authorities in establishing thresholds for acceptable levels of risk, not by simple and straightforward enforcement of explicit rules, but by the execution of discretionary powers based on relatively vague and goal-oriented provisions.[11]

And indeed, the new requirements regarding rest, restitution and night work occupied much attention as a 'compliance issue'. In order to meet the requirements, a number of measures had to be considered: rebuilding cabins, hiring flotels, accommodating schedules according to capacities, planning of different types of activities and so forth, all of which would potentially generate large compliance costs. And as indicated, the issue initiated a call from the industry for a more systematic use of cost-benefit analyses in the regulatory processes. Although some actors interpreted the industrial responses as acts of active non-compliance, the official argument was rather that the interpretation provided in the guidelines and in the enforcement policy did not have the proper legal mandate, as these had not been explicitly presented in the formal regulatory process.[12]

[11] The interpretations as noted were endorsed by the ministry and possibly also originated from political interference. Rumors had it that the ministry (led by a Labour Party minister) had yielded to strong union pressure.

[12] See: http://www.olf.no/hms/retningslinjer/?26481

The report also brought up more general concerns regarding enforcement practices. It was argued that comparable cases were not always treated equally, depending on both disciplinary area and individual agency officials. The issues related to work hours, rest and restitution were specifically mentioned as unduly warranted. As the regulations generally required much discretionary judgement in order to establish compliance, the process and method of making such judgements were all the more critical. Some thus voiced the opinion that authority judgements had to follow approximately the same methodology for risk evaluation as that which was codified in the industrial standard for risk analysis.[13] It was also suggested that many companies often saw the agency and the ministry as 'externally voicing the same views', possibly causing a reluctance to appeal. The relatively high proportion of appeals supported by the ministry was suggested as an indication that complaints were often justified, and the Norwegian Oil and Gas Association argued more generally that the right to appeal should be used more actively by the industry.

The Norwegian Oil and Gas Association also commissioned expert reports from two different law professors, each supporting their own objections. The first report was technically based on the argument that the regulatory process had not provided proper information about the consequences, which in this case were considerable (even if contested) in terms of economic costs. Because the parties had not been properly involved in the formal hearing process, the requirement was to be considered legally invalid. An explicit reference to this report was made in the 'Norwegian Oil and Gas Association guidelines for rest and restitution', indicating that the companies were expected only to consider the relevant solutions with reference to the general wording of the provision, and that no application for 'exemption' from the regulation would be considered necessary if this requirement was met. The next expert report appeared in 2007, a year after the Norwegian Oil and Gas Association report, and was commissioned to specifically address the legal status of the night work restrictions. The report itself took a broader perspective, however, going back to the justifications for the special exemptions given to the petroleum industry regarding working hours. These exemptions originated from 1977 and the birth of the new Working Environment Act (occupational health and safety) and were justified by the need for more flexible requirements adapted to the specific operational concerns in this industry; these new interpretations, it was argued, was a reversal, in effect eliminating or equalising the formerly established differences. Furthermore, within the purpose-based legal framework, allowing much industrial discretion in implementing regulatory goals, and warranted in higher level regulations (and policy documents), interpretations at agency level would not have sufficient status as a legitimate legal standard. Any reinterpretation of formally processed regulatory

[13] NORSOK standard for risk analysis (Z-013, see http://www.standard.no/imaker.exe?id=244).

framework would have to take place through new specified regulations, following proper regulatory procedures.

External legal scrutiny did not end here, however. Two critical review papers were to appear from the Institute of maritime law at the University of Oslo Law School (Logstein 2007, 2009).[14] These reviews adopted a balanced but essentially formal legal perspective, providing a more comprehensive legal analysis of critical regime properties and again probing the working arrangement regulations. The extensive use of non-formalised rules was examined within the spectrum of regulatory practices of the PSA: in general guidelines to the regulations, in statements and letters of interpretation, in identical letters to the regulatees and even in statements given in the different collaborative arenas (such as the Safety Forum). Measured against established principles of legality and due process, these regulatory practices were seen to lack proper status as formal law. The interpretive privilege of the PSA was acknowledged to some extent, but only insofar as it enacted the regulator's informative tasks, as codified in general administrative law. This task would be limited to informing about alternative ways of complying with the general and purpose-based rules, not of limiting the range of options beyond what would strictly follow from these rules. The case thus triggered the enrolment of new actors (see Chapter 12 in this volume), particularly introducing more rigorous legal scrutiny in combination with a call for more dedicated assessments of the risk-cost-benefit balance.[15]

However, the PSA appeared firm in their position regarding the risks involved if workers were not sufficiently rested and alert. Obviously there was a double pressure, however, as the unions also protested against what was seen as more lenient enforcement practices in terms of interpreting the legal content against the diversity of real-world situations to which these could be applied. How strict should the clearly ambiguous phrase, 'normally alone', be interpreted? What kinds of activities should be allowed to take place during night shifts? The amount of discretion involved in these cases was considerable, and apparently, some practical compromises and accommodations were made in the process.

Union arguments can be summarised as follows. Companies had no regard for the risks involved and lacked a proper 'fatigue management' program. Platform overcrowding was often a result of bad planning. Activities were 'renamed' and

[14] This engagement from the law school was indeed prompted by external funding from the Norwegian Oil and Gas Association, in itself, of course, not compromising the professional quality of the reviews.

[15] It should be noted, however, that the companies did not respond uniformly to the requirement; some adopted a more pragmatic and timid stance. Partly, this variety would reflect very different compliance costs, but that would hardly be a sufficient explanation. Some even took pride in making these investments, both as a (possible) contribution to risk reductions and as a 'gesture' to the welfare of the workers. Indeed, one company proudly presented these investments as a sign of their general generosity and positive attitude in all welfare and HSE matters. In another company, it was also suggested that a too lenient compliance policy could be taken as a lack of 'solidarity' with their industrial fellows.

categorised strategically in order to provide compliance. Only clear and precise rules could ascertain real compliance, and delegation to local collaborative committees had to be avoided because this would place undue pressure on the local safety and union representatives to accept facility- or company-specific agreements. There were even examples of bilateral contract schemes where workers willing to 'hotbed' were offered (substantial) economic compensation. Although attractive for the individual, these schemes certainly compromised and possibly undermined the legitimacy of the 'safety argument'.

The controversy has indeed been an ongoing issue to this day, also involving the ministry in a more active role in the regulatory process. As for cabin sharing, a recent regulatory amendment now specifies more precisely under which conditions 'hotbedding' can be allowed (although not simultaneously), namely during (1) restoring of physical barriers and in other acute situations, (2) turnaround/revision stop and (3) hook-up and start-up. Still, compensating measures shall be considered and discussed with employee representatives. This amendment was adopted by the ministry against the advice of the PSA, and represented an unusual case of the agency (and of course the unions) being overruled by the ministry.

Most important from our point of view, however, is how this case challenged the 'conditionality' of the trust balance and exposed potential fragilities when critical issues about the socio-economic distribution of risk were at stake. The risk-cost-benefit considerations that were thus introduced involved rather complex questions: What was really the magnitude of the risks involved? How much risk reduction would be achieved by the variety of solutions available in different operational situations? What were the compliance costs, and did they justify the (contested) risk reduction compared to alternative allocations of risk-reducing efforts? And, of course, what was really the impact of 'welfare concerns' for the workers in these requirements?

The latter issue was even extended to the societal level, as the costs generated were juxtaposed with the amount of welfare that could be produced in other sectors (e.g. elderly care) dependent on public funding, the availability and standards of which also depended on revenues and taxes from the petroleum industry. Industry representatives would occasionally suspect the welfare aspects, in practical terms, of the 'single cabin' requirement to be the 'real' motivation, and consequently to regard the risk arguments as a smoke screen. This was really a luxury demand, forced through the decision-making process by active lobbying at the political level, disregarding due processes, formal hearings and, if not disrupting, at least unduly bypassing the legitimate tripartite collaboration processes that had preceded the regulatory reform. Finally, it deprived the industry of the possibility granted through the regulatory system to apply local knowledge and discretionary judgements adapted to the operational requirements on a case-by-case basis.

Beside those arguing substantially for the factual and risk-based justifications, proponents within the unions, at the shop-floor level and at other levels, would occasionally support the welfare argument, but would also refer to substantial revenues generated in the industry as the justifying factor, claiming that this was something they really could afford with the current oil prices. Company managers themselves would not hesitate to sleep in luxury hotels on their business trips. Such arguments were of course also applied more generally in relation to HSE costs, welfare costs, wage levels, working hours and so on.

11.4.2. *Case II: Regulating Culture*

The 2002 regulations also introduced a new provision requiring the companies to develop a 'good and sound HSE culture' that was to generate much discussion and controversy and expend much interpretive energy. The background for incorporating the culture provision in the regulatory framework can be related to the increasing scholarly and industrial preoccupation with safety culture, in its dual meaning as both a cause of accidents and substandard safety conditions and as a possible avenue for improving the organisational management of risk. Furthermore, it can be related to an increasing concern about deteriorating safety within the industry in the late 1990s (Ryggvik 2000), and a perceived inadequacy and insufficiency of formal risk management systems in coping with the situation as presented by Rosness and Forseth (Chapter 12) in this volume. The regulation thus reflected an increasing awareness of how the more deep-seated organisational properties and processes did or did not contribute to the proper management of risks. The controversies about the risk level did also reflect a concern that parts of the industry had developed an overreliance on their own safety performance; that self-complacency and hubris had crept into the company souls, and that treatment 'in depth' was necessary.

But the culture provision was also seen as a clear continuation of the already established principles of enforced self-regulation as embedded within the tripartite context. It was thus meant to generate processes and a framework for dialogue as to its meaning and impact in order to penetrate further into the self-regulatory and self-improving mechanisms of the industry. The white paper preceding the regulatory reform specified that a good culture implied 'continuous improvement and learning' and that HSE aspects should be integrated as part of the common values of the regulated organisations, to be reflected in established attitudes, competencies and behaviours.

As could be expected, the provision soon generated substantial and interpretive challenges which were reflected in supervisory practices and enforcement as well as its role in the overall regulatory strategy. The context of ordinary audits appeared inadequately fit for identifying and addressing safety culture, and the status of the provision in terms of regulatory enforcement was subject to much uncertainty. The most substantial contributions in the follow-up process were various seminars and conferences on the topic and the production of a thematic booklet. Both of these initiatives received much attention and interest. The former served more as deliberative and partly confrontational forums than as arenas for presenting the digested PSA 'culture policy'. The latter was largely based on ideas emerging from the High Reliability Organisation research (Roberts 1990, 1993; Weick and Sutcliffe 2001) and James Reason's interpretations of these. Safety culture was thus seen as an informed culture, further specified as being flexible, just and fair, based on open reporting and on organisational learning abilities (Reason 1997). In addition, the importance of the tripartite and participative framework was stressed.

Critical factors in the implementation process can roughly be related to (primarily external) substantive and (primarily internal) interpretive challenges. The former involved critical and controversial risk management issues particularly materialised through a diversity of 'cultural' safety programs that proliferated within the industry. A number of such safety programs have been launched during the years, many of which have been comprehensive 'behavioural' programs addressing large portions of operational personnel (Lindøe and Engen 2007; Ryggvik 2008). They were often referred to as 'culture programs', with the explicit or implicit intent of promoting proper attitudes and safe behaviour at the sharp end.

The programs generally reflected a preoccupation with avoiding injuries and fatalities by keeping lost-time incidents (LTI) figures low, including also the controversial idea that there was a proportionate relationship between minor errors and major accidents and that the total number of errors therefore had to be reduced, also referred to as the 'Iceberg Theory' (Heinrich et al. 1980). The programs thus served as triggers for accentuating some hotly debated risk management controversies. This triggered questions about the location of responsibility within the 'causal chains' as well as the efficacy of the programs in terms of their risk-reducing potential: Were the investments devoted to avoiding injuries justifiable and/or effective against the larger risk picture? Were the addressees properly selected in terms of justifiable and/or effective intervention strategies? As the programs were so closely related to the implantation of a 'safety culture' within the companies, the concept – and the provision – thus came to carry the burden of handling some very critical and contested issues about risk and its (mis)management.

It was notoriously difficult, however, to link HSE culture to major accident risks and process safety in any clearly communicated manner. In the fall of 2004, a serious gas blowout occurred on Statoil's Snorre Alpha platform. A catastrophic outcome was avoided only by chance, and by a hazardous rescue operation by the remaining crew. Poor HSE culture appeared as an important contributing factor in several post-event accounts, including a thorough investigation report commissioned by Statoil, as well as in statements from the minister of labour and from the CEO of Statoil, although the latter used the diagnosis for voicing the view that the cultural problem was 'local' and confined to the facility rather than being company-wide. The PSA itself refrained from using the label in their investigation (PSA 2005). This reflects the growing reluctance from the agency to use the provision in the enforcement processes, which leads to the interpretive and communicative challenges thus confronted.

How could the concept of safety culture be meaningfully applied to the socio-technical complexities of the industry? What was the 'meaning of the word' in operational terms? The regulatory context itself exacerbated these conceptual problems, requiring as it did some minimum level of clarity and transparency as to how the provision could be implemented in terms of regulatory interventions. A considerable degree of restraint and modesty was clearly evident among agency officials and managers as to how this should be done. Even managers responsible for regulatory affairs questioned its applicability, with reference to the difficulties involved in verifying the conditions for its fulfillment: What was 'good enough'?

Two interrelated conceptual problems were particularly salient. First, some initial attempts to use the provision in regulatory encounters had raised the question about its reference to organisational scope and context. The targeting of relevant organisational units appeared as one important challenge, in particular as these had become increasingly transient and blurred over the years, with new forms of more dynamic and flexible organisational forms. Were there systematic differences between companies or installations, and which were the most appropriate units delineating and identifying cultures? Furthermore, what was the impact of long- and short-term changes in such cultures, once identified? As one PSA supervisory coordinator lamented: 'I think it is a very demanding and challenging word to use. It opens a discussion about its meaning. They will say 'we have it', or it may be that parts of the company 'have it' and other parts does not. So it turns into a question about the use of words – if the "whole company has a bad HSE culture"'(see Kringen 2009, p. 313).

The second problem was related to the status of HSE culture in relation to the conceptual and explanatory schemes already available, such as could be found in the large body of reporting systems, investigation methodologies and other classificatory systems devised for ordering the diversity of causal factors available on the explanatory menu. Interpretive challenges were thus exacerbated by the fact that no systematic analysis was provided of how HSE culture was related to the large and comprehensive body of regulatory requirements and risk management models already in operation. To some extent, they all had gatekeepers and institutionalised 'protections' and practices supporting and perpetuating their survival. Thus, there were several 'competing' approaches, frameworks and methodologies that had obvious but unexplained linkages to HSE culture.

The dimensions of HSE culture outlined in the PSA booklet reflected the broad scope of the provision: good HSE culture required a series of trade-offs and considerations along a number of salient dimensions: safe behaviour versus external frameworks, economy versus HSE, formalistic risk bureaucracies versus 'cultural' risk organisms, the value of statistical monitoring versus experience-based and local knowledge and so forth. But questions appeared about whose behaviour and whose culture were really addressed, and what relation culture had to other shapers if these were not also themselves really a part of culture, such as rewarding systems/KPIs, contract conditions and schedules – in short, framework conditions laid down in decisions made far from the shop floor level. How could the 'culture' label (or its composite properties) be attached in any meaningful manner to the highly complex and transient operational processes and organisational structures within the industry? Was culture predominantly to be found and 'engineered' in the attitudes and behaviours of workers within the constraints of the workplace context? Were these local workplace contexts shaped by decision-making processes at higher echelons, also deserving to be addressed in terms of culture? Was culture to be revealed in the overall priorities and trade-offs made in the board rooms or licence committees? Was culture the residual factor that either 'escaped' the formal systems of risk management or indeed filled in the human and organisational software needed for making them work in practice? Was compliance with procedures good culture only if procedures were good? How could indicators of culture be interpreted and aggregated in order to justify cultural diagnoses?

Given the hyper-functional and highly abstract nature of the concept in terms of its potential meanings and possible contexts of application, the linkages between means and ends were correspondingly blurred. At the other end of the scale, somewhat reified and essentialising notions of culture would regularly appear, neither of which could easily mobilise any viable consensus. Attempts to reduce and specify the concept in terms of quantifiable indicators or precise templates for action were largely resisted but gradually adopted as a general agency policy.

Taken together, the conceptual and the substantial problems related to HSE culture appeared mutually reinforcing: it was of great importance to dissociate the proper implications of the provision from the behaviour-based safety programs, and this in itself occupied much time and energy. HSE culture was so much more than just behaviour at the sharp end. But the holistic approach also had costs and augmented the difficult task of conveying the practical applications of the ever-expanding and increasingly abstract contents of the cultural 'black box'. The translation process was all the more complex. Was good HSE culture just another word for any given 'state of the art' risk management, in effect producing a tautology?

Although interviews with managers and officials in the agency did reveal some common themes and associations relating culture to the promotion of informed organisational learning, it also questioned what was really the added value of introducing culture as just another label for this. And even if the dominance of engineers in the agency had not hampered the proliferation of human and organisational approaches to risk management, it was still an internal 'culture' with a taste for concrete and immediately operational concepts and tools. This leads to a final note on the internal organisational challenges involved in translating HSE culture to a viable template for coordinated engagements with the industry. It clearly presupposed a lot of organisational patience, curiosity and engagement across disciplinary boundaries, and probably a fair expectation that the effort would be rewarded. A decline in interest and attention was gradually apparent, largely reflecting the difficulties involved in developing conceptual understandings of HSE culture that could be sufficiently shared and appreciated within the agency. There had also been some critical organisational discontinuities in the process. Key officials in the regulatory reform process had left the agency when the new regulatory framework was finally launched, and to some extent, the content and significance of HSE culture had to be 'reinvented' by relatively recently employed officials, partly recruited for the purpose. The follow-up was organised as a project outside the rank and file and thus not deeply entrenched within the rest of the organisation. The evidence shows that the substantial value of the 'cultural dialogue' with the industry was largely the privilege of the few dedicated officials affiliated with the HSE culture group. Instead of executing a well-founded regulatory strategy, they appeared as organisationally marginalised, but still encumbered with the burden of having to interpret the implications of the provision and to communicate their insights, not only externally but also internally.

11.5. DISCUSSION

In sum, the regulatory system opens a large space for discretionary judgement, negotiation and potential conflict where thresholds for and forms/degrees of intervention may be unclear and unpredictable. The parties are involved in continuous arguments about the balancing of economic and safety-related issues, and also trade-offs between competing and contested risk concerns. Under normal conditions, however, some regulatory uncertainty seems to be a price willingly taken by all parties, even though it implies that enforcement practices in effect contain a considerable degree of standard setting.

The enforcement options and practices reflect what has been referred to as a 'tit-for-tat' strategy (Ayres and Braithwaite 1992). That is, measures taken comprise a diversity of informal and formal interventions applied on and contingent upon a broad range of circumstances. Although the gravity of the observed transgressions is important, the use of instruments will also be influenced by historical compliance records and estimated effect of the specific instrument in the specific situation. The availability of instruments within the enforcement pyramid allows great flexibility and reliance on accommodative modes. Having ultimately the power to influence the chances of being granted access to the attractive markets, the PSA is able to 'speak softly with big sticks'. As observed by Ayres and Braithwaite (1992: 19): 'Paradoxically, the bigger and more various are the sticks, the greater the success regulators will achieve by speaking softly'. Importantly, however, this also gives the agency much power not clearly apparent from the formal enforcement record. Regulatory compliance is simply achieved by the mere presence of possible reputational costs and sanctions. Although enforcement occurs through bilateral control of industrial actors, the regulatory system also involves and requires more generalised treatment of critical issues across domains and disciplines and beyond single agency-company encounters. Collaborative arenas are established for dealing with such issues within the tripartite framework.

Although the regime enjoys widespread legitimacy in its basic features, a diversity of problems must be continually dealt with and negotiated. Critical questions recur with varying force and intensity about what contributes to safety and how certain and/or significant is the causal contribution of any given measure or intervention. The complex and highly technical nature of the industry makes legal and legalistic approaches difficult to apply in very transparent and clear-cut ways, and lawyers on both sides appear to succumb to the supremacy of the problem-solving engineers.

The two cases reviewed earlier in the chapter illustrate in their specific ways the legitimacy and accountability challenges that can arise from such purpose- and systems-based regulatory regimes, specifically the challenges involved in translating overall principles and vaguely stated purposes to operational templates for regulatory action. The challenges are exacerbated as the regulatory context implicitly or explicitly involves an assessment of compliance, or at least some justified evaluation of how certain practices correspond with or deviate from regulatory provisions and purposes. Both cases also clearly involve the institutional setting of the regime. The tripartite framework relies on deep involvement through negotiations, and seeks consensus. The overall agreement on the legal and institutional framework for discourse stands in some contrast to the occasionally strong and real conflicts. In this participative regulatory processes the regulator faces many dilemmas in manoeuvering between powerful stakeholders and navigating within the contested terrain of substantially competing interests.

The issue of regulatory legitimacy involves several dimensions, but can largely be distinguished as degrees of regulatory intervention related to regulatory purpose and forms of intervention related to means and methods. Vulnerabilities are related to several factors, including the reasonableness and legitimacy with which such requirements are followed up in implementation policies and practices. Norm-setting does not provide a given standard but will be subject to continuous negotiation, and the relationship between statutory rules and enforcement is fluent and flexible.

The most obvious difference between these cases is the degree to which legal assessment and formal procedures were involved in the regulatory process. Following the four-fold criteria for assessing accountability and legitimacy issues introduced earlier, some key points are briefly summarised in Table 11.1. Importantly however, the accountability elements do not represent stages in the regulatory process, as originally suggested by May. Indeed, all elements are of relevance to all stages within this regulatory system. For example, legal accountability does not apply only to the production of rules, but to the whole process, because lawmaking is actually taking place continually.

Case I clearly caused allegations of regulatory unreasonableness. Within principles/purpose-based regimes, however, the very concept of unreasonableness will in itself trigger basic questions about several aspects of legitimacy because the legal mandate ultimately involves considerations of risk tolerance and degrees of regulatory intervention. Or perhaps more precisely, in the absence of precise and democratically established normative determinants, the questions of reasonableness and unreasonableness will be transferred into new domains of governance which will involve new accountability issues related to proportionality and evidence-based justifications of causal links between fluid and contested instruments – as well as targets.

Professional discourse (and legitimacy) is intrinsically linked to the governmental and legally mandated role – in this case the norms and logic of the legal system encountered in the world of risk regulation and management, the latter predominantly populated with engineers and a practical problem-solving agenda, but also involving deep uncertainties and controversies regarding risk acceptance, effectiveness and proportionality. Against this background, the strategy and response from the Norwegian Oil and Gas Association could be seen primarily as a case of 'principled disagreement' (Kagan and Scholz 1984) more than one of calculated non-compliance (Ayres and Braithwaite 1992). Furthermore, the case also illustrates how the pyramidal apex in the enforcement hierarchy may not be the point where expected, desired and legitimate compliance is achieved, but rather a source of non-transparent resignation or industrial self-discipline.

Case II illustrates first of all how the challenges facing regulators in non-prescriptive and purpose-based regimes move far beyond the simple command-and-control scheme and compliance monitoring. Discretionary powers and judgemental requirements are magnified, ultimately implying the difficult tasks of evaluating the risks at hand and the organisations' ability to cope with them. Ardent investigation into the deeper and complex chains and configurations of causes serves as a leading theme in these proactive regulatory strategies (Hopkins 2007). On the one hand, addressing 'culture' may appear as a quick fix and as a shorthand encapsulation of a largely undefined set of dimensions. On the other hand, cultural approaches

TABLE 11.1. *Analysis of the two cases*

Legitimacy and accountability elements	Normal state	Case I	Case II
Legal	Described as legally void, lacking precise and predictable rules and seldom externally scrutinised, such as in the courts. Largely adopted by lawyers inside the regime, but questioned by external professionals.	Triggered and involved the entry of legal professionals and assessment and considerations of due process in regulatory, standards, processes and outcomes.	No legal assessment involved, but lack of legal status, testing, and indeed testability still led to questioning of the impact and relevance of the provision as such.
Bureaucratic	High degree of delegation and regulatory discretion awarded at agency level. Regulators occasionally use the threat of prescriptive regulation as a remedy for failure in tripartite negotiations.	Questioning discretionary powers and assessments. Lack of transparency. Unclear mandates and proper channels of law enforcement. Risk-cost-benefit assessments called for.	Organisationally marginalised project. Enforcement policy not clearly stated.
Professional	Often unclear and contested understanding of risk, assessments of means-end mechanisms. Practical and problem-solving engineers still often agree on a case-by case basis.	Scientific basis unclear. Means-end relationships questioned (principled disagreement). Proportionality of interpretations and intervention questioned.	Scientific basis unclear. Means-end relationships questioned. Triggered substantial risk management controversies but also serious interpretive challenges.

Political/ democratic	The regime as such enjoys a high degree of support, from top to bottom. Regularly endorsed through ministerial reports and parliamentary decisions. But levels of risk tolerance and acceptable management practices are only vaguely discussed and defined as political templates for risk governance.	Involved negotiation, lobbying and strategic effort from top to bottom. Enrolled new actors and activated ministerial interference.	Some political involvement in initial phases, and culture appears as an attractive addressee in the political rhetoric. Otherwise largely unchallenged.

may potentially address safety-critical factors vital for organisational performance (Turner and Pidgeon 1997).

Anyhow, the PSA had to confront a number of difficult tasks, making sense of the complexities of HSE culture, making it operational within the framework of their regulatory strategy, and at the same time facing industrial initiatives, programs and 'cultural interpretations'. The provision triggered substantial and conceptual problems, both for the regulators and for the industrial actors. Substantial problems were related to basic risk management controversies, such as the conflict between behaviour-based and holistic and systemic strategies. Conceptual problems were related to the meaning of 'culture' and its semiotic and substantial location within the plethora of existing conceptual frameworks and schemes, such as those reflected in risk management systems, safety programs and investigation models. The experiment thus seriously challenged the diagnostic capabilities of the regulators, operating as it did on the frontiers of the avenues available for regulatory intervention.

Both cases involve challenging professional as well as normative considerations. Expert-based knowledge and criticism entered into the controversies and discourses. In the former case, these ranged from the estimation of the risk-producing effect of temporary and 'non-optimal' sleeping conditions to the overall societal welfare costs and benefits of the requirements. The total welfare outcome would range from state revenues contingent on cost-effective extraction and production of petroleum resources to welfare contingent on offshore working conditions, the inherent risks related to the latter ambiguously lurking in the background as a contested means-end issue (e.g. hotbedding equals fatigue equals major accident risk). The knowledge-based effect of the culture provision was no less clear. Clearly, the assumption about a causal convergence between types of incidents has been subject to much controversy. It is well known that major industrial accidents can occur in spite of exceptionally low LTI figures, such as in the Esso Longford and the BP Texas refinery accidents (Hopkins 2000; 2008). Hopkins employs the term 'mono-causality' in a critique of the assumption that accidents can be prevented (only) by reducing the total number of incidents and errors, large or small. Hale (2001) refers to this idea as an 'urban myth' that has haunted several industries for years. But these ideas appeared only as perverted and perhaps unpredictable and uncontrollable translations of the 'cultural project'. On the normative side, the culture discussions also raised critical normative questions regarding overall priorities. Even if behaviour-based strategies may have effect under certain conditions (Hale and Boys 2013), there will always be room for discussing to what extent, for instance, risk-producing technologies should be contained by risk-adaptive behaviour, or to what extent the technologies themselves should be designed for allowing fallible behaviour. There is no straightforward procedure for calculating the cost-benefit ratio of investments in 'forgiving technologies' as against the behavioural programming schemes. And some of the estimations involved would in any case be highly contestable, such as the value of a flexible and 'error-friendly' working environment.

Clearly, legitimacy issues shift in character as regulations become more systems oriented. Some are concerned with how these new regulatory approaches imply a transfer of responsibility from statutory institutions and government agencies to non-governmental bodies and regulatees (Black 2008). Using examples from different regulatory domains, May examines accountability shortfalls following from prescriptive regulations being substituted by systems-based and performance-based regimes. Typically, such regimes face challenges regarding all aspects of accountability, in particular because regulators at some point and in some form are expected to specify and legitimately demonstrate how certain practices can be regarded as unlawful. In the case of working conditions, the outcome was to specify in more detail the conditions for allowing certain activities. The industry appeared successful to some extent, and a relatively unusual resumption of steering power was assumed by the ministry. As noted by May (2007, p. 23), 'Any regulatory regime must confront a fundamental issue of how tight controls should be in seeking consistency versus how much discretion should be granted in promoting flexibility and innovation'. Accountability and legitimacy shortfalls may be encountered in both systems. It might be considered a coincidental paradox that the outcome in the first case favoured the party most interested in the latter while achieving the former.

The nature and impact of rules and their implementation cannot be fully grasped without taking the institutional and contextual regime dimensions into account. This involves the broader structures of rules, norms, interests, actors and stakeholders that constitute and shape the way in which regimes deal with the risks subject to regulation. There are blurred boundaries between these overall system features, such as between rulemaking and enforcement practices. The cybernetic distinction between standard-setting and behaviour modification becomes unclear, as the range of standard-setting 'acts' extends well beyond those confined to the traditional rulemaking processes. This triggers the intrusion of legal considerations and accompanying demands for legal certainty, accountability and due process.

11.6. CONCLUSION

Societal risk management is a multi-level phenomenon ranging from legislative and government-imposed rules and institutions to the management and execution of single-work operations on petroleum facilities. Key issues related to regulatory robustness involve concrete elaborations of critical components in purpose- and systems-based regulations, and how these are manifested within the larger regulatory context, involving strong tripartite traditions and a climate for interventionist state action. Risk regulation involves the whole process of evaluating the risks, their magnitude and character, understanding the often complex and intersecting causal chains involved in 'producing' the risks, and likewise, the kind of 'management' required for preventing the risks from materialising as real losses of social objectives and values deserving public protection. Although the present system enjoys a high degree of legitimacy among the parties involved, it is also vulnerable because of disparate interests and contested means, combined with vague and purpose-oriented legal standards. Despite being trust-based, trust among parties is highly conditional, with unclear boundaries and thresholds for disruption and conflict.

The existence, and potential application, of formal enforcement instruments still serves as a basic framing condition against which these dialogical encounters takes place. And some degree of 'existential uncertainty' on the part of the industry provides a convenient and perhaps also necessary condition for enacting the regulatory role with a corresponding degree of 'existential certainty'. After all, they face the most powerful companies, in both international and national terms; to have some golden shares in their reputational assets may be considered legitimate even at the cost of reduced legal accountability. The 'empty legal space' may thus also serve some higher purpose, given that the regulatory powers are used with self-discipline and continuous awareness of the various legitimacy challenges that must be coped with within the context of enlightened 'dialogue'.

Using precise legal criteria and quantified and comparable risk-cost-benefit estimates could ultimately undermine this dialogue. Regulations based on vaguely stated principles, purposes and system requirements have qualities that leave much room for regulatees to adapt to expectations and demands in accordance with internal operational and technological considerations and changes.

Clear rules can regulate the simple and the stable with consistency and certainty, whereas the complex and changing rules need broader principles (Hale and Borys 2013). Braithwaite has argued that principles should be backed by non-binding rules, and furthermore that 'binding principles backing non-binding rules are more certain still if they are embedded within institutions of regulatory conversation that foster shared sensibilities' (2002, p. 75). Now, this is exactly how the system is designed to operate, and normally with a considerable degree of success. A legitimate institutional system for participative collaboration must still be able to handle real and 'unsolvable' conflicts of interest, and indeed the robustness of the system is tried and tested in just these types of cases.

To some extent it can be argued that many competing positions within this regime are beyond settlement based on empirically substantiated evidence as to how agreed-upon purposes should translate into accepted or sufficiently safe practices. Rather, they represent different forms of 'equilibriums' of risk management, each of which might in theory contribute equally to achieving regulatory outcomes. For instance, behaviour-based approaches may in some instances be as effective as the provision of forgiving technologies or comfortable scheduling and resource planning. Different measures may thus be functional equivalents in terms of their safety-producing effect, but certainly not in terms of legitimacy. One is thus reminded of the argument put forward long ago in the Royal Society publication on risk from 1992, outlining several dimensions in risk management as constituting unsolvable dilemmas, including several of the issues reviewed her, such as between no-blame versus high-blame approaches and between positions arguing for and against the necessity of trade-offs between economy and safety. These dilemmas were later referred to as representing 'recurrent sets of opposing views in risk management' that partly reflected 'competing world views' (Hood and Jones 1996, p. 9; Royal Society 1992).[16]

[16] It was not assumed that all positions clustered around the extreme ends, but was rather seen as a way of ordering the debates and some of their associated assumptions. To this could be added that the dimensions also represented an attempt to allow for a more 'relativist' and less 'objectivist' discourse in the risk debates, not assuming that opposing views could be 'settled by science' or reduced to a question about scientific certainty.

In this context, law and risk governance do not constitute two distinct and separate normative structures. Law is what provides the overall context for legitimate public intervention in the private sphere, but at the same time the national cultural context and tripartite framework blur and complicate the boundaries. Legal rules embody general normative concerns codified in a number of ways, and integrated as institutionalised templates within the regulatory domain, which thus appears to substantiate both the legal and social paradigms in the regulatory and socio-legal literature. While basically embracing the social paradigm, whereby legal rules should be seen as 'only one set of norms competing with others that derive from other systems' within any given institutional and formative context, Black argues that it is still important 'not to lose sight of law' (Black 1997, pp. 52–53). The cases presented here focus in particular on the interaction between such partly competing normative foundations, and also illustrate the changing visibility of law in these contested terrains.

REFERENCES

Ayres, I. and Braithwaite, J. (1992) *Responsive Regulation. Transcending the Deregulation Debate*. Oxford University Press: Oxford.

Baldwin, R. and Cave, M. (1999) *Understanding Regulation. Theory, Strategy, and Practice*. Oxford University Press: Oxford.

Bardach, E. and Kagan, R. (2002): *Going by the Book: The Problem of Regulatory Unreasonableness*. Second Edition. Transaction Publishers: New Brunswick, NJ.

Black, J. (1997) 'New institutionalism and naturalism in socio-legal analysis: Institutionalist approaches to regulatory decision making', *Law & Policy*, 19 (1): 51–94.

Black, J. and Baldwin, R. (2010) 'Really Responsive Risk-Based Regulation', *Law and Policy*, 32 (2): 181–213.

(2008) 'Constructing and contesting legitimacy and accountability in polycentric regulatory regimes', *Regulation & Governance*, 2: 137–164.

Braithwaite, J. (2002) 'Rules and principles: A theory of legal certainty', *Australian Journal of Legal Philosophy*, 27: 47–82.

Braithwaite, V., Braithwaite, J., Gibson, D. and Makkai, T. (1994) 'Regulatory styles, motivational postures, and nursing home compliance', *Law and Policy*, 16 (4): 63–94.

Coglianese, C. and Mendelson, E. (2010) 'Meta-regulation and self-regulation', in R. Baldwin, M. Cave and M. Lodge (eds.): *The Oxford Handbook of Regulation*. Oxford University Press: Oxford, pp. 146–168

Dahle, T.G. (1994) 'The Norwegian approach to safety in offshore petroleum activity', *Journal of Loss Prevention in the Process Industries*, 7 (4): 379–381.

Difi (2012) *Graves det dypt nok? Om utredningsarbeidet i departementene*. Difi rapport 2012: 8.

Dølvik, J. E. (2007) 'Introduksjon' & 'Konklusjoner', in *Hamskifte: Den norske modellen i endring* (The Norwegian Model in Transition). Gyldendal: Oslo.

Gilad, S (2010) 'It runs in the family: Meta-regulation and its siblings', *Regulation and Governance*, 4 (4): 485–506.

Gustavsen, B. and Hunnius, G. (1981) *New Patterns of Work Reform. The Case of Norway.* Universitetsforlaget: Oslo.

Hale, A. R. (2001) 'Conditions of Occurrence of Major and Minor Accidents'. Paper presented at the seminar *Le risqué de défaillance et son contrôle par les individes et les organizations*, Gif sur Yvette, France.

Hale, A.R. and Borys, D. (2013) 'Working to rule or working safely', in Bieder and Bourrier eds: *Trapping Safety into Rules.* Ashgate.

Heinrich, H.W., Petersen, D. and Roos, N. (1980) *Industrial Accident Prevention: A Safety Management Approach.* Fifth Edition. McGraw Hill: New York.

Hood, C. and Jones, D.K.C. (eds.) (1996) *Accident and Design: Contemporary Debates on Risk Management.* UCL Press: London.

Hood, C., Rothstein, H. and Baldwin R. (2001) *The Government of Risk. Understanding Risk Regulation Regimes.* Oxford University Press: Oxford.

Hopkins, A. (2000) *Lessons from Longford. The Esso Gas Plant Explosion.* CCH: Sydney.

(2006) 'What are we to make of safety programs?', *Safety Science*, 44 (7): 583–597.

(2006b) 'Studying organizational cultures and their effects on safety', *Safety Science*, 44 (10): 851–952.

(2007) 'Beyond compliance monitoring: New strategies for safety regulators', *Law and Policy*, 29 (2): 210–225.

(2008) *Failure to Learn: The BP Texas Refinery Disaster.* CCH: Sydney.

Hovden, J. (2002) 'The development of new safety regulations in the Norwegian oil and gas industry', in B. Kirwan, A. Hale and A. Hopkins (eds.): *Changing Regulation. Controlling Risks in Society.* Pergamon: Amsterdam.

Hovden, J., Nilsen, M.R., Steiro, T. and Sten, T. (2000) *Utfordringer for arbeidet med helse, miljø og sikkerhet (HMS) i norsk petroleumsbransje* (HSE Challenges in the Norwegian Petroleum Industry). SINTEF rapport STF38 A00404: Trondheim.

Kagan, R.A., and Scholz, J.T. (1984) 'The Criminology of the Corporation and Regulatory Enforcement Strategies', in K. Hawkins and J.M. Thomas (eds.): *Enforcing Regulation.* Kluwer-Nijhoff: Boston.

Kringen, J. (2009) *Culture and Control. Regulation of Risk in the Norwegian Petroleum Industry.* Faculty of Social Sciences, University of Oslo: Oslo.

Lindøe, P.H. and Engen, O.A. (2007) 'Behaviour based safety and the Nordic model', in T. Aven and J.E. Vinnem (eds.): *Risk, Reliability and Societal Safety: Proceedings of the European Safety and Reliability Conference.* Taylor & Francis: London, pp. 1705–1712.

Lodge, M. (2004) 'Accountability and transparency in regulation: Critiques, doctrines and instruments', in J. Jordana and D. Levi-Faur (eds.): *The Politics of Regulation. Institutions and Regulatory Reforms for the Age of Governance.* Edward Elgar: Cheltenham, pp. 124–144.

Logstein, H.S. (2007) 'Myndighetenes forvaltning av sikkerhetsregelverket i petroleumsvirksomheten' (Government regulation of safety in the petroleum sector) *Marlus*, 350: 63–79.

(2009) 'Styring ved ikke formaliserte regler ved norsk sokkel' (Governance through non-formalized rules on the Norwegian shelf) *Marlus*, 371: 187–224.

May, P. (2007) 'Regulatory regimes and accountability', *Regulation and Governance*, 1 (1): 1–21.

Ministry of Labour (2002) Report No. 7 to the Storting (2001–2002) *Health, environment and safety in the petroleum activities.*

(2006) Report No. 12 to the Storting (2005–2006) *Health, environment and safety in the petroleum activities.*
(2010) Report No. 29 to the Storting (2010–2011) Joint responsibility for a good and decent working life.
Norwegian Oil and Gas Association (2006) *Kost/nytte-utfordringer i HMS-regelverksregimet på norsk sokkel* (Cost-Benefit Challenges in the HSE Regulations for the Norwegian Shelf). Norwegian Oil and Gas Association -rapport 24.03.2006. Stavanger.
OECD (2010) *Risk and Regulatory Policy. Improving the Governance of Risk.* OECD Reviews of Regulatory Reform. OECD Publishing: Paris.
OECD (2011) *Regulatory Policy and Governance. Supporting Economic Growth and Serving the Public Interest.* OECD Publishing: Paris.
Parker, C. (2006) *Meta-Regulation: Legal Accountability for Corporate Social Responsibility.* University of Melbourne Legal Studies, Research Paper No. 191. Melbourne.
PSA (2005) 'Investigation of gas blow-out on Snorre A, Well 34/7–31/A, 28 November 2004. http://www.ptil.no/English/Helse+miljo+og+sikkerhet/Tilsyn+og+raadgivning/5_granskinq_gassutblaasing_snorre-a.htm.
Reason, J. (1997) *Managing the Risks of Organizational Accidents.* Ashgate: Aldershot.
Roberts, K.H. (1990) 'Some characteristics of one type of high reliability organizations', *Organization Science,* 1: 160–176.
(1993) 'Cultural characteristics of reliability enhancing organizations', *Journal of Managerial Issues,* 5: 165–181.
Rogowski, R. and Wilthagen, T. (1994) *Reflexive Labour Law.* Kluwer: Dewenter.
Rossness, R. (2001) 'Slank og sårbar. Om verdien av organisatorisk redundans' (Lean and Vulnerable. On the Value of Organisational Redundancy). SINTEF-Rapport STF38 A01413. Tröntheim.
Rossness, R., Håkonsen, G., Steiro, T. and Tinmannsvik, R.K. (2000) 'The vulnerable robustness of high reliable organisations: A case study report from an offshore oil platform'. Paper presented at the 18th ESReDA Seminar on Risk Management and Human Reliability in Social Context, Karlstad, Sweden.
Royal Society (1992) *Risk: Analysis, Perception, Management.* The Royal Society: London.
Ryggvik, H. (2000) 'Offshore safety regulations in Norway: From model to system in erosion', *New Solutions,* 10 (1–2): 67–116.
(2008) 'Sikker atferd i historisk perspektiv' (Safe Behaviour in Historical Perspective), in R.K. Tinmannsvik (ed.): *Robust arbeidspraksis. Hvorfor skjer det ikke flere ulykker på sokkelen?* (Robust Work Practices. Why are There no More Accidents on the Shelf?) Tapir Akademisk Forlag: Trondheim, pp. 183–197.
Ryggvik, H. and Smith- Solbakken, M. (1997) *Norsk oljehistorie (III) Blod, svette og tårer* (Norwegian Oil History (III) Blood, Sweat, and Tears). Ad Notam Gyldendal: Oslo.
Scott, C. (2004) 'Regulation in the age of governance: The rise of the post-regulatory age', in J. Jordana and D. Levi-Faur (eds.): *The Politics of Regulation. Institutions and Regulatory Reforms for the Age of Governance.* Edward Elgar: Cheltenham.
Shaver, K.G. (1985) *The Attribution of Blame: Causality, Responsibility, and Blameworthiness.* Springer: Heidelberg and New York.
Shrader-Frechette, K. (1991) *Risk and Rationality.* University of California Press: Berkeley.
Tinmannsvik, R.K. (ed.) (2008) *Robust arbeidspraksis. Hvorfor skjer det ikke flere ulykker på sokkelen?* (Robust Work Practices. Why Are There No More Accidents on the Shelf?) Tapir Akademisk Forlag: Trondheim.

Turner, B.A. and Pidgeon, N.F. (1997) *Man-Made Disasters*. Second Edition. Butterworth-Heinemann: Oxford.

Vaughan, D (1996) *The Challenger Launch Decision: Risky Technology, Culture and Deviance at NASA*. The University of Chicago Press: Chicago.

Vaughan, D. (2005) 'Organizational rituals of risk and errors', in B. Hutter and M. Power (eds.): *Organizational Encounters with Risk. An Introduction*. Cambridge University Pres: Cambridge, pp. 33–66.

Vinnem, J.E. (2006) *ALARP prosesser. Utredning for Petroleumstilsynet. Sluttrapport fase 2. Gjennomgang av selskapenes dokumentasjon og praksis* (ALARP Processes. Review of Documentation and Practices in the Companies). Preventor: Stavanger

Weick, K.E. and K.M. Sutcliffe (2001) *Managing the Unexpected: Assuring High Performance in an Age of Complexity*. Jossey-Bass: San Francisco.

12

Boxing and Dancing

Tripartite Collaboration as an Integral Part of a Regulatory Regime

Ragnar Rosness and Ulla Forseth

12.1. INTRODUCTION: A REGULATORY REGIME THAT WORKS IN PRACTICE BUT NOT IN THEORY?

When discussing the regulatory regime of the Norwegian petroleum industry with researchers from non-Scandinavian countries, we have noticed that they are often perplexed by "the Norwegian model". From the vantage point of established juridical and economic perspectives on regulation (e.g. Baldwin and Cave 1999; Kaasen, Chapter 5 this volume), the industry regulations may seem too open-ended, the inspections too few (Ryggvik, Chapter 15 in this volume), and the reactions to non-compliance too soft to produce acceptable health, safety and environment (HSE) performance. This resembles the paradox faced by students of high-reliability organisations: these organisations seemed to be working in practice but not in theory (LaPorte and Consolini 1991). Tripartite collaboration between trade unions, employers' federations and the political and regulatory authorities is often emphasised as a key to understanding the Norwegian model (Levin et al. 2012; Bang and Thuestad, Chapter 10 in this volume), but it may be challenging for outsiders to understand how this collaboration works in practice (Hart 2007).

The metaphors of "boxing" and "dancing" have been used to characterise shifting patterns of adversarial and co-operative modes of industrial relations/social partnerships (Huzzard et al. 2004). We shall use these metaphors as a point of departure for exploring the robustness of the regulatory regime in the Norwegian petroleum industry and its relationship to the tripartite collaboration on HSE. In this chapter we analyse collaboration and conflict between trade unions, employers' federations and the political and regulatory authorities in a period around the year 2000, when these interactions were particularly visible in the media. This analysis complements the outline of tripartite collaboration in Chapter 10 by providing a more dynamic account. We employ a theoretical framework that allows us to broaden the view of "who regulates" and to capture some of the power games related to the regulation of HSE. We argue that the Norwegian regulatory regime achieved robustness through its capacity to enroll new actors or to redefine the role of current actors when faced with internal disturbance or changes in the environment.

12.2. A MAJOR ACCIDENT THAT NEVER HAPPENED

Just before the millennium shift, a controversy concerning the safety of petroleum exploration on the Norwegian continental shelf threatened to disintegrate the existing tripartite collaboration on HSE. Union representatives claimed that HSE conditions had eroded after a period of intensive cost-cutting, whereas industry representatives claimed that HSE conditions had never been better. After the intervention of the political and regulatory authorities, a more co-operative climate gradually emerged from mid-2000. The tripartite collaboration was revitalised and several new tripartite arenas were established. A major research project was initiated to help to build a common perception of the risk level in the industry (Ryggvik 2003; Moen et al. 2010).

The historian Helge Ryggvik (2003) characterised the outcome of the HSE controversy as "a major accident that never happened". The safety system that had been established during the 1980s in response to numerous accidents, including the blowout at Ekofisk Bravo in 1977 (NOU 1977, p. 47) and the capsizing of *Alexander L. Kielland* in 1980 (NOU 1981, p. 11), had eroded during the late 1990s. The number of major hydrocarbon leaks had increased, and the industry could be seen as heading for a new major accident. However, Ryggvik suggested, this accident never materialised, owing to the revitalisation of tripartite collaboration and safety work in general from the second half of 2000.

By the "robustness" of a public regulation regime we refer to its capacity to fulfil the intentions of the legislators in the face of changes in the regulatory environment (Rosness et al. 2012). The revitalisation of tripartite collaboration and the accompanying revitalisation of safety work reflected the robustness of the Norwegian regulatory regime, with its reliance on tripartite collaboration between authorities, employers and employees. In order to understand this process, we conceptualise the regulatory regime as an actor-network and use discourse analysis to explore how the actors went about enrolling other actors for their purposes.

12.3. CONCEPTUALISING A REGULATORY REGIME AS AN ACTOR-NETWORK

The term "risk regulation regime" is typically used to compare the ways in which risk is regulated in different policy domains or in different countries (e.g., Hood et al. 2001). In accordance with this purpose, Hood et al. (2001) see risk regulation regimes as relatively bounded systems with some degree of continuity over time. Viewing regulatory regimes as relatively bounded and stable systems may be a precondition for comparing regulatory regimes in a straightforward manner. Other conceptualisations may, however, allow for a richer analysis of how regulatory regimes achieve robustness, for instance when faced with changes in the political environment.

Actor-Network-Theory (ANT) is a perspective or an analytic approach according to which phenomena such as organisation, knowledge, agency and power are conceived as effects generated by patterned networks of heterogeneous materials (Law 1992; Latour 2005). The actors that constitute such networks may be individuals, groups and organisations, but also non-human entities such as oil rigs, markets, laws, regulations and oil reservoirs. The director of a regulatory agency will no longer be a director if we remove her computer, her office, her telephone, her budgets, the laws and regulations, her inspectors and the regulated organisations and facilities. In this sense, an actor is also a network. In everyday life we usually take the effects of familiar networks for granted and thus disregard or "black-box" these extensive networks – a simplifying process often referred to as punctualisation.

The patterning of actor-networks requires ordering efforts called *translations*. The networks that make up a punctualised actor are "borrowed, bent, displaced, distorted, rebuilt, reshaped, stolen, profited from and/or misrepresented" to generate effects such as agency, organisations or power (Law 1992, p. 6). Such ordering efforts may face resistance from actors who strive to build a different network. A central task of ANT is to analyse such ordering struggles.

In accordance with ANT, we explore the implications of viewing the regulatory regime in the Norwegian petroleum sector as *a potentially fragile constellation of actors who may have their own agendas*. Applying this theoretical approach to the conflicts on safety in the Norwegian petroleum industry around 2000, we are led to ask the following research questions: What actors did the three parties (trade unions, employees and authorities) try to enroll during the different phases of the conflict? How did they go about enrolling these actors? Was the restoration of tripartite collaboration associated with the establishment of a new and more robust actor-network – a network that adapts to changes but retains its capacity to fulfil the intentions of the legislators? If so, what made this new network robust?

12.4. TRACING TRANSLATION EFFORTS THROUGH DISCOURSE ANALYSIS

Actor-networks can be identified by tracing the efforts made by actors to enrol other actors or to resist being enrolled. Our main tool for tracing these efforts is discourse theory and analysis (Foucault 1972). The concept of discourse highlights how certain interpretations become fixed so that they are taken for granted, naturalised or normalised. The same discourse, the same characteristics of the way of thinking and speaking, will then appear across a range of texts. There is a constant struggle between different discourses in order to define and categorise phenomena in our world. A particular discourse gains hegemony by being unchallenged and naturalised, so that its premises are taken for granted.

Discourse analysis covers a wide range of multidisciplinary perspectives (Wetherell, Taylor and Yates 2001), and it comprises a complete "package" that includes philosophical premises, theoretical models, methodological guidelines and specific techniques for analysing language (Jørgensen and Phillips 1999). Critical discourse analysis, as formulated by Fairclough (1995), advocates a text-oriented analysis to capture the ways social and political domination are reproduced by text and talk by joining together three different strands of analysis: (1) text, (2) discursive practice and (3) social practice. The starting point is a close reading and analysis of a specific text such as a feature article in the media or an interview transcript. Discursive practice concerns how the writer/storyteller draws on existing discourses and genres to produce a text, and how the receiver relies on existing discourses when interpreting the text. Social practice has to do with how discourse both reproduces and changes knowledge, identities and social relations, including power relations, and simultaneously is shaped by other social practices and structures (Fairclough 1992).

12.5. DATA AND METHODS

The research design for this work relies on the analysis of documents and "texts" in a broad sense. In addition to being sources of data, these texts were important "actors" in the dispute on safety in the Norwegian petroleum sector. We draw on a strategic sample of public documents addressing the controversy, such as feature articles and news stories, trade union periodicals, pamphlets, Web sites, white papers and research reports. The collection of data and analysis proceeded as an iterative process. Analysis consisted of multiple readings of the sampled texts and identification of events and actors. We then screened the documents in order to determine the substance of the views regarding the dispute on the safety level. In the next step, we selected and coded texts, text elements and quotes that portrayed the main standpoints of the stakeholders. Finally, we analysed how the actors made use of existing discourses and rhetorical means, use of metaphors and images, transitivity (how processes and events were connected or not connected to subject and object) and modality (the tones and styles of a text, such as use of humour and sarcasm). Our concern was with the way language was being used to describe and make sense of a phenomenon (regulation of safety), and the rhetorical devices employed in translation processes to enrol new actors. After the first phase of analysis, we did a new literature search for additional text to supplement important parts of the analysis and add texts on the period after the dispute.

Our analysis is mainly restricted to the actors that featured in our sample of texts. We thus have not captured actions and actors operating "backstage", such as informal contacts between trade unions and politicians. Relations between oil companies and contractors were beyond the scope of this work as this is a complex issue that merits an analysis in itself.

12.6. RESULTS

The major part of this section will be devoted to episodes in 2000 and 2001 that illustrate how some actors tried to enrol other actors during the controversies. First, however, we outline the historical context of tripartite collaboration in the Norwegian petroleum sector.

12.6.1. *Tripartite Collaboration in the Norwegian Petroleum Industry: Historical Context*

The international oil companies resisted the early attempts of national trade unions to establish themselves on Norwegian production installations and promoted the establishment of local employee organisations (OLF, n.d.). However, in 1977, three local employee organisations joined to form The Federation of Offshore Workers' Trade Unions (Oljearbeidernes Fellessammenslutning, or OFS). The Norwegian Confederation of Trade Unions (Landsorganisasjonen, or LO) responded by forming the Norwegian Oil and Petrochemical Workers' Federation (Norsk Olje- og Petrokjemisk Fagforbund, or NOPEF). The employers' federation (Norske Operatørselskapers Arbeidsforening, or NOAF) was established in 1981. This organisation was a precursor to the Norwegian Oil Industry Association (Oljeindustriens Landsforening, or OLF) which later changed its name to the Norwegian Oil and Gas Associations.

Trade unions and employers' federations in Norway have a long tradition of negotiating improvements related to health and safety (Engen and Lindøe 2008; Øyum et al. 2010). The Industrial Democracy Programme (Emery and Thorsrud 1969; Gustavsen 1992), inspired by ideas from the Tavistock institute in London, was launched in Norway in the late 1960s. It was supported by the Confederation of Norwegian Enterprises (Norsk Arbeidsgiverforening, or NAF), the LO and the government, based on the expectation that industrial democracy and worker participation would lead to increased productivity and an improved work environment, and thus benefit both workers and employers. In 1978, the research programme "Safety Offshore" was launched. A significant share of the resources was devoted to research on "The

Human and Organisational Aspects of Safety", which to a large extent continued the approach and ideology of the Industrial Democracy Programme (Qvale 1985).

An important outcome of the Industrial Democracy Programme was the Norwegian Work Environment Act of 1977. This law prescribed close collaboration between workers and employers with regard to HSE issues, and it involved significant restrictions on the management prerogative of the employers (Ryggvik 2008). One such feature established the right of the safety representatives to halt dangerous work without being held liable for the economic consequences (Forseth, Torvatn and Andersen 2009). The Work Environment Act thus made tripartite collaboration an integral part of the regulatory regime for HSE. The practice of collaboration between the workers' union and management at the enterprise level developed into a comprehensive system of bi- and tripartite collaboration. This is sometimes referred to as "the Nordic model," whereas others prefer the term "the Norwegian model" to emphasise specific national features (Dølvik and Stokke 1998, Levin et al. 2012).

According to Hovden (2002), the regulatory regime of the petroleum industry has been through three main phases:

1. In the 1970s, the industry was characterised by the meeting between a "wild west" approach to safety and a regulatory regime based on traditions from the onshore industry and ideas from the nuclear and chemical industries. The Norwegian authorities were dependent on international oil companies to explore the oil resources, but their explicit ambition was to promote a national Norwegian oil industry and to make the international companies adapt to Norwegian HSE standards.
2. During the 1980s, the Norwegian Petroleum Directorate (NPD) had established a moral authority and the necessary power base to introduce a regime of enforced self-regulation ("internal control") based on new risk- and performance-based criteria (goal-based or functional regulations; see Chapter 10 by Bang and Thuestad in this volume). According to the government, the petroleum industry was at the forefront of the Norwegian industry in the area of HSE during the 1980s and into the 1990s, and the unions were an important actor in this work (St.meld. nr. 7 2001–2002; Ryggvik 2003). These developments were accompanied by a marked reduction in fatal accident rates.
3. During the 1990s, the industry experienced reduced profit as a consequence of lower oil prices. The NORSOK project ("Norsk Sokkels Konkurranseposisjon") was launched, with an official ambition to reduce the costs of extracting oil without compromising HSE. New standards for coordination, control and construction were developed by the industry itself (Hovden 2002). The development during the 1990s was characterised by extensive changes: restructuring of organisations, workforce reductions, introduction of new models of

TABLE 12.1. *Significant events related to tripartite collaboration on HSE in the Norwegian petroleum industry around 2000.*

Time	Event
1977	The Working Environment Act establishes a legal framework for tripartite collaboration
1978	The "Safety Offshore" research programme is launched
1986	Internal control of HSE regulations is implemented in the Norwegian petroleum sector
1995	The NORSOK forum submits proposals for reducing the cost and time of new field developments by 40–50% while maintaining Norways leading position with regard to HSE
8 September 1997	Two pilots and ten passengers are killed on their way to the oil production vessel *Norne* in a helicopter crash
13 September 2000	NPD issues a letter to the licensees on the Norwegian shelf requesting new actions to improve safety
Autumn 2000	Safety Forum is established as a new tripartite arena
24 December 2000	An offshore worker is killed in an accident at the production platform Oseberg Øst
2000/2001	Working Together for Safety is established
April 2001	The report from the pilot project "Risk Level on the Norwegian Shelf" is issued
December 2001	The Ministry of Work and Administration issues the white paper, "About health, environment and safety in the petroleum sector", St.meld. nr. 7 (2001–2002)

> cooperation and major changes in the organisation of construction projects. Parts of the industry went through major mergers (St.meld. nr. 7 2001–2002). The capacity to influence the assignment of new development fields had been an important part of the power base of the NPD. This capacity became less relevant because the new fields were less prosperous. Against this background the trade unions raised concerns about the safety level in the industry.

We have provided a timeline summarising what we see as important events and actors related to the dispute on the safety level on the Norwegian continental shelf. Table 12.1 gives an overview of events related to the erosion and revitalisation of tripartite collaboration in the Norwegian petroleum sector around 2000.

12.6.2. *The Norne Helicopter Accident*

On 8 September 1997, a Super Puma helicopter on its way to the floating production, storage and offloading unit (FPSO) at the Norne oilfield crashed just before arrival (AAIB/N 2001). The accident caused the death of the two pilots and all ten passengers. The immediate cause of the accident was technical, starting with a fracture of the power transmission shaft between the starboard engine and the main gearbox. However, the Norne accident could also be construed as a consequence of the new construction practices that were introduced with the NORSOK standards. A central ambition was to reduce the time to design and build offshore installations. In accordance with an ambitious project schedule, the operator, Statoil, carried out a lot of completion work after the FPSO had been positioned at the offshore oilfield. Many workers had to shuttle by helicopter to and from the FPSO on a daily basis, and they were thus exposed to increased accident risk and exhausting workdays. Union representatives raised this issue, and the Norne accident was also followed up by the Norwegian parliament, in a written query to the Minister of Petroleum and Energy. The NPD made the following comment in their Annual Report for 1997:

> Transportation of personnel to and from the installations is considered to be air traffic, and thus does not directly fall under the petroleum authorities' sphere of responsibility. However, the incident has contributed to bringing the spotlight to bear on sufficient living quarters capacity on the installations as an alternative to helicopter transportation. Time pressure in the development projects has in several cases resulted in installations being installed on the field before they are completed. This leads to a significant need for labor which must commute on a daily basis because the living quarters are not designed for a temporary increase in manning. ... Therefore, the Norwegian Petroleum Directorate has taken the initiative of contacting the involved parties in order to evaluate the problems involved in daily commuting to and between offshore installations. ... From the employee's perspective, it has been pointed out that daily commuting constitutes a "special risk" which some employee groups are exposed to, as it is primarily personnel from contracting companies who are subjected to daily commuting. (Norwegian Petroleum Directorate 1998, p. 84)

The NPD adopted the position of the unions that there was a connection between the shuttling practice and the Norne accident. Through this translation the accident was enrolled as an actor in the broader discussion concerning the NORSOK regime and its influence on safety. By contacting the involved parties in order to evaluate the problems, the NPD also used the accident as an occasion to maintain and reconfirm tripartite collaboration on HSE.

12.6.3. *Clashing Views on HSE in the Petroleum Industry*

During the late 1990s, a polarisation developed between the trade unions and the petroleum industry. The trade unions claimed that safety offshore was deteriorating, whereas the industry claimed that safety was better than ever. We examine two expressions of these positions.

On 23 August 2000, the editorial in *Oljeindustrien*, the journal of the Norwegian Oil and Gas Associations, stated their views on the safety level:

> The oil industry has always given safety high priority. There are no incentives at all for not doing that. This applies least of all to the oil companies that have to compete for new licences and are dependent on safe and efficient operation of their production facilities. I have worked for so long in public administration that I know what emphasis is put on safety statistics and safety evaluations when new licences are granted. It is important that the authorities continue this practice. The oil industry spends enormous resources on safety work, and statistics show that there has been radical and continuous improvement in the safety level in the industry throughout the 80s and 90s. It is therefore very unfortunate and totally wrong when it is claimed that safety was better in the past. Safety has never been better than it is now. It has also been claimed from certain quarters that safety work has been given lower priority or neglected in conjunction with the reorganizing and restructuring which the petroleum industry had to carry out to strengthen the competitive edge and the profitability. This is not the case. It is the opposite that is happening. The focus on safety is particularly high in times of great changes. (quoted by the Norwegian Petroleum Directorate 2000, p. 121)

The message is clear in this text – "Safety has never been better than it is now" – and it is presented as an objective truth. There is no use of passive tense in the text, and the oil industry stands out as the primary agent and is repeated several times. ("The oil industry has always given safety high priority".) The industry is thus presented as taking responsibility. The author gives the authorities credit for using their power and paying attention to safety performance when granting new licenses. The Norwegian regulatory regime is thus enrolled as a guarantee that the oil industry gives safety high priority. There are a lot of reinforcing adverbs ("always", "enormous", "very" etc.). Indeed, the word "safety" is repeated nine times in order to strengthen the message. There is no room for doubt or recognition of dilemmas between efficient operations and safety. The argument about lower priority on safety is said to come from "certain quarters", indicating that these sources are biased and dubious. The argument that safety has never been better is based on "facts" and statistics – what can be counted and measured. These underlying premises that the safety level is reflected in what has been counted or measured seem to be taken for granted within this discourse, so that other ways of talking about and interpreting what is going on seem irrational. There is no allowance for the possibility that the assessment of safety is a sense-making process that opens up different interpretations and judgements. There is also no reference to how employees in the oil industry perceive their own safety and working environment. This editorial draws on and reinforces an *economic-operational discourse* by using phrases such as "strengthen the competitive edge" and justifies the need for "reorganizing and restructuring".

How did the trade unions promote their views on safety in the petroleum industry? In a news release from the Norwegian News Agency (NTB) on 14 November 2000, a report from the consulting firm Kommuniké AS, containing harsh criticism of the oil companies and The Norwegian Petroleum Directorate, was cited and used as evidence: "'Oil-Norway' [i.e. the Norwegian oil industry] has developed a culture of greed which may lead to major losses of human lives and damages to the environment". Two trade unions, Lederne (The Confederation of Leaders) and OFS, backed the report. The Organisational Secretary of the latter, Roy Erling Furre, was quoted by the news agency giving his evaluation of the report:

> There are harsh statements in the report, and serious accusations are directed against the petroleum industry. But the use of such language is fully justified, because the safety on the shelf is in decline. … During the recent years we have got an increasingly risk willing and aggressive industry, which weakens the emergency preparedness for economic reasons and denies the problems. On previous occasions we have been told by the industry that we pursue unserious criticisms and that safety in the North Sea has never been better, but we can observe at close range that this is not the case", says Furre, who wants to see responsible politicians on the arena. He claims that in this issue, OFS is on line with their organisational rivals in the North Sea, Norwegian Oil and Petrochemical Worker's Union (NOPEF) (News release, NTB, 14 November, 2000).

In this case, trade unions enrolled a report from a consultant company as evidence for their own point of view. The comment opened with sentences in the passive tense ("serious accusations are directed against the petroleum industry"). The absence of an explicit subject conveys neutrality and independence on the part of the consultants who wrote the report. Then there is a change of mode and agents and events are linked: A "we" is introduced in a way that seems to imply that the oil industry has become a problem for the whole Norwegian society. The industry is portrayed as "increasingly risk willing and aggressive", and consequently the trade unions draw a line between "us" and "them". We get to know that there is a disagreement between the collective voice of the trade union(s) and the industry when it comes to the safety level on the Norwegian continental shelf. This is also an operational discourse from the inside, but with strong emphasis on protection of the workers and the environment. We will label this an *operational/protective* discourse. There is little use of metaphors in this excerpt, but the industry is characterised as "aggressive". The arguments are clear without necessarily revealing specific sources for the interpretation. One exception is when the representatives from the trade union mention their own observations ("observed at close range"). We hear about "harsh statements" and "accusations", and there is little use of hedges, that is, words modifying the authority of the statement.

We can thus observe the struggle between different discourses. On the one hand, the industry drew on a discourse proclaiming "safety in the North sea has never been better". On the other hand, the trade unions formulated a competing discourse drawing on their observations ("but we can observe at close range that this is not the case"). Each party tried to establish their view as the truth. However, there are also several similarities between the competing discourses. The industry representatives bolstered their views with statistics, and the union representatives mobilised consultant reports as "objective" and "independent" evidence. Representatives from the industry strengthened their discourse by labelling the critique from the unions as "unserious", whereas the union representatives characterised the industry as "risk willing and aggressive". Messages were repeated, and there is little use of hedges. The parties were boxing, but they seemed to be playing the same game.

12.6.4. *The Establishment of Safety Forum*

In the middle of June 2000, the trade union NOPEF, acting through the LO Industry, took an initiative to establish a safety council for the oil industry (Næss 2000). The union leaders from NOPEF drew on a safety discourse and referred to the concerns of the NPD about reduced safety level offshore. They also criticised their counterpart, Norwegian Oil and Gas Associations, by stating that they had "no trust that Norwegian Oil and Gas Associations would intensify the work to prevent serious accidents" as they "constantly failed to include the employees" in decisions concerning safety. On July 3, the Minister for Local Government and Regional Development, Sylvia Brustad, announced her support for the initiative. The new arena of collaboration, Safety Forum, was to be administered by the NPD. President of Norwegian Oil and Gas Associations, Tharald Brekne, made the following comment: "We welcome every initiative that can contribute to increased safety and better emergency preparedness. … [If] such a safety forum is administered by the NPD, then I think it will be OK. Besides I have no further comments" (Næss 2000, p. 11).

It was hardly an option for Norwegian Oil and Gas Associations to refuse to participate in this forum. The president did not seem overly enthusiastic and he underscored the importance of the NPD as the administrator. With this move, the trade unions, supported by the political authorities, had enrolled a new actor, a *tripartite arena* where the unions, the employers' federations and the authorities could exert a significant influence on the agenda. This step also created psychological *commitment* to cooperation among the parties (Pfeffer 1992). Once they were enrolled, they had committed themselves to contribute to a process leading to less antagonism and more intensive co-operation. Even as the parties continued to draw on the same discourses in their dispute about the safety level, they took tangible action to revitalise tripartite collaboration. Now they were boxing and dancing.

12.6.5. *The NPD Enrols the Goal-Based HSE Regulations*

According to the Norwegian newspaper *Aftenposten* (October 6, 2010), the NPD threatened to revert to more detailed HSE regulations:

> The authorities threaten with more detailed regulation if the oil companies do not increase their efforts.
>
> The good collaboration that previously characterised the relationship between the industry and the authorities, no longer functions in a constructive manner, the NPD claims. The companies increasingly challenge the authorities' framework conditions. Such a development could lead to an increased degree of detailed regulation by the authorities.

The threat to revert to more detailed regulations (e.g., Baram, Chapter 7 in this volume) may be interpreted as a necessary precaution to keep safety from eroding in a situation where the companies challenge the authorities' interpretations of the goal-based rules. One may also read this threat as a display of power. The goal-based regulations give the industry freedom to choose the solutions they consider optimal as long as they comply with the goals stipulated by the functional requirements. The NPD threatened to remove this freedom. More detailed regulations might force the industry to select solutions they perceive as suboptimal and detailed regulations may hamper technological and organisational development. In this manner, the NPD attempted to make an association between the goal-based HSE regulations and industry loyalty to the authorities' framework conditions.

12.6.6. *The Offshore Accident on Christmas Eve*

On Christmas Eve of 2000, a fatal accident occurred at the platform Oseberg East, operated by Norsk Hydro on the Norwegian continental shelf. A service engineer, aged thirty-two, employed by one of the contractors, died after being hit by two tons of production pipes from a crane lift on the pipe deck. The mother of the victim, Lisbeth Olgadatter Wathne, became an activist for increased safety levels offshore. After the accident she made several appearances in the media. She met the Minister of Petroleum and Energy and the director of the NPD and was invited to give speeches on several occasions at trade union conferences. The following excerpts are taken from her opening speech at the annual meeting held by one of the trade unions at Statoil, Statoilansattes Forening, on 14 June 2001.

> I am the mother of Rune Grønningen, who was killed at Oseberg Øst on Christmas Eve 2000. He was the victim of WORK ENVIRONMENT CRIME, which describes a situation where one has not complied with the laws and regulations that apply. One can be convicted for this. ...
>
> One would expect that an operator who is directly the cause of a fatal accident, would spend large resources to investigate how to help the bereaved. However, that is not the case.... Common decency would call for the highest-ranking executive, the managing director, to make the effort to call us ... and simply apologise on behalf of the company that something like this could happen. Just that. ... Until now, the managing director has had nearly three months on his hand to make this phone call, but no ... he has still not found it appropriate. At any rate, I have not heard that Norsk Hydro has taken the full responsibility.... On the contrary, Myklebust [managing director of Norsk Hydro] sets off primarily to defend his own company. I see that he is quoted in Aftenposten [national newspaper] on February 11 this year stating that "it is very misleading" how I have argued publicly because people may come to believe that it is dangerous to work offshore! What arrogance! What a total lack of social antennas! Is he not capable of reading statistics? Does he not understand that there is a misfortune behind every reported accident?

In her speech Watne positioned herself as a mother who has lost her son. She characterised the fatality as the result of a crime, and thus introduces a moral and juridical frame. The personal and moral orientation persisted when she recounted how Norsk Hydro, represented by its managing director, treated her after the accident. Wathne also contrasted the formal safety systems, with their rules and reports, with the working conditions, as the workers experienced them:

> I did not speak about rules and reports. I spoke about the work environment .... How is it to work at a place where the workforce is constantly replaced? ... It is under these difficult working conditions that extremely dangerous work operations are carried out. ... Every single petroleum worker I have spoken to had examples of accidents – and near misses to tell about. How is this possible?

Wathne thus launched a new actor, the event of "work environment crime". Her criticism, however, went beyond a single company and she accused those in power positions within the industry, "the offshore clergy" (Wathne 2002, p. 12) for whom "the pursuit of profit was a supreme sovereign", for cultivating a culture of violating procedures. In this way, a single company, the petroleum industry, managers in powerful positions within the industry and the pursuit of profit were all enrolled in the network. She joined forces with the workers in the industry, retelling their accounts of incidents and accidents from the "inside". She claimed that accident statistics were misinterpreted and emphasised the necessity of looking beyond official figures searching for the incidents and stories behind the numbers. Thus, workers and their storytelling were also mobilised.

To get her message through, Wathne used several stylistic devices: personalising the message, clear and emotional language, sarcasm, repetition of core sentences, accusations, provocative questions and exclamation marks, metaphors and storytelling. She laid bare her personal situation and was fearless in her accusations and the naming of stakeholders. Wathne thus introduced a new discourse that emphasised ethical/judicial aspects and the personal experiences of petroleum workers (Forseth and Rosness 2010). She challenged the existing discourse where the line of arguing was based on economy, statistics and third-party reports. From an actor-network perspective, this fatal accident can be seen as an actor because it has aroused strong emotions and media attention and has brought new actors centre stage.

12.6.7. *The Unions Adopt the New, Personalised Discourse*

Wathne's introduction of an ethical discourse found resonance with the unions' struggle for a safe work environment offshore. This is evident in an article entitled "Safety out of control?" written by Roy Erling Furre from OFS on 10 January 2001, shortly after the fatal Christmas Eve accident:

> A mis-culture of hiding injuries has developed in order to ensure nice figures on the so-called zero-level statistics [Lost Time Accident]. This is exactly what Johan Petter Andresen was a victim of when he broke his finger and was dismissed because he did not accept to hide the injury. We know about more cases where people have been asked to cover up for such injuries. A girl who had injured her finger did not have her injury reported – not until she had to amputate due to an infection; then it was recorded in the statistics. A high degree of manipulation does not lead to a good level of safety and this is the reason why we do not trust the Norwegian Oil and Gas Associations. (Furre, 2001)

In this article Furre made use of some of the same stylistic devices as Wathne. He illustrated his arguments with two stories in order to present the fates behind the statistics. He identified a male offshore worker who had been injured and stated his name and details about the injury, whereas the female worker is anonymous and referred to as "a girl".

Furre also challenged the validity of injury statistics as a safety indicator. His stories illustrate how statistics are subject to interpretation and manipulation. In this quote, Furre used sarcasm when he talked about the "so-called zero-level statistics". He also used strong words such as "manipulation" and "cover up" and concluded that trust in Norwegian Oil and Gas Associations was low.

12.6.8. *Working Together for Safety*

In 2001, a new major collaborative project called Working Together for Safety was launched. This project was initiated by the industry led by Norwegian Oil and Gas Associations and involved many of the major actors in the tripartite constellation. It had five aims (Working Together for Safety 2001):

1. increased safety in the petroleum industry offshore;
2. reduced risk of personal injury and major accidents;
3. improved trust of the industry among its employees and their families;
4. strengthened trust and cooperation among the actors of the industry; and
5. improved reputation for the industry.

Working Together for Safety established itself as an important arena for collaboration on specific tasks within the industry, such as development of common work permit procedures. Working Together for Safety has thus complemented Safety Forum, which has functioned as an arena for setting the HSE agenda and negotiating a shared understanding of HSE issues. Another distinction was that NPD administrated Safety Forum, whereas their role in Working Together for Safety was that of an observer. The roles of Safety Forum and Working Together for Safety are also discussed by Bang and Thuestad in Chapter 10 of this volume.

12.6.9. *Enrolling the Research Community in Consensus Building: The RNNP Project*

As a response to the controversy about the risk level the NPD launched the "Risk level on the Norwegian Shelf" (Risikonivå på Norsk sokkel, or RNNP) pilot project. As described by Bang and Thuestad (Chapter 10, Section 10.6) in this volume, the project was performed in collaboration with Norwegian research institutions, and its expressed purpose was to establish a methodology to assess trends in the risk level and safety work on the Norwegian continental shelf (St.meld. nr. 7 2001–2002).

The pilot project report, which was published in April 2001, included (1) a survey of various indicators of the development of the risk level, (2) interviews with some of the key actors in the industry, and (3) proposals for further work (Norwegian Petroleum Directorate 2001). These proposals included a questionnaire to be distributed to all employees working on offshore installations. By including interviews and questionnaires, the RNNP project legitimised the opinions of offshore workers and union representatives as valid and necessary input to an assessment of safety in the petroleum sector.

The pilot study report explicitly supported the claims of the trade unions and the NPD that safety was deteriorating (p. 3):

> Based on the data material and the indicators that have been used in this project, the risk level seems to be increasing. Admittedly there are a limited number of indicators in the pilot project, but the negative trend during the past 2–3 years is judged to have general validity.
>
> It is characteristic for the total picture that there are no indicators that display a substantial improvement.

The RNNP project was a means to obtain trustworthy information and analyses, but also a means to build consensus between the parties and to secure commitment to future efforts to improve safety. The following paragraph from the preface to the pilot project report illustrates this point:

> Objectivity and trustworthiness are key words when one is to form a considered opinion about safety. The results from the project have been presented for Safety Forum, where the trade unions and the employers' federations are represented. The comments thus far have been positive and constructive with expectations that this will contribute to a common platform for safety improvement.

The double nature (scientific and political) of the project report is reflected in the list of authors, which includes NPD and the consultant and the research institutions that participated in the work. It is also reflected in the conclusions, as illustrated by the following excerpt (p. 3): "A clear need to improve the collaborative climate between the parties and to reach a more similar perception of reality has been demonstrated". The call for improved collaboration is presented here as a result of an empirical scientific study.

The significance of RNNP was related to its position in a greater actor network, in particular its relations to Safety Forum and NPD. The RNNP studies attained legitimacy and attention from its links to Safety Forum. For NPD, the RNNP was an instrument for providing knowledge and for influencing the agenda for safety work in the industry.

12.6.10. *The White Paper on Health, Environment and Safety in the Petroleum Activities*

On 9 June 2000, the Ministry of Petroleum and Energy issued a white paper on oil and gas activities (St.meld. nr. 39 1999–2000). One of the ten chapters in this document was devoted to HSE issues. The white paper referred to the concerns of the NPD that the risk level appeared to be increasing. However, this document did not signal a high level of ambition with regard to HSE:

> While the challenges formerly consisted in bringing risk and injury frequencies down to an acceptable level, the challenge in the years to come will primarily be

> to prevent that the development causes the risk and consequently accidents and injuries to increase. (p. 19)

On 3 July 2000, two members of the parliament presented a proposal that the government should issue a white paper on HSE conditions on the Norwegian continental shelf (Dokument nr. 8:90 1999–2000). The Ministry of Labour and Government Administration issued a new white paper on health, environment and safety in the petroleum activities on 14 December 2001 (St.meld. nr. 7 2001–2002). This white paper signalled a higher ambition level than the previous one:

> The Government wants the petroleum sector to be a leading industry with strong focus on health, environment and safety at all levels of the activities, and which has "continuous improvement" and "precaution" as fundamental principles. (p. 9)

The white paper took a clear stand with regard to the controversy concerning the risk level:

> It has been documented that the risk level in the petroleum sector is increasing. One of the causes of the negative trend is the far-reaching change processes that were carried out during the nineties. Technological, operational and organisational efficiency improvement measures were implemented without the industry paying sufficient attention to the consequences with regard to health, environment and safety. (p. 9)

The white paper also made tripartite collaboration and employee participation a precondition for the goal-based regulations:

> Collaboration between the parties and employee participation are important preconditions for the inspection and regulation model that applies to health, environment and safety in the activities. This follows, not least, as a consequence of the authorities having abandoned detailed regulation and control. The authorities have, instead, made the parties responsible for finding the best solution by themselves within the boundaries for prudent activity drawn by the authorities. Feedback shows, however, that collaboration between the parties and participation has not been sufficiently taken care of during the later years. There is consequently a need for renewed and strengthened attention to these issues. (p. 10)

The white paper was an important policy statement and thus reinforced the link between tripartite collaboration and the goal based regulations.

12.6.11. *Boxing and Dancing Together for Safety*

What was the long-term impact of the series of events discussed earlier in this chapter? In a series of interviews published on the homepage of the Petroleum Safety

Authority, Norway, five years later, some of the main protagonists expressed their views on Safety Forum. The director of the Safety Department of the NPD *and* chairman of Safety Forum, Magne Ognedal, expressed his views on the process from controversy towards a more collaborative climate between the parties: "One of the most important tasks of the Safety Forum was to work for a climate and a culture of mutual respect. We struggled a little bit in the beginning, but this has increasingly improved" (Ognedal 2005).

Two of the union representatives in Safety Forum underscored the importance of a formal arena for interaction:

> To us this is one of the most important arenas for communication with all the stakeholders in the industry. Several of the oil companies are rather rigid in their views on HSE. Indeed, their 'world-wide' focus takes away the focus from many important processes – and leads to an 'Americanisation' of important issues – something that contributes in a very narrow focus on HSE. (Furre 2005)

> The Safety Forum is a very demanding arena – especially for the employers. This is not the place for delaying answers, exhaling problems, trying to avoid responding or to evade challenges regarding HSE. ... The road from an effective and conflict solving organ to a farce is not necessarily long! (Karlsen 2005).

After stressing the importance of Safety Forum as a tripartite arena, both union representatives were more concerned with, and critical of, the oil companies than reflecting on their own role and contribution. Karlsen spoke frankly about the issue of power and underscored that the earlier power tactics of the oil companies were not as easy to deploy in the face-to-face setting in the forum. Furre, for his part, raised the question of who has the power to define and operationalise the concept of safety, and the influence of cultural differences in this respect. The two quotes demonstrate that the unions were in favour of Safety Forum but they still went on attacking the oil companies.

The newly appointed director of HSE and operations of Norwegian Oil and Gas Associations, Knut Thorvaldsen, entered the arena with the following statement: "Safety Forum and Working Together for Safety ... have both been successful. Both arenas are central in achieving continuous improvements in HES and for making the petroleum industry a leading industry in matters of HES" (Thorvaldsen, 2005). The quote is a positive policy statement where the success of Safety Forum and Together for Safety as arenas for collaboration was emphasised and there was no critique of the unions or other stakeholders.

These quotes indicate that tripartite collaboration was still very much alive five years after the events that initiated it. It had not developed into a nirvana where all controversies have been surmounted. The parties were still boxing and dancing together for safety.

12.6.12. *Tripartite Collaboration and HSE Performance*

Is it possible to identify any connection between the developments in tripartite collaboration and the HSE performance of the Norwegian petroleum industry? A visual inspection of results from the RNNP project indicates that the trends for three important HSE indicators changed shortly after 2000. The total indicator for major accidents on the Norwegian shelf, normalised against work hours, showed a rising trend between 2002 and 2004, and a declining trend in the following years (PSA 2012, p. 112, figure 79). The number of injuries related to work hours showed a steady level from 1996 to 2000 and a declining trend in the following years, which gradually levelled out around 2004 (PSA 2007, pp. 130–131, figures 117 and 118). The number of *serious* injuries showed a rising trend from 1996 to 2000, a declining trend from 2001, and levelled out around 2005 (PSA 2007, p. 132, figure 119). These results are compatible with a causal connection between the developments in tripartite collaboration and HSE performance. However, the results do not constitute a rigorous proof that such a causal connection exists, because the possibility remains that the trend shifts may have been caused by other factors.

12.7. DISCUSSION

12.7.1. *What Actors Did the Three Parties Try to Enroll during the Different Phases of the Conflict?*

We have identified a number of efforts by the different actors to enrol new actors during the different phases of the controversy concerning the risk level, for instance:

- The oil companies, Norwegian Oil and Gas Associations, the trade unions and the NPD enrolled statistics and research reports to underpin their views.
- The trade unions enrolled the political authorities in their efforts to establish a new arena for tripartite collaboration (Safety Forum).
- NPD enrolled the goal-based HSE regulations through its threat to revert to more detailed regulations.
- NPD tried to enroll the oil companies through their letter to the licensees.
- The mother of a an offshore worker who lost his life offshore enrolled his fatal accident, the mass media and the reputation of Norsk Hydro, its managing director and the Norwegian petroleum industry.
- NPD enrolled the research community by initiating the RNNP project.
- The ministry and the NPD enrolled the other political parties by issuing a white paper on HSE in the petroleum industry.

The controversy and the efforts to revitalise tripartite collaboration led to an expansion of the network of actors involved in the discourse on offshore safety. For instance, the government and the research community were mobilised. We suggest that this mobilisation of new actors was a prerequisite for the revitalisation of tripartite collaboration. This implies that the capacity and willingness to enroll new actors was a crucial feature of the robustness of the regulatory regime in the Norwegian petroleum sector, with its reliance on tripartite collaboration.

Our analytical approach was not sensitive to actors leaving the regulatory actor-network or losing significance within it. Consequently we are not in a position to determine whether some actors were "lost" during the controversy. Moreover, the analysis only captured translation efforts that could be discerned from the public documents.

12.7.2. *How Did the Parties Go about Enrolling New Actors?*

The disputing parties used a wide range of discursive techniques in their effort to enroll new actors, convince them of the correctness and advantages of their views and overcome resistance. In the first phase, the unions, the employer's federation and the authorities were drawing heavily on accident and incident statistics as "objective facts" but interpreting them differently. Additionally, experts and researchers were called on to strengthen and support their views. Typically, the actors argued in a dialectical manner challenging the positions of their counterparts. In this process the union representatives and the victim's mother presented other types of data than statistics, such as events, stories and narratives from the field. Consequently the employees' own experiences and viewpoints were voiced and became more valid.

We have illustrated how the conflicting combatants tried to persuade and mobilise others by using different styles and tones. Modality and transitivity have been identified as tools in critical discourse analysis (Fairclough 1995). Modality has to do with the way actors connect with their statements. Modality was expressed by style, intonation, humor and satire. At the beginning there was a fight about numbers and statistics. In text excerpts from Norwegian Oil and Gas Associations, claims were presented as facts ("the oil industry has always given safety high priority", "the focus on safety is particularly high in times of great changes") and the content was reinforced ("always", "enormous"). There was no room for doubts or recognition of dilemmas between efficient operations and safety. There was little use of hedges by the actors, words such as "a little" and "maybe", that would undermine the message and signal that the world is complex and ambiguous. Transitivity concerns how processes and events are connected or not connected to subject and object. The argument about deteriorated safety level was rejected by the Norwegian Oil and Gas Associations and said to come from "certain quarters" indicating that these sources were dubious and not trustworthy. The mother of the victim in the fatal accident offshore articulated a counter-discourse with strong emotions, vivid metaphors and harsh critique, personification and personal accusations of the "offshore clergy" manifested in the persona of the managing director of Norsk Hydro. She used stories and narratives to convey meaning and evoke emotions. Traces of this kind of discourse were later found in the discourses of the union representatives and the NPF. The statements and arguments of the conflicting parties were thus explicable on the basis of the different discourses that we had identified in a previous analysis of the controversy (Forseth and Rosness 2010): (1) *protective/operational* (the authorities and the trade unions), (2) *economic/operational* (employers' federation) and (3) *moral/judicial* discourse (articulated by the mother of a victim of a fatal accident offshore).

12.7.3. *The Restoration of Tripartite Collaboration*

Enrolling actors through translation involves building power and overcoming resistance. However, power relations are not stable; they must be recreated. A recurrent issue in ANT studies is how actor-networks attain durability. According to Latour (2005), many actor-networks achieve durability by enrolling non-human actors. A durable limitation of speed on a road is often achieved by introducing artificial bumps, so that drivers have to reduce their speed unless they are willing to risk damaging their cars. An organisation may standardise the performance of certain tasks by introducing data-processing systems that restrain the possible ways the task may be performed.

None of the actors that were captured in our analysis had the capacity to physically restrain the behaviour of other actors in the way artificial bumps restrains speed on a road. However, several of the actors that were mobilised had the capacity to influence or restrain action in more subtle ways. The introduction of new arenas for collaboration, such as Safety Forum, influenced the patterns and style of interaction among the parties. When NPD issued a public letter to all licensees with the message that safety was eroding and that there was a need to revitalise tripartite collaboration, it became problematic for the industry *not* to respond in a constructive manner. The unions' idea of a safety forum was picked up by the director of the NPD, who succeeded in persuading the Minister of Local Government and Regional Development that this could serve their common interest in promoting and improving safety in the petroleum industry. The parties could then invite the employers' federation with an offer that was hard to refuse. By fixing power relations through this forum, however, the parties could not push their interests as solo players, but had committed themselves to discussing and sharing standpoints. On the other hand, the parties represented in this forum became an *obligatory passage point* (Clegg 1989; Clegg et al. 2006) for the construction of facts and policy statements.

The NPD had succeeded in creating a regulatory web where the actors were embedded and interests fixed.

The RNNP pilot project later enrolled the research community with the goal of establishing a methodology and a regular mapping of the safety levels. A reference group for the project was established in which all the stakeholders were invited to participate. However, the NPD published the report in its own name and had the power to comment and edit the content. In retrospect, one might say that the RNNP was an important step in the formulation of a new discourse on safety, but it can also be described as a *hybrid* of research and politics. The NPD is responsible for inviting tenders for the collection and analysis of the survey data, and it selects the appropriate topic for the qualitative part of the project. Statistics take the lead, and these statistical indicators tend to become fixed in order to be able to follow trends in the safety levels. This illustrates how power is inscribed into actors and material artifacts. The concerted actions to establish new arenas for collaboration and to build common perceptions of the state of safety may have helped to build a durable actor-network.

Many of the new arenas that were created, such as Safety Forum and the reference groups of RNNP and other research projects, tended to recruit participants from a small population of dedicated persons in trade unions, employers' federations, oil companies and the NPD. As a consequence, these people met each other frequently and had the opportunity to get familiar and develop informal and pragmatic styles of interaction.

12.7.4. *Converging Sense-Making and Reinforcement of Organisational Structures*

Karl Weick's (1993) study of the deterioration of organisational structure and sense-making in a smokejumper crew during the Mann Gulch forest fire provides a clue for understanding *the combined impact* of the new arenas for collaboration and the RNNP project. In the referred case, the smokejumper crew was trapped in a canyon with only a slim chance of escaping the rapidly approaching fire front. Their foreman improvised a new survival technique by lighting a fire and ordering the crew members to lie down in the area that had burned. No crew members obeyed the order. The foreman survived because the burnt-out area created a hole in the passing fire. Only two of the fifteen smokejumpers who tried to run away from the fire survived.

Weick suggested that the organisational structure of the smokejumper crew disintegrated as the crew members were unable to make sense of the orders they received. At the same time, sense-making processes were affected as the group structure disintegrated. The crew members' failure to obey the order of the foreman was thus the consequence of a process of mutual destruction of sense-making and structure in the group.

The revitalisation of tripartite collaboration in the Norwegian petroleum sector may be regarded as the opposite process to the mutual destruction of sense-making and structure that Weick (1993) found in the Mann Gulch disaster. The regulatory authorities promoted a convergence of sense-making processes by initiating the RNNP project. At the same time, they reinforced the structures for collaboration by establishing Safety Forum. The industry responded by establishing a new tripartite arena, Working Together for Safety. The revitalisation of tripartite collaboration may thus be viewed as a process where converging sense-making and the development of organisational structures for collaboration mutually reinforced each other.

12.7.5. *Who Regulates?*

Regulation may be conceived as *an activity that exerts control or influence on the regulated industry, technology, system or activity on behalf of society* (Baldwin and Cave 1999; Hopkins and Hale 2002). In a formal sense, the Norwegian state regulates HSE in the petroleum sector. However, ANT invites us to seek a more complex answer by thinking of regulation as an effect produced by a patterned network or constellation of actors. To answer the question "who regulates", we need to identify the actors that contribute to the *effect* that we may call "regulation of risk": What actors contribute to achieving the *control* or *influence* on HSE conditions and HSE performance that the legislators seek to attain?

We have identified numerous efforts by some actors to enroll other actors during the controversies concerning the risk level and the ensuing efforts to revitalise tripartite collaboration. All these actors should be considered candidates when we try to determine "who regulates". Moreover, the answer does not have the form of a list of isolated actors. Rather, the answer should be sought in the dynamic and patterned network of *relations* that are created, maintained and changed by the actors.

According to this logic, the parliament, the ministry and the NPD still regulate, but they are no longer alone. The industry regulates through its participation in the development of new regulations, but also through its compliance with, negligence of or resistance to the regulations. The trade unions regulate, for example, through their lobbying activities to establish Safety Forum. Thus the tripartite collaboration, with its episodes of conflicts, negotiations and joint action, regulates HSE in the Norwegian petroleum industry.

The exact delimitation of "who regulates" – that is, what actors and relations contribute to regulation – is open to controversy and alternative interpretations. It is not possible to determine the causal impacts of each action by each actor by way of experiments or other "objective" methods. According to Latour (2005, p. 46), "to use the word 'actor' means that it's never clear who and what is acting when we act, since an actor on stage is never alone in acting. … By definition, action is *dislocated*. Action is borrowed, distributed, suggested, influenced, dominated, betrayed, translated." Conceptualising regulation as an actor-network thus raises an uncertainty about *who* regulates. The ANT perspective also implies that the regulatory regime no longer has obvious boundaries. We are not allowed to stick to a preconceived delimitation of a regulatory regime, for example, that "the state" or the NPD regulates. Rather, we have to trace the dynamics of actor-networks by tracing translation efforts and their effects. In times of confrontation, some actors may drop out of the networks, and others may be enrolled. The regulatory regime is not necessarily a relatively bounded system with a high degree of continuity over time. Stability has to be demonstrated and explained rather than taken for granted or implied by definition.

12.7.6. *ANT and Discourse Analysis as a Research Approach*

The choice of research approach has extensive implications for the identification of research issues, the conduction of data collection and analysis and the interpretation and application of results. By choosing a research approach that combined ANT and discourse analysis, we committed ourselves to "following the actors", be they human or non-human, without regard for predefined boundaries to concepts such as "regulatory regimes". This allowed us to capture the extensive mobilisation of new actors during the controversy and to reconsider the issue of "who regulates". The approach helped us to highlight the significance of ambiguities and rhetoric in power games. We were also able to capture both the emerging conflict and the ensuing revitalisation of tripartite collaboration in a coherent account.

This analysis approach does not provide simple answers to questions such as "what made the parties resume collaboration"? Our best answer to such questions is not a single action or condition, but the whole account, including the dynamic relations between the actors. In this way, the analysis pays respect to the complexity of the processes studied, but it does not provide for simple generalisations or straightforward predictions. Some readers may find this frustrating, whereas others may see such complexity as a hallmark of social processes. The latter may find that the logic of ANT, with its focus on enrolling new actors through translation, resembles the thinking of effective strategists facing a need to build a network of alliances to accomplish their goals.

12.7.7. *Implications for the Regulation of Risk*

The actor-network perspective and discourse analysis may help us to understand how regulatory regimes can achieve robustness when faced with changes in their environment. The HSE regulatory regime in the Norwegian petroleum sector managed to reverse or halt a negative trend in HSE conditions and tripartite collaboration by enrolling numerous new actors. The robustness of a regulatory regime is therefore not just a matter of keeping the current actor-networks intact. It is also a matter of enrolling new actors or redefining the roles of current actors as the environmental conditions facing the industry and the regulatory authorities change. The recent industry focus on corporate governance and corporate social responsibility, for instance, points to new actors which conceivably may be enrolled to back up the regulatory regime, but which may also take on their own agendas in ways that run counter to the HSE objectives of the authorities.

Tripartite collaboration can be described in terms of formal rights and obligations and associated practices. We found, however, that the actions of the parties went far beyond the exercise of formal rights and obligations. The trade unions took a broad range of actions to reverse what they perceived as an erosion of HSE conditions. For instance, they mobilised the political authorities. The NPD also went beyond a narrow, formalistic approach, for instance, when they addressed the licensees with their concerns about the HSE conditions or when they threatened to introduce more detailed regulations. Norwegian Oil and Gas Associations also went far beyond formalism when they initiated Working Together for Safety. The actions that went beyond the exercise of formal rights and obligations were crucial for the revitalisation of HSE work and tripartite collaboration.

We have argued that the regulatory system is not stable – new actors are enrolled, while others may resist the current order. Does this imply that regulatory regimes, in the broader sense outlined here, need to be managed? In the present case, the political authorities did intervene. The Minister for Local Government and Regional Development responded positively to an initiative from a trade union (NOPEF) to establish a safety council for the oil industry. Our analysis suggests that this apparently minor intervention may have been a crucial step in the efforts to revitalise tripartite collaboration on HSE. Other interventions include questions raised by parliamentarians to the responsible minister. This suggests that the role of the political authorities in creating a robust regulatory regime is not limited to top-down management through instructions or prescriptions. They may, on some occasions, need to respond to proposals or challenges and allow themselves to be enrolled into actor-networks created by other actors.

Conceiving regulatory regimes as actor-networks also has implications for proposals to transfer elements of a regulatory regime to a new context. It is hard to predict what happens when a small segment of an actor-network is transferred to a new setting, outside its original actor-network. We do not deny that adopting elements from other regulatory regimes can lead to favourable results, but one should be prepared for surprises when a new set of actors "twist and bend" the elements of the regulation regime to make them suit their own purposes. This reservation also applies to generalisations from the present study. We do not claim that a blueprint of tripartite collaboration in accordance with "the Norwegian model" can be transferred to other countries to achieve a similar robustness of the regulatory regime (Hart 2007). The capacity of the parties to fight their way towards revitalised collaboration depended on skills and mutual expectations that have been developed through many decades of boxing and dancing (Levin et al. 2012). It also depended on a balance of power and dependencies where no party could afford to ignore the actions of the other parties. Regulatory regimes in other countries and cultures may draw on other resources to give a voice to HSE concerns, join forces, create shared visions across stakeholder groups and establish new practices.

Are there "good" and "bad" (or functional and dysfunctional) discourses when it comes to the governance of risk? To answer this question, we should realise that power is embedded in discourses (Foucault 1972; Fairclough 1995). A hegemonic discourse may support a particular worldview to the extent that it is taken for granted and becomes invisible to the people participating in the discourse. Alternative worldviews are excluded as unthinkable, irrelevant or irrational. In the context of risk governance, a discourse may become dysfunctional to the extent that it excludes other relevant discourses. Clashes between competing discourses, as we saw in an early phase of the controversy on the risk level, may be the first step in a process of breaking out of a hegemonic discourse and opening the field to alternative worldviews. From this point of view, the most destructive discourses may be those that are totally self-contained and immune to challenges raised by alternative discourses or contradicting facts. Discourses that allow for interpretive flexibility, that do not claim monopoly and that recognise contradicting facts and contradicting worldviews would appear more functional (Bråten 1983; Schulman 1993).

We should also recognise that the "functionality" of a given discourse depends on the context. The highly personal discourse introduced by the offshore worker's mother after the fatal accident, with its focus on ethical/judicial aspects, may have speeded up the process towards a revitalisation of the tripartite collaboration. In other contexts, similar discourses may contribute to scapegoating or create distrust.

How do the results of this study relate to the generic risk governance framework presented by Renn in Chapter 1 of this volume? The participants in the conflict did not themselves refer to the generic risk governance framework or a similar normative framework. A mapping of the controversy to the risk governance framework is thus an interpretation imposed on the actors and their actions by the researchers. With this reservation in mind, we may notice that the conflicts analysed in this chapter did not concern the acceptability of the activity (oil production on the Norwegian continental shelf) as such. Rather, the conflict concerned the continued management of risks that the parties considered tolerable – that is, acceptable to the extent that the risks were properly managed and adequate risk reduction measures were implemented. The conflict illustrates that the challenges related to the stages of "interdisciplinary risk estimation", "evaluation", "management" and "monitoring and control" are not resolved once and for all. Controversies occurred as the oil industry changed its practices in response to environmental conditions such as fluctuating oil prices and less prosperous oilfields. New arenas for risk communication and participation (e.g., Safety Forum, Working Together for Safety) were important means to revitalise tripartite collaboration and build new trust, but so was a new regime for interdisciplinary risk estimation (RNNP) as well.

Normative models of risk governance regimes may, intentionally or unintentionally, convey the idea that risk governance processes should be masterminded by an actor that stands outside the controversies. This is not what happened when the tripartite collaboration on HSE was revitalised. Rather, the controversies as well as the steps to revitalise collaboration were initiated by the stakeholders themselves. The political and regulatory authorities did intervene, but this was mainly in response to being enrolled by the stakeholders. The regulatory regime, with tripartite collaboration as an integral part, thus displayed an amazing capacity for self-correction. We propose that this capacity for self-correction is an important contribution to the robustness of the regulatory regime in the Norwegian petroleum industry. Both the ability to engage in a conflict when HSE was under pressure, and the capacity to subsequently join forces and revitalise collaboration, both boxing and dancing, were crucial aspects of this robustness. Controversies such as those discussed by Kringen (Chapter 11) in this volume are not necessarily symptoms of a regulatory regime in deterioration. On the contrary, prolonged absence of controversy in a field with strong and real conflicts of interest should be taken as a warning sign that the regulatory regime may be losing its capacity for self-correction.

12.8. CONCLUSIONS

We have analysed the erosion and subsequent revitalisation of tripartite collaboration on HSE in the Norwegian petroleum sector around 2000. The regulatory regime was conceptualised as an actor-network, and we used a discourse analysis approach to pinpoint translation efforts associated with the enrolment of new actors in the controversy. This approach allowed us to make sense of a historical development that would be difficult capture if we had restricted ourselves to viewing the regulatory regime as a relatively bounded and stable system of rules, incentives, monitoring and sanctioning of deviations.

The controversy concerning the risk level and the subsequent efforts to revitalise tripartite collaboration led to a rapid expansion of the actor-network dealing with safety in the petroleum sector. We suggest that this capacity and willingness to enrol new actors was a prerequisite for the revitalisation of tripartite collaboration. The actors used a variety of rhetoric devices to promote their positions during the controversy. We identified different discourses that embedded contrasting epistemological and ethical premises.

The durability of the revitalised tripartite collaboration was achieved mainly through mechanisms that facilitate, stimulate or provoke certain action patterns, rather than by enrolling actors that physically restrained action. Arenas such as Safety Forum, for instance, provided a setting where the parties could challenge each other and where they had the opportunity to develop informal and pragmatic styles of interaction. By establishing arenas such as Safety Forum and at the same time promoting the formation of common perception of the state of safety in the petroleum sector, the authorities promoted a process where converging sense-making and the development of organisational structures for collaboration mutually reinforced each other.

The question of "who regulates" does not have a simple answer when we conceptualise a regulatory regime as an actor-network. The parliament, the ministry and the NPD regulate, but a number of other actors also contribute to the effects that we collectively label HSE regulation. Thus the tripartite collaboration, with its episodes of conflicts, negotiations and joint action, also regulates HSE in the Norwegian petroleum industry.

The robustness of a regulatory regime depends on its capacity to enrol new actors or to redefine the roles of current actors when faced with internal disturbance or changes in the environment. Transferring elements of a regulatory regime to a new context involves transferring a small segment of an actor-network to a new setting, where a new set of actors may "twist and bend" the elements in ways that produce unexpected effects.

We propose that the capacity for self-correction demonstrated in this case is an important contribution to the robustness of the regulatory regime for HSE in the Norwegian petroleum industry. The ability to engage in a conflict when HSE was under pressure and the capacity to subsequently join forces and revitalise collaboration were equally important aspects of this robustness.

REFERENCES

AAIB-N (2001) *Report on the Air Accident 8 September 1997 in the Norwegian Sea, approx. 100 NM West North West of Brønnøysund, involving Eurocopter AS 332L1 Super Puma, LN-OPG, Operated by Helikopter Service AS*. Report 47/2001, Air Accident Investigation Board, Norway. Retrieved 2012–07–20 from http://www.aibn.no/ln_opg_eng_total-pdf?lcid=1033&pid=Native-ContentFile-File&attach=1

Baldwin, R. and Cave, M. (1999) *Understanding Regulation. Theory, Strategy, and Practice.* Oxford University Press: Oxford.

Bråten, S. (1983). *Dialogens vilkår i datasamfunnet.* [The Conditions for Dialogue in the Computer Society.] Universitetsforlaget: Oslo.

Clegg, S.R. (1989) *Frameworks of Power.* Sage: London.

Clegg, S.R., Courpasson, D. and Phillips, N. (2006) *Power and Organizations.* Sage: Thousand Oaks, CA.

Dokument Nr. 8:90 (1999–2000). Forslag fra stortingsrepresentantene Karin Andersen og Hallgeir H. Langeland om å fremme en tilleggsmelding til St.meld. nr. 39 (1999–2000)... Retrieved 2012–03–14 from http://www.stortinget.no/no/Saker-og-publikasjoner/Publikasjoner/Representantforslag/2000–2001/dok8–200001–090/

Dølvik, J.E. and Stokke, T.A. (1998) "Norway: The Revival of Centralized Concertion", in A. Ferner and R. Hyman (eds.): *Changing Industrial Relations in Europe.* Blackwell Publishers: Oxford, pp. 118–145.

Emery, F. and Thorsrud, E. (1969) *Democracy at Work.* Martinus Nijhoff: Leiden.

Engen, O.A. and Lindøe, P.H. (2008) "Atferdsbasert sikkerhet tilpasset norske forhold" [Behaviour-Based Safety Adapted to Norwegian Conditions], in R.K. Tinmannsvik (ed.): *Robust arbeidspraksis. Hvorfor skjer det ikke flere ulykker på sokkelen?* Tapir: Trondheim, pp. 199–211.

Fairclough, N. (1992) *Discourse and Social Change.* Polity Press: Cambridge.

(1995) *Critical Discourse Analysis.* Longman: London

Forseth, U. and Rosness, R. (2010) "Arguments about Safety – Discourse Analysis", Paper presented at the 5th International Conference *Working on Safety*, 7–10 September, Røros, Norway.

Forseth, U., Torvatn, H. and Andersen, T.K. (2009) "Stop in the Name of Safety – The Right of the Safety Representative to Halt Dangerous Work", in S. Martorell, C. Guedes Soares and J. Barnett (eds.): *Safety, Reliability and Risk Analysis*, Vol. 4. Taylor and Francis: London, pp. 3047–3054.

Foucault, M. (1972) *The Archeology of Knowledge.* Routledge: London. Translated from L'Archéologie du Savoir. Gallimard: Paris.

Fune, R.E. (2001) *Safety out of Control?* Retrieved 13 March 2012 from http://www.safe.no/index.cfm?id=272129

(2005) *Frå golvvask til fagleg oppvask.* Interview, 15 March 2005. Retrieved 2 February 2012 from http://www.ptil.no/nyheter/fraa-golvvask-til-fagleg-oppvask-article1862–24.html

Gustavsen, B. (1992) *Dialogue and Development.* Van Gorcum: Assen, The Netherlands.

Hart, S. (2007) "Industry, Labour and Government in Norwegian Offshore Oil and Gas Safety: What Lessons can we Learn?" *The Workplace Review*, November 2007.

Hood, C., Rothstein, H. and Baldwin, R. (2001) *The Government of Risk. Understanding Risk Regulation Regimes.* Oxford University Press: Oxford.

Hopkins, A. and Hale, A. (2002) "Issues in the Regulation of Safety: Setting the Scene", in B. Kirwan, A. Hale and A. Hopkins (eds.): *Changing Regulation. Controlling Risks in Society.* Pergamon: Amsterdam, pp. 1–12.

Hovden, J. (2002) "The Development of New Safety Regulations in the Norwegian Oil and Gas Industry", in B. Kirwan, A. Hale and A. Hopkins (eds.): *Changing Regulation. Controlling Risks in Society.* Pergamon: Amsterdam, pp. 57–77.

Huzzard, T., Gregory, H. and Scott, R. (eds.) (2004) *Strategic Unionism and Partnership: Boxing and Dancing?* Palgrave Macmillan: Basingstoke.

Jørgensen, M.W. and Phillips, L. (1999/2008) *Diskursanalyse som teori og metode* [Discourse Analysis as Theory and Method]. Roskilde Universitetsforlag/Samfundslitteratur: Roskilde.

Karlsen, K. (2005) *Engasjert nettverksbygger*. Interview, 15 March 2005. Retrieved 2 February 2012 from http://www.ptil.no/nyheter/engasjert-nettverksbygger-article1861–24.html

LaPorte, T.R. and Consolini, P.M. (1991) "Working in Practice But Not in Theory: Theoretical Challenges of 'High-Reliability Organizations'", *Journal of Public Administration Research and Theory*, 1:19–47.

Latour, B. (2005). *Reassembling the Social. An Introduction to Actor-Network-Theory*. Oxford University Press: Oxford.

Law, J. (1992) *Notes on the Theory of the Actor Network: Ordering, Strategy and Heterogeneity*. Centre for Science Studies, Lancaster University: Lancaster (Revised in 2001 and 2003). Retrieved July 26, 2012, from http://www.google.no/url?sa=t&rct=j&q=john%20law%20notes%20actor%20network%20theory&source=web&cd=1&ved=0CFkQFjAA&url=http%3A%2F%2Fwww.lancs.ac.uk%2Ffass%2Fsociology%2Fpapers%2Flaw-notes-on-ant.pdf%2F&ei=sSkRUPWAPNP24QTHhIDoCQ&usg=AFQjCNHogijPLfdMjV5RWeY25uEBYQuJwA

Levin, M., Nilsen, T., Ravn J.E. and Øyum, L. (2012) *Demokrati i arbeidslivet – Den norske samarbeidsmodellen som konkurransefortrinn* [Democracy in Working Life – The Norwegian Model as Competitive Advantage]. Fagbokforlaget: Bergen.

Moen, A., Blakstad, H.C., Forseth, U. and Rosness, R. (2010). "Disintegration and Revival of Tripartite Collaboration on HSE in the Norwegian Petroleum Industry", in R. Briš, C. Guedes Soares and S. Martorell (eds.): *Reliability, Risk and Safety. Theory and Applications. Vol.* 3. Taylor & Francis: London, pp. 2191–2198.

Næss, A. (2000) Brustad sier ja til sikkerhetsråd. [Brustad Says Yes to a Safety Council]. *Dagens Næringsliv*. Oslo, 4 July, p. 11.

Norsk Telegrambyrå (NTB) (2000). Grådigheten truer liv og miljø i Nordsjøen. [Greed Threatens Life and Environment in the North Sea]. *Norsk Telegrambyrå*. Stavanger. 14 November.

Norwegian Petroleum Directorate (1998). *Annual Report* 1997. Stavanger.

Norwegian Petroleum Directorate (2001). *Utvikling i risikonivå – norsk sokkel. Stavanger. Sammendragsrapport*. [Development in Risk Level – Norwegian Shelf. Summary Report]. Norwegian Petroleum Directorate: Stavanger.

Norwegian Oil and Gas Associations (n.d.) *Olje- og gasshistorien*. [The Oil and Gas History]. Retrieved 15 March 2012 from http://www.olf.no/no/Faktasider/Oljehistorie/

NOU (1977) *Ukontrollert utblåsning på Bravo 22. april* 1977 [Uncontrolled Blowout at Bravo on 22 April 1977]. Green Paper No. 47. Universitetsforlaget: Oslo.

NOU (1981) *Alexander L. Kielland-ulykke* [The *Alexander L. Kielland* Accident]. Green Paper No. 11. Oslo: Universitetsforlaget.

NOU (1999) *Analyse av investeringsutviklingen på kontinentalsokkelen*. [Analysis of the Development of Investments on the Continental Shelf]. Green Paper No. 11. Statens forvaltningstjeneste: Oslo.

Ognedal, M. (2005) *Helhetsfokus på jobb – millimeterfokus i fritiden*. Interview, 15 March 2005. Retrieved 2 February 2012 from http://www.ptil.no/nyheter/helhetsfokus-paa-jobb-millimeterfokus-i-fritiden-article1858–24.html

Øyum, L., Skarholt, K., Ravn J.E. and Nilssen T. (2010) "The Industrial Relations of Safety – Differences in Tripartite Collaboration in Norwegian Industries", in R. Briš, C. Guedes

Soares and S. Martorell (eds.): *Reliability, Risk and Safety: Theory and Applications*. Vol. 2. Taylor & Francis: London, pp. 1277–1284.
Pfeffer, J. (1992) *Managing with Power. Politics and Influence in Organizations*. Harvard Business School Press: Boston, MA.
PSA (2007) *Utvikling i risikonivå – norsk sokkel. Fase 7 hovedrapport 2006*. [Development of Risk Level – Norwegian Shelf. Phase 7 Main Report 2006]. Report Ptil-07–01, Petroleum Safety Authority – Norway.
PSA (2012) *Risikonivå i petroleumsvirksomheten. Hovedrapport, utviklingstrekk 2011, norsk sokkel*. [Risk Level in the Petroleum Industry. Main Report, Trends 2011, Norwegian shelf]. Petroleum Safety Authority – Norway.
Qvale, T.U. (1985) *Safety and Offshore Working Conditions. The Quality of Work Life in the North Sea*. Universitetsforlaget: Stavanger.
Rosness, R., Blakstad, H.C., Forseth, U., Dahle, I.B. and Wiig, S. (2012) "Environmental Conditions for Safety Work – Theoretical Foundations", *Safety Science*, 50 (10): 1967–1976.
Ryggvik, H. (2003). *Fra forvitring til ny giv. Om en storulykke som aldri inntraff?* [From Disintegration to New Initiative. About a Major Accident That Never Happened]. TIK working document. 26/2003. University of Oslo: Oslo.
(2008) "Verneombudet, kollektivet og sikkerheten" [The Safety Delegate, the Collective and the Safety], in R.K. Tinmannsvik (ed.): *Robust arbeidspraksis. Hvorfor skjer det ikke flere ulykker på sokkelen?* Tapir: Trondheim, pp. 149–164.
Schulman, P.R. (1993) "The Negotiated Order of Organizational Reliability", *Administration & Society*, 25 (3): 353–372.
St.meld. nr. 39 (1999–2000) *Olje- og gassvirksomheten*. [Oil and Gas Activities.] White Paper. Ministry of Petroleum and Energy: Oslo.
nr. 7 (2001–2002). *Om helse, miljø og sikkerhet i petroleumsvirksomheten*. [About Health, Environment and Safety in the Petroleum Sector]. White Paper. Ministry of Work and Administration: Oslo.
Thorvaldsen, K. (2005) *Erfaren nykommer i OLF*. Interview, 15 March 2005. Retrieved 2 February 2012 from http://www.ptil.no/nyheter/erfaren-nykommer-i-olf-article1857–24.html
Wathne, L.E. (2002) "Dødsulykker må aldri skje" [Fatal accidents must never happen]. *Dråpen*, no 1, 2002, 12–15.
Weick, K. (1993) "The Collapse of Sensemaking in Organizations: The Mann Gulch Disaster", *Administrative Science Quarterly*, 38: 628–652.
Wetherell, M., Taylor, S. and Yates, S.J. (2001) *Discourse Theory and Practice*. Sage: London.
Working Together for Safety (Samarbeid for Sikkerhet). (2001). *På vei mot en felles HMS-kultur*. [On the Way Towards a Common HSE-Culture]. Pamphlet.

13

Emergent Risk and New Technologies

Ole Andreas Engen

13.1. INTRODUCTION

This chapter discusses how the Nordic model or Nordic practices have been redefined through technological transformations and how cost efficiency programmes have challenged the institutional mechanism that keeps the tripartite system intact. The discussion concentrates on the technological transformation process that took place on the Norwegian shelf during the 1990s and show how the NORSOK programme, which was built partly on the institutions of the Nordic model, actually threatened to destroy that same model from within. The technological transformation process connected to the NORSOK programme created the premises for the risk regulation of the Norwegian petroleum industry we see today, but revealed vulnerabilities and dysfunctional elements of the system which are still present. On a more general level, the chapter highlights how legitimacy of safety regulation is being put under pressure when one set of values, namely economics, are being given more relative weight than other sets of values, namely environment and safety.

The concept of Nordic model has different connotations depending on perspective and level of analysis. According to Moene et al. (2009), the Nordic model is a way of organising society combining social security and capitalist dynamics in a distinctive way. From a petroleum industrial point of view, the Nordic model implies specific ways of organising production offshore, chief amongst them the instructions that the petroleum industry has to accept the same rules of the game as the rest of the Norwegian industries. During the 1970s and the 1980s, the increased acceptance and influence of the Nordic model coincided with the establishment of certain technological pathways offshore. This meant a harmonising of the workplace environment, which implied that the working conditions offshore were subjected to the same legal framework as for the industrial workers onshore (Ryggvik 2008). It also unified different groups concerning technological choices that created a basis for agreement on how to organise very large development projects offshore and to utilise them in a broader socio-economic perspective (Engen 2009). The practice of the Nordic model thus mirrors the establishment of specific technological pathways that dominated the Norwegian petroleum industry from the 1980s to the beginning of the new century (Engen et al. 2012).

Rosness and Forseth (Chapter 12) in this volume describe how the Norwegian petroleum industry at the turn of the century was characterised by a deep controversy between the union representatives and industry representatives concerning the HSE conditions on the Norwegian continental shelf. The atmosphere between the different groups was tense and threatened to destroy the tripartite collaboration. However, a revitalisation of the Nordic model took place and led the tripartite collaboration back on track, which, according to Rosness and Forseth, revealed the robustness of the safety practices within the Nordic model. In this chapter we argue in favour of two additional reasons why the tripartite collaboration was threatened during the 1990s. The first is attributable to the liquidation of the industrial paradigm connected to the Condeep technology, and the second is attributable to an extensive cost-cutting programme – NORSOK. Both reasons are deeply interwoven because the established technological methods had to be changed in order to reach the goal of NORSOK.

The chapter is organised as follows. First we introduce such key notions as governance structures, institutional mechanisms and technological pathways. This is followed by a conceptualising of the tripartite collaboration (Figure 13.1), which illustrates how the Nordic model presumes a collaborative structure on regulative, industrial and working-life levels as well as at different workplaces and individual levels. In this section we also briefly explain the two main concepts, namely the Condeep technology and the NORSOK programme. Thereafter we go through the main phases of the NORSOK programme and discuss its technological consequences and how it affected the HMS conditions on the continental shelf. Through this discussion we give a justification of why the tripartite collaboration deteriorated during the 1990s, and present the Nordic model as a robust regulatory regime. Finally we comment on the vulnerability of trading off one set of values against another set of values within the tripartite collaboration

13.2. GOVERNANCE STRUCTURE, INSTITUTIONAL MECHANISMS AND REGULATION

We argue that the regulatory practice of the tripartite system and technology can be understood in a dialectical relationship. Arguments for a closer relationship between technology and regulatory practices are also underlined in definitions of technology which comprise regulatory elements. According to Hughes (1987), technological systems include technical devices, the organisational routines and procedures, legislative artifacts and scientific and other knowledge elements such as skills, rules of thumb and

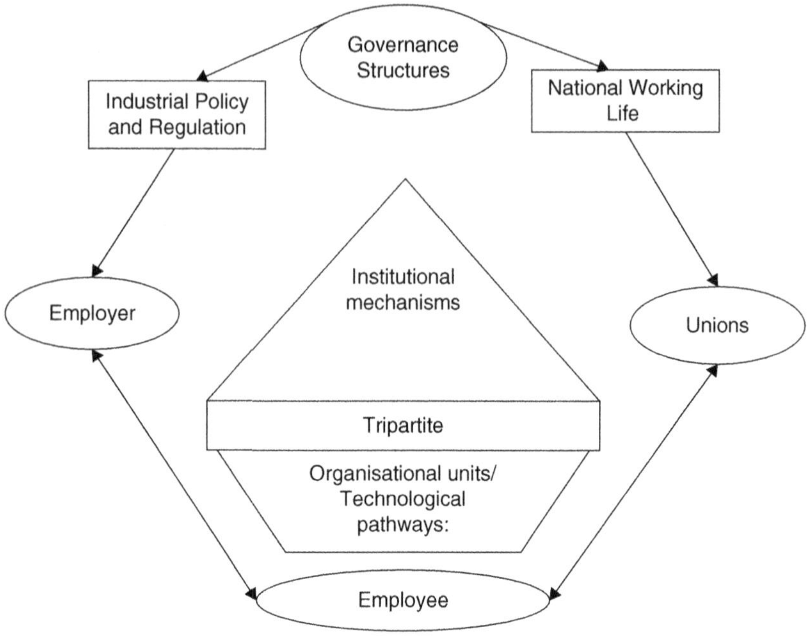

FIGURE 13.1 The socio-technical framework of governance, institutions and practices of the Nordic model.

norms for the handling of the technology. These elements are not distinctive or autonomous factors, but form a seamless web that constitutes technological pathways.

There have been many efforts to build complex models of risk regulation on the basis of socio-technical theory and system theory (Rasmussen 1997). Accordingly, in this chapter, risk regulation as practice derived from the tripartite collaboration is considered to be shaped by institutional structures and arrangements as well as social processes. All actors, as well as risk regulators, within the context of technological systems are influenced by institutional procedures, principles, expectations and norms encountered in cultural and historical frameworks. In Hughes's definition of technological systems, all activities directly coupled with the petroleum production in the North Sea could thus be regarded as a technological system (Engen 2002; Olsen and Engen 2007) aimed at producing, transporting and selling crude oil and natural gas in a safe manner. Within this technological system, a technological pathway could be regarded as an integrated unit with a limited purpose supporting the overall objective, for example, a platform design.

In our context – the socio-technical system of the Norwegian Petroleum Industry – institutional aspects of technology and the Nordic model can be expressed through the following figure (Figure 13.1).

Figure 13.1 conveys a broad institutional perspective on industrial policy, regulation, working life and technological pathways. The integrated organisational unit – department, installation, plant, and the like – follows traditional socio-technical models of risk regulation where the organisational units combine the technical production system with the safety management system (Olsen and Lindøe 2009). The diamond-like figure in the middle illustrates feedbacks where information and effects from industrial policy and risk regulation are being handled through a complex set of institutional mechanisms. These institutional mechanisms consist of actors i.e. companies, unions, employers associations and governmental organisations, and are constituted by their mutual relationships.

Governance structures and institutional mechanisms concerning safety are composed of a hierarchy of regulations imposed by, amongst others, legislators, regulatory agencies and field inspectors (Hood et al. 2001). In Norway this role is partly taken up by the Petroleum Safety Agency (PSA; see Textbox III). In general, governance structures refer to acts of public agencies and corporations in order to implement regulations, routines and procedures to enhance safety performance in a socio-technical system. By institutional mechanism we refer to prevailing norms, scripts and guidelines which are embedded in and between organisations and governmental institutions. The institutional mechanism filters and influences the behaviour of the heterogeneous group of organisational actors which all have different capacities in their influence on safety, environmental issues and so forth. In case of major accidents, the institutional mechanisms may serve as social amplifiers of the feedback effects where media pressures, investigation committees and interactions between safety agencies and organisations create a web of relations which may affect technical choices and technological design processes (see Textbox I). Major accidents followed by external pollution may therefore influence technological design and make it more robust because such accidents affect the society at large and attract immediate negative attention of the market and the public opinion.

13.3. THE INDUSTRIAL POLICY AND REGULATION IN THE CONDEEP ERA

13.3.1. *The Condeep*

The technology that came to dominate the oil and gas production on the Norwegian continental shelf until the end of the 1990s was large integrated platform constructions – the Condeep (abbreviation of a "concrete deep water structure"). Condeep refers to a type of gravity-base structure for oil platforms. A Condeep usually consists of a base of concrete oil storage tanks from which one, three or four concrete shafts rise. The Condeep itself always rests on the sea floor, and the shafts rise to about 30 metres above sea level. The platform deck itself is not a part of the construction. This gravity-base structure for a platform was unique in that it was built from reinforced concrete instead of steel, which was the norm up to that point. The platform was made for the heavy weather conditions and the great water depths found in the North Sea.

The Condeep design was also closely connected to the Norwegian State Oil company, Statoil. Statoil became the main instrument of the state in developing Norwegian petroleum competence and shaping a particular technological pathway, relying on the Norwegian concession system to strengthen its dominance. Agreements regarding training and transfers of knowledge and technology from other companies were negotiated, and Statoil itself took the role of intermediary in delegating tasks to the Norwegian industry. As the Norwegian oil companies and main suppliers gradually built up their competence in the shelter of a protectionist policy, the public administrative apparatus for oversight of the oil and gas industry matured.

The decline in oil prices during the 1980s was a shock to the whole petroleum industry (Engen 2002). The main industrial actors were strongly encouraged to co-operate in order to develop new cost-saving technologies by forming NORSOK (an abbreviation of *NOR*sk *SO*kkels *K*onkurranseposisjon [The Competitive Situation of the Norwegian Shelf]), an industrial program for development of new technologies and standards, organisational development and new contractual relations, regulations and new initiatives for co-operation and negotiations between oil companies and their suppliers. The main objective was to reduce average costs of oil and gas extraction by as much as 50 per cent. The program was inspired by the similar British initiative, CRINE (*C*ost *R*eductions *In* a *N*ew *E*ra).

NORSOK represented a historical and institutional break with the protectionist practices of the Norwegian concession system and the technology policy of the 1970s and 1980s. This also implied a political shift from an active and interventionist oil policy to a more passive one that sought to link various actors rather than dictating terms to them. NORSOK, which began in 1993, introduced a process that allowed both oil companies and main suppliers to enjoy greater freedom when choosing technological concepts, subsuppliers, location of bases, headquarters and so on.

13.3.2. *Condeep and the Nordic Model*

The Condeep era in the 1970s and 1980s coincided with the institutionalising of the Nordic model in three ways. First, it created workplaces offshore on a larger scale than it did before. Second, the Condeep design unified the interests of industrial groups and the government because it joined together political goals of building up the Norwegian workforce and competence with financial goals of increased revenues from the large offshore fields. Third, the platform design satisfied regulatory requirements concerning safety and risks. Indirectly we may say that the Condeep regime was responsible for launching the two most important institutional improvements in the Norwegian working life after the Second World War: the working environmental act and the internal control reform. Both reforms endeavoured to modernise the Norwegian working life in general and to improve safety performance in particular, and both reforms significantly stimulated and challenged the tripartite collaboration on the Norwegian shelf.

The Bravo blowout in 1977 and the *Alexander L. Kielland* accident in 1980 had created an enormous pressure on all participants to reorganise the safety work and to develop technological arrangements in order to reduce the risk level and at the same time increase and adjust the production in accordance with the governmental and corporate goals (see Chapter 8). The safety department in the Norwegian Petroleum Directorate (NPD; see Textbox III) experienced an increase in self-confidence when it achieved a certain level of influence on the technological design at Statfjord B. During platform's construction, the NPD instructed the operating company (Mobil) to build the living quarters as an isolated unit beside the production quarter. After intensive negotiations, Statoil and Mobil gave in, and the NPD could satisfactorily declare that safety regulations definitely could have an impact on technological designs.

TEXTBOX I

In the 1970s, the NPD was accused of having written the most expensive letter in Norwegian oil history, when in November 1976 it said no to the construction of the Statfjord B platform as a copy of the Statfjord A. The NPD believed that platform A was built as a gas stove and should have living quarters on a separate platform for safety reasons. The opposition was massive. Mobile, which at that time was the operating company, feared it would be forced into dual solutions elsewhere in the world as well. The solution was one platform with a clearer separation between residential and processing systems. However, this led to a one-year postponement in construction. The NPD was subjected to severe pressure from all sides, not least because the shipyards along the Norwegian coast were experiencing an acute shortage of orders. The media, politicians, representatives and business leaders were all criticising the decision.

The story of the living quarters on Statfjord B is the most spectacular historical example of safety regulations directly shaping technological designs on the Norwegian continental shelf, where the NPD actively managed to utilise governance structures and institutional mechanisms to force the oil companies to change the technological choices. More important perhaps is the role that the NPD was assigned during the 1980s, as a main coordinator with other regulators in order to develop an integrated and unified set of regulations concerning safety, working environment, natural environment and occupational health. Regulating the rotation schemes and the introduction of the barrier philosophy became part of the same multi-purpose approach taken by the NPD. From a socio-technical point of view, the role played by NPD illustrates how such a regulatory agent may utilise governance structures and institutional mechanisms on different levels to shape technological decisions processes, as described by Bang and Thuestad (Chapter 10) in this volume.

13.3.3. *Working Life – Technology and WEA*

The implementation of the Working Environmental Act (WEA) in 1977 marks a breakthrough of the Man-Technology-Organisation (MTO) perspective concerning regulation of the workplace conditions in accordance to health, safety and environment in Norway. The MTO perspective demands that the technology should be adjusted to the individual worker and not the opposite. According to Figure 13.1, the WEA redefines governance structures, has an impact on the relationships between public agencies and private companies and alters the institutional mechanisms affecting interaction and norms of co-operation between the parties. Several studies in the 1980s and 1990s confirmed how the implementation of the WEA shaped the conditions of the workplace by adjusting and adapting the technical environment. Increased focus on workplace ergonomics and "the appropriate technology movements" significantly illustrate increased efforts to satisfy the requirements of the MTO perspectives on both offshore and onshore facilities (Ryggvik 2008).

TEXTBOX II

The purpose of the Norwegian Working Environment Act of 4 February 1977 is to ensure a work environment that gives employees full safety from physical and mental harm with a protective technical, industrial hygiene and welfare standards at all times in accordance with the technological and social developments in society. Moreover, the law ensures safe employment and meaningful work for the individual employee. Finally, the law provides a foundation for businesses to solve their problems working in partnership with the social organisations and the control and supervision of public authorities.

Hence, the WEA was a milestone for the socio-technical thinking on risk and safety regulation in the Norwegian petroleum sector. According to Ryggvik (2008), the WEA was not only an institutional and organisational innovation concerning safety management in general, but also a final condemnation of management principles derived from more behaviourally oriented safety philosophies. The behaviour-based safety (BBS) approach has its roots in the early stages of safety management and regulation in United States. Combinations of psychological, engineering and Taylorism-inspired organisational studies gave birth to an intricate system of instructions to make the individual worker internalise behavioural patterns that were considered safe. A few studies also confirm positive effects of such safety regimes (Krause et al 1999). Ryggvik (2008), however, argues that BBS neglects the complex relationships between the individual and the technological system and transfers the responsibility for safety from the managers to the individual workers, thus creating scapegoats when incidents and accidents happen. According to Ryggvik, BBS thus represents the opposite of the basic values of the WEA.

Although the WEA in hindsight is well known for encouraging, initiating and implementing democratic norms and institutions in the workplaces, Ryggvik (2008) also emphasises how the act underlines the systemic approach and unambiguously puts focus on socio-technical factors in order to develop a safe working environment. In principle this means that risk regulation should put human conditions before technological advances. Ryggvik (2008) also confirms how technological development in the 1980s followed the main principles of the risk regulatory system led by the NPD. In most cases the technological conditions in individual workplaces on the platforms were formed and shaped with the help of concrete sections of the law. Technical devices that were constructed in order to minimise noise, vibration and other strains are just a few of many examples. This was also the period when technology and working life became a new field for psychologists and other social scientists.

The systemic approaches of the WEA support our perspective of compatibilities between technology and tripartite collaboration during the 1980s. Ryggvik (2008) confirms that regulative aspects in certain periods of the development of the Norwegian safety regime had consequences for the technological development. However, it is important to point out that this mechanism is not direct or causal. It is more accurate to say that the WEA, through how it changed the institutional settings, altered and affected the technological decision-making procedures. According to an institutional perspective, it is thus reasonable to argue for a certain relationship between the co-operative norms of the Nordic model, the Condeep technology and risk-regulating aspects of the WEA (Karlsen and Lindøe 2006; Lindøe and Engen 2012).

13.4. THE NORSOK PROCESS AND THE TECHNOLOGICAL TRANSFORMATION OF THE NORTH SEA

13.4.1. *The NORSOK Program*

At the turn of the century, the socio-technical character of the Norwegian petroleum industry had changed completely. The Condeep paradigm had been liquidated,

Statoil had been privatised, great merger processes had taken place amongst main suppliers and three oil companies were reduced to one when Norsk Hydro bought Saga and thereafter merged its petroleum division with its own. On a general level, the governance structures were altered. Statoil went public, and the Norwegian government withdrew to a more passive owner position, to a certain degree freeing the company of political influence (Engen et al. 2012). Such abdication of power had direct consequences for the technological pathways and thereby more indirectly influenced production, workplace design and safety regulation. The changes of the 1990s visualise the contradiction in terms that existed within the socio-technical system of the Norwegian petroleum industry. On the one hand, the tripartite collaboration constituted institutional possibilities of cost-efficient change management such as the NORSOK program. One the other hand, the cost-efficient changes threatened the pillars on which the tripartite collaboration was based. The NORSOK program thus illustrates the strength but also the vulnerability of the tripartite system.

NORSOK was launched by the Norwegian government in order to improve the competitive abilities of the Norwegian petroleum industry. The backcloth was the oil price decline of the mid-1980s and the economic recession that followed. Both investment rates and expenditures connected to exploration and drilling fell dramatically. In the beginning of the 1990s there was therefore a general agreement amongst the Norwegian government, the largest oil companies and main suppliers about the need to encourage implementation of new technology and organisational designs, new contractual arrangements and improved procedures for accomplishing and launching development projects. The main goal was to reduce the general cost level by 50 per cent.

The architect of NORSOK was the Ministry of Oil and Energy. At the end of the 1980s and beginning of the 1990s, negative price and investment trends persisted. In order to meet the complex and comprehensive challenges that followed in the wake of the new economic conditions, the NORSOK program became organised in seven priority areas: cost analysis, standardisation, new contractual relations, documentation and information, restructuring and transportation, more cost-efficient safety, health and environment management and finally an extensive area discussing framework conditions.

The recommendations from the NORSOK committees may be summed up in the following two points:

> To meet the future challenges facing the Norwegian petroleum industry, effective interaction between all the groups within the sector is required. The government, the oil companies and the suppliers have thus to undergo "a cultural change", meaning a prioritized focus upon increased added value. (Engen 2002)

> The cost increasing impact of the prevailing contractual relations between the suppliers and operating company has to be removed. Effective co-operative organizations between the actors and technical simplifications are going to be established through new contract models and simplifications of the political and economic external conditions. (Engen 2002)

The objective of the first NORSOK organisations (The Committee of Development and Operation and the NORSOK Secretariat) was to unite industrial groups and interests within the socio-technical system. The NORSOK Secretariat and the Collaboration Committee were assigned the tasks of further encouraging and persuading the actors to behave in accordance with the NORSOK recommendations. NORSOK was thus an overall program of action formalised through different organisations assuming a broad co-operation between all relevant petroleum groups in Norway.

13.4.2. *Activating the Tripartite System*

A relevant issue is how it was possible to organise such a program and at the same time attain committed co-operation across organisational and political boundaries. The answer, as argued earlier, is to be found in the institutional character of the socio-technical system at that time. The Condeep paradigm, the protectionist Norwegian policy concerning utilising the petroleum industry for economic development and the increase in employment rate had induced the unions to take an active position in order to involve themselves in industrial development. After a period marked by heavy strikes during the 1970s, the Confederation of the Norwegian Enterprises (NAF) and the Norwegian Confederation of Trade Unions (LO), after pressure from the conservative ministry, had entered into an agreement to further harmonise working life (Ryggvik 2008), which on a more general level was formalised by the Petroleum Activity Act in 1985 (see Chapter 8 in this volume). This also encouraged a stronger collaborative commitment from LO, which at the end of the 1980s accepted a general agreement of reducing the wage and price growth in Norway – the so-called Solidarity Act. It is important, however, to emphasise that LO did not represented all the petroleum workers. During the 1980s and 1990s, the other major union, SAFE, chose to reject these agreements, which to some extent has placed it on the sidelines of the collaboration movement that was taking place.

The new governance structures united most different industrial groups. Governmental actors, especially from the Ministry of Oil and Energy, had realised that the inefficiency of the socio-technical system threatened development of future oilfields and thereby future tax income. The suppliers and oil companies feared lower profit margins resulting from inefficient organisational procedures connected to development projects, unfair sharing of risk premiums and low innovative ability and capacity both amongst the oil companies themselves and amongst the suppliers. LO and the Confederation of Norwegian Enterprises together were equally worried about the economic outlooks for the Norwegian petroleum industry. Accordingly there were common interests of utilising the governmental structures and institutional mechanisms in order to change the entire socio-technical system. Even though the NORSOK program is not a project based on the tripartite system, it was made possible because of the same institutional mechanism, which guaranteed a broader degree of participation and commitment from the actors in the working life than similar initiatives taken on the British continental shelf (i.e. the previously mentioned CRINE).

The NORSOK program significantly challenged risk and safety regulations. Increased cost efficiency has a downside in safety, and the recommendation from NORSOK to change technical and organisational procedures implied new risk profiles of the development projects. In the wake of the Bravo and *Alexander L. Kielland* accidents, the regulatory safety regime took centre stage. NORSOK implied the need to screen all routines, procedures, organisational models and practices. As mentioned earlier, the safety regulation on the continental shelf was based on the WEA, which also was deeply embedded in the ideology of the tripartite collaboration. For certain NORSOK advocates from the employer side it was not obvious that the WEA principles favoured the best economic solutions or the best practices offshore. LO for its part demanded that the restructuring process not on any level should deteriorate HSE regulations or threaten safety offshore. Every time this issue was touched upon during the process, the employee organisations threatened to withdraw from the program. This illustrates that NORSOK had been built on a vulnerable foundation and that a tripartite system requires a certain degree of consensus and power balance between the employee's and employer's sides of the working life. In order to keep such a balance, the government undertook the position of a mediator in critical phases of the program. Accordingly, the employee organisations associated with LO chose to stay in NORSOK until the end of the program in 2001.

13.4.3. *From Integration to Disintegration*

In 1995, NORSOK published seven reports which together suggest a completely new framework of inter-organisational routines and procedures on the Norwegian shelf. At the time the reports were released, all industrial groups seemed satisfied. In particular, the main suppliers dominated by large Norwegian industrial corporations Aker and Kvaerner expressed enthusiasm and great expectations in gaining a larger share of future development contracts in the wake of the NORSOK program. The LO unions were also moderately satisfied because the NORSOK report concerning health and safety issued guarantees that all HSE standards and levels should be kept despite technical, organisational and institutional changes. Practical realities, however, soon removed such optimism and revealed that the interrelated networks of the socio-technical petroleum system are also characterised by strategic interests, alliances and unequal distribution of power – not least when it comes to closely connected issues such as production, financial risk and safety.

One of the most important goals of NORSOK was to introduce new and efficient organisational models handling large development projects – the so-called EPC(I) contracts. Briefly described, such contracts imply that one company is assigned one large contract from the oil company, which includes all phases in a development project: engineering (E), procurement (P) and construction (C). In some cases the contract includes installations (I) as well. The new contractual models required a reorientation amongst the main suppliers, and the large Norwegian suppliers Aker and Kvaerner answered by restructuring their organisations via uniting nationally based fabrication yards, engineering and installation departments. The objective was to gather and fine-tune the competencies in one organisation and to minimise the interfaces between the different departments. The main suppliers believed such reorganising of the contractor market would benefit the fabrication yards along the western coast of Norway, strengthen competence in engineering and contribute to innovations concerning petroleum-based technical solutions.

The main suppliers were disappointed. After the reorganisations there were only three groups of suppliers and two specialised system suppliers of subsea installations that were able to handle the new contracts. Few actors, equal competence and a limited amount of projects created a tight market and induced destructive competition in the supplier market. This favoured the oil companies. Whereas price and delivery schedules are measurable variables, a comparison of technical and organisational competence is difficult. Accordingly the suppliers who could offer the most favourable prices and tight delivery schedules were preferred. To the oil companies this implied lowering the budgets and tightening the deadlines, while to the main suppliers it implied an overestimation of their technical, commercial and organisational capacity. EPC contracts gave the suppliers larger contracts and increased control and responsibilities of the development projects, but also increased risk. The oil companies took a more passive role in projects, keeping general surveillance of the projects and left it to the suppliers to handle unforeseen events and incidents.

The NORSOK program, however, assumed that the oil companies and main suppliers should develop "interaction patterns" in the contractual relations that would result in efficient and mutually beneficial routines and procedures. Enthusiasm from NORSOK and faith in individual technical and organisational abilities led the suppliers to push their capacities to the extreme limit. Marginal price-setting and tough delivery schedules further led to increased internal costs, and even though the suppliers produced at full capacity, the profit margins were decreasing. The oil companies answered by utilising the market situation and further pushing the prices and demanding shorter delivery time. Both actions were inconsistent with "the gentleman's agreement of NORSOK" and threatened safety and existing HSE standards.

While the oil companies extended their hegemony, the suppliers in their eagerness to increase market share ended up in destructive competition. This situation threatened the foundation of the NORSOK program. The tight supplier market created a wide range of negative spin-offs. Within "the Condeep paradigm", the oil companies had organised the sub-suppliers in close and transparent networks. The purpose was to develop relevant competence, ensure sufficient quality of the products and establish predictable procurement procedures. The safety and regulation regime was also a part of this structure. The oil companies were obliged to ensure that their suppliers operated according to the principles of the WEA. The new contractual relations assumed that the main suppliers were given the responsibility for those networks. The new role of the main suppliers therefore also implied coordinating the versatile competence of the heterogeneous group of sub-suppliers and efficiently utilising their competence in the development projects. In addition, they were also given responsibility to maintain the safety and HSE standard. However, these obligations became difficult to fulfil.

13.4.4. *Economic Pressure*

To the small and medium-sized enterprises (SMEs) acting as sub-suppliers offshore and onshore, NORSOK became a painful experience. As a result of the new market conditions, the main suppliers gradually proclaimed that they did not possess a sufficient degree of freedom within the development projects to undertake their obligations towards the SMEs. Hence, the pressure from the oil companies against the main suppliers were forwarded down through the hierarchy of contractual relations and caused further displeasure with the NORSOK program. During the second half of the 1990s, the unions reported several violations of the health and safety conditions and of the working time regulations (Engen 2002).

The picture, however, was far from unambiguous. Surveys conducted in 1997 and 1998 demonstrated that capacity suppliers were more negatively affected than the so-called system suppliers. The explanation is that most capacity suppliers were standardised producers and possessed few technical and organisational capabilities regarding handling environmental challenges. Technically specialised suppliers seemed to be better equipped to utilise environmental changes, improve their products and sophisticate their organisational capabilities. Accordingly they could also generate better financial results. There is, however, considerable uncertainty as to the results of these surveys. Later studies have shown that underreporting and other sources of errors indicated that the HSE problem connected to the NORSOK program was underestimated (Ryggvik 2012).

Whether the HSE conditions were attributable to the way the main suppliers handled the new contracts, to the difficulties in adjusting to changed market conditions, or simply represented an expression of harder competition at all levels is difficult to determine exactly. One way or another, the new inter-organisational relations created instability in the socio-technical system. Many of the companies were in some periods left without predictable contractual agreements and were facing turbulent and risky technical and financial alterations. Building up mutually beneficial procedures takes time and has to rely on well-functioning formal and informal relations. The main suppliers and the small and medium-sized companies lacked the inter-organisational apparatus necessary for maintaining the procurement networks. They also lacked the right instruments to develop and incorporate reliable HSE conditions in the contracts.

More surprising was that the common efforts to simplify routines, procedures and inter-organisational interactions further encouraged the antagonism between the relevant industrial groups and also negatively influenced safety regulations. The reduction of documentation and controlling bureaucracy and the introduction of technical and organisational standardisation constituted the concrete attempts to break down inefficient organisational structures. But because of the extensive application of ISO and CEN certifications, the new NORSOK standards actually signified more bureaucratic red tape instead of simplified procedures. For most suppliers this implied new requirements concerning HSE in addition to those already existing.

The consequence was heavy organisational pressure, and the extensive use of overtime to comply with the time schedules and the new requirements created considerable displeasure in LO unions. As previously mentioned, the willingness of LO to co-operate has to be viewed in the context of the so-called Solidarity Act, whereby moderate wage agreements and cost reductions would improve the national competitiveness and secure employment. The conflict, however, also arose when the employer side consequently referred to NORSOK as an justification for dismissals, organisational rationalising and reformulation of the safety regulations of working life. To LO this was at variance with the co-operative principles of the NORSOK organisations and a general break with the rules of the game in economic life, that is, the Nordic model.

The growing antagonism between the oil companies and the main suppliers, the disappointment of the SMEs and the pressure on the unions further complicated cooperation in NORSOK. The struggle for hegemony therefore prevented a successful finalisation of the NORSOK program. Despite the harmonic point of departure when establishing the NORSOK organisations, it seemed that the program created winners and losers instead of strengthening the balance between the main actors. This is further illustrated by the claim from the suppliers that 75 per cent of the cost reductions attributable to the NORSOK program devolved on the government through the taxation system, and 20 per cent on the oil companies. Accordingly, the suppliers felt they were both losing the benefits and paying the price for the changes that took place.

The NORSOK program was indeed characterised by contrasts. On the one hand, the technical and organisational development paths formed new competitive technological trajectories which replaced the Condeep paradigm. New technological solutions such as floating installations and subsea technologies and methods were introduced – new technologies which both satisfied cost-efficient requirements and were better adapted to new contractual relations and inter-organisational models. On the other hand, the inter-organisational relationships also encouraged positioning, bargaining and strategic behaviour. The relationship between the NORSOK program and organisational performance concerning cost efficiency and HSE thus seemed inevitable. Expectations towards the NORSOK program concerned its role as a contributor to stable and regular interrelated activities which balanced safety and new cost-efficient technical and organisational solutions were broken. Accordingly, new technological pathways threatened to destabilise the petroleum industry and thus the foundations of the Nordic model. At the beginning of the new century, the consequences of NORSOK challenged the institutions that had been developed by the governmental agencies and the unions in accordance with the internal control reform and the WEA. NORSOK did not remove these principles, but the systemic makeover induced by NORSOK seriously threatened the legitimacy of the basic pillars of the Nordic model.

13.4.5. *Risk and Technology*

Rosness and Forseth (Chapter 12) in this volume do not give an explicit reason why the controversies concerning safety increased dramatically at the turn of the century. This chapter is an attempt to complete the picture and to discuss more generally the relationship between the practice of safety regulation, the Nordic model and trends in the socio-technical development. NORSOK was basically a cost-efficient program, and HSE was not the main issue for the different industrial groups working with the program. It is thus empirically difficult to trace the incidents and deteriorated HSE practice directly to recommendations and implementation from the NORSOK committees. However, interviews conducted with key informants in the unions during 1998 and 1999 confirm that NORSOK became an excuse for changing HSE practices in the companies, even though the report did not recommend such organisational changes. Utilising NORSOK in order to squeeze the local shop stewards was common. NORSOK was a lever for the employers in local negotiations and contributed to deteriorate the relationship between the employers and employees both in the oil companies and the supplier industry. According to union representatives, many shop stewards felt obliged and pressed to follow the managers in order to improve cost efficiency at the expense of the HSE level. Even though the official intentions of the NORSOK program were not to deteriorate HSE at any level, both modifications of safety regulations and related incidents related to them gave NORSOK a bad reputation (Engen 2002).

The incidents and number of accidents increased during the 1990s. According to Hovden (2002), several trends in the risk picture could be identified: (1) an increase in major hazards related to problem with gas leakage, and (2) increased risk and uncertainty regarding safety levels for floating installations. This was mainly related to new technology. The Norne accident in 1998, as described by Rosness and Forseth (Chapter 12) in this volume, illustrated very clearly the high-risk arena of shuttle traffic by helicopters. There were also increased accidents related to the activities of service vessels and increased vulnerability caused by increased use of ICT systems, automation and reduced manning on the installations. However, most studies showed low and stable frequencies of occupational accidents.

The exact reasons for these incidents are far from unambiguous. Some are attributable to lack of maintenance, but a large part can also be traced back to new technology and contractual relations. The investigation report after the Norne accident discusses how tighter schedules and increased use of rotating teams increase the risk of such type of accidents. Norne is a floating installation with reduced capacity with respect to living quarters, and therefore dependent on shuttling people more frequently. New contractual relations implied that the oil companies became less responsible for design and accomplishments of the development projects and also less responsible for general HSE. The increased economic pressure combined with lack of knowledge and experiences in risk and safety management constituted new vulnerabilities and reduced robustness in the socio-technical system. Accordingly, the basis that characterised the governance structure and institutional mechanism of the Nordic model was put under pressure. NORSOK implied shifts in the technology and organisation, and the regulatory system – as exemplified by Norne – was not able to follow up on the requirements of new technology and working procedures in the socio-technical system. In some respect we see an institutional inertia, where risk regulation and safety management were sacrificed on the altar of new cost-efficient technological solutions.

Accordingly, NORSOK is a relevant case to understand the vulnerability of the Nordic model. The NORSOK program reduced the significance of the government in the socio-technical system and changed the governance structure by giving more degrees of freedom to the oil companies and the main suppliers. However, the financial winners in this social-technical transformation were the Norwegian government and the oil companies – the government by increasing the basis for tax income because of cost reduction and the oil companies by increasing the profit rate. The losers were the suppliers and the unions. The main suppliers increased their responsibilities in handling and conducting large development projects. But larger responsibilities implied a larger part of the financial risk. In most cases the main suppliers lacked competence and the financial capacity to bear the projects. The result was a heavier pressure on the SMEs and the deteriorating conditions for their workforces.

New technology, new contractual relations and changing working conditions lead to the NPD losing the initiative concerning upgrading the risk-regulating framework. In pre-NORSOK perspective, the regulatory authorities had to admit that oil prices and economy had overshadowed safety issues. The challenge for the NPD/PSA, therefore, was to win back the power and reinitiate safety in a proactive manner. This further underlines the irony of the NORSOK program where the institutional prerequisites for establishing NORSOK – that is, the tripartite collaboration that characterises the regulation regime of the North Sea – were the principles that were most adversely affected during the NORSOK period. The pressure against HSE regulations not only threatened to impair the Nordic model amongst the unions and employers on the Norwegian shelf, but also was one of the key pillars and facilitators in the tripartite collaboration, namely PSA/NPD. Worth noticing is that NPD/PSA actually participated in NORSOK steering committee by the resource director, which further underlines the main objectives of the program: effective resource allocation and cost efficiency. After the NORSOK period, the NPD/PSA initiated several programs and regained the power. How this regaining process proceeded is described more thoroughly by Bang and Thuestad (Chapter 10) and Rosness and Forseth (Chapter 12) in this volume.

13.5. FINAL REMARKS

The NORSOK program visualised the dialectical relationship between the technological development and the regulatory practice of tripartite system in the sense that new technologies, new contractual relations and organisational forms challenged the co-operativeness of the parties in the Norwegian petroleum industry. The governance structures which are basically built upon the concessions laws, the jurisdiction of the formal authorities and the legal framework were not changed significantly by the NORSOK program. At the outset it was decided that general frameworks such as the petroleum act and the tax laws were not subject for revisions. The NORSOK program, however, changed the procedures and routines of the oil companies and suppliers and co-operation and interactions amongst the key industrial actors, and thereby altered the market conditions. The NORSOK program also challenged the power balances in the Nordic model and, on a more general level, laid new premises for prevailing institutional mechanisms. One consequence was the relationship between oil companies and the suppliers, which ended up with giving the oil companies such as Statoil and Hydro more power to squeeze the suppliers through the contracts. Another consequence was the increased struggle between the employer and the employee organisations and an increased division between the main labor organisations on the Norwegian shelf. As mentioned earlier, OFS decided to stay out of the formal process while the LO unions decided to participate in order to "keep one hand on the wheel", as one LO leader put it (Engen 2002).

In hindsight, it is reasonable to interpret the NORSOK program as an attempt to challenge not only the Nordic model as it had developed through the 1970s and the 1980s, but also the general Norwegian petroleum policy. NORSOK was a concrete expression of a new type of industrial policy. Anti-protectionism, liberalism and non-intervention were policy instruments in favour of which the conservatives in Norway had argued for years, and when the labor party established the NORSOK program, this ideology was further promoted by key actors in the NORSOK groups who worked openly in favour of a petroleum industry freed from governmental interventions and regulations. As mentioned previously, the liberalisation movement won many of the battles on the NORSOK arena, but this was also attributable to the fact that the social democratic politicians who had established NORSOK also worked for a revision of the practices of the Nordic model. The infant industry policy which had created the Condeep paradigm, expensive technological pathways and costly effects of the WEA and the internal reform also had its critics in the Labor Party. Accordingly, NORSOK became an arena where arguments for cost efficiency and technological transformation dominated the agenda, and where alternatives were not discussed.

TEXTBOX III

On 1 January 2004, the Norwegian Petroleum Directorate became divided. The part of the agency that safeguarded the interests of safety and working environment in petroleum operations was separated as an independent regulatory body – the PSA. The PSA reports to the Ministry of Labor and Government Administration and had been supplemented with resources from the former Directorate for Fire and Electrical Safety (DBE) and the Labor Inspectorate. PSA ensures overall supervision of the health, safety and working environment in the petroleum activities, regardless of whether the activities take place on the continental shelf or on land.

Accordingly, the NORSOK program created fundamental prerequisites for the Norwegian petroleum policy after the millennium shift and for how the institutional mechanisms become confronted by the NORSOK groups. This was also true for institutions such as NPD/PSA. As mentioned earlier in the chapter, the resource department of the former NPD was in favour of most of the ideas behind the NORSOK program. The safety divisions in the former NPD were far more skeptical. The decision to split the NPD into a new NPD and an independent PSA had little to do with NORSOK, but rather was a result of an increased awareness that optimal and cost-efficient resource allocation and safety regulations present so many contradictions in terms that it is impossible to handle them within the same organisational framework (Lindøe and Olsen 2009). Nonetheless, it were the same contradictions that became visible in the NORSOK program.

The PSA redesigned the institutional mechanism which implied establishing new arenas for tripartite collaboration, surveys and surveillance of the risk level onshore and more sophisticated regulatory framework concerning new technologies onshore. However, the consolidation of the tripartite collaboration was also challenged by an organisational restructuring amongst the suppliers and oil companies. In order to handle the new contracts and develop competence that could handle large development projects, the main suppliers merged and finally ended up as one huge company, Aker Solution. Similar development took place amongst the oil companies. In 2001, Statoil was privatised and in 2007, the merger between Statoil and Hydro Oil & Gas finalised the Statoil's long struggle to get rid of the label as a state-governed company. The merger increased the company's position of near-complete dominance on the Norwegian shelf and enhanced its international ambitions. The merger of Statoil and Hydro created new momentum of the Norwegian petroleum industry. The "new Statoil" represents challenges for the regulatory governments – not least the increased power the company now possesses vis-à-vis the governmental institutions.

The tripartite relationship is based on trust between the parties, but of equal importance to the tripartite relationship is the balance of power. Ryggvik (Chapter 15) in this volume and Lindøe et al. (2012) underline how the Nordic model implies strong unions, strong and competent oil companies and strong competent governmental institutions. The merger of StatoilHydro usurped the power and thus upset the power balance in the tripartite collaboration. The merger also increased the domination Statoil possessed over the technological pathways and towards the regulatory authorities concerning risk regulation and safety. The NORSOK program changed the socio-technical character of the Norwegian petroleum industry by challenging the practices of the Nordic model. These practices survived, however, despite new technological directions and alterations in governance structures and institutional mechanisms. Even though the principles of the WEA and the Nordic model are being challenged continuously by different types of safety regulation systems, new technologies and economic pressure, it seems that the basic principles remain intact. The discussion of the NORSOK program shows how the explanations lie in the interdependency and balance of power between the parties. The strength of the Nordic model is that collaboration is a very potent instrument when it is necessary to change the socio-technical system. At the same time collaboration will be undermined if the balance of power is upset or certain actors dominate the process and gain power at the expense of other actors.

REFERENCES

Engen, O.A. (2002) *Rhetoric and Realities. The NORSOK Programme and Technical and Organisational Change in the Norwegian Petroleum Industrial Complex*. Dissertation dr. polit. University of Bergen: Bergen.

(2009) "The Norwegian petroleum innovation system. A historical overview", in J. Fagerberg, D. Movery, and B. Verspagen (eds.): *Innovation, Path Dependency and Politics*. Oxford University Press: Oxford, pp. 179–207.

Engen, O.A., Langhelle, O. and R. Bratvold (2012) "Is Norway really Norway?", in B. Schaffer and T. Ziyadov (eds.): *Beyond the Resource Curse*. University of Pennsylvania Press: Philadelphia, pp. 259–283.

Hood, C., H. Rothstein and R. Baldwin (2001), *The Government of Risk*, Oxford University Press, Oxford.

Hovden, J. (2002) "The Development of New Safety Regulations in the Norwegian Oil and Gas Industry", in Kirwan, Hale And Hopkins (eds.) *Changing Regulations*, Pergamon: Oxford.

Hughes, Thomas P. (1987) "The Evolution of Large Technological Systems," *In The Social Construction of Technological Systems: New Directions in the Sociology and History of Technology*, Eds Wiebe E. Bijker, Thomas P. Hughes, and Trevor J. Pinch. The MIT Press: Cambridge, Mass.

Karlsen, J.E. and Lindøe, P. (2006) "The Nordic model at a turning point?" *Policy and Practice in Health and Safety*, 4: 17–30.

Krause, T.R, Seymor, K.J. and Sloat, K.C.M. (1999) "Long-term evaluation of a behaviour-based method for improving safety performance: A meta-analysis of 73 interrupted time-series replications", *Safety Science*, 32: 1–18.

Lindøe, P. and Engen, O. A. (2012) "Offshore safety regime – A contested terrain". Paper presented at the Conference *Working on Safety*. Sopot, Polen.

Lindøe, P. and Olsen, O. E. (2009) "Conflicting goals and mixed roles in risk regulation. A case study of Norwegian Petroleum Directorate", *Journal of Risk Research*, 2 (3–4): 427–441.

Lindøe, P., Baram, M. and Patterson, J.R. (2012) "Risk regulation – an assessment of US, UK and Norwegian approaches". Paper presented at *ESREL*. Helsinki, Finland.

Moene, K.O., Mehlum, H., Barth, E., Lind, J.T. and Krüger, I. (2009) *Den skandinaviske modellen og økonomisk ulikhet*. Fordelingsutvalget, Finansdepartementet 2009: Oslo, pp. 328–351.

Olsen, O.E. and Engen, O.A. (2007) Technological change as a trade-off between social construction and technological paradigms", *Technology in Society*, 29 (4): 456–468.

Olsen, O.E. and Lindøe, P. (2009) "Risk at the ramble: The international transfer of risk and vulnerability", *Safety Science*, 47: 743–755.

Rasmussen, J. (1997) "Risk Management in a Dynamic Society: A Modelling Problem", *Safety Science*, vol. 27: 183–213.

Ryggvik, H. (2008) *Atferd, Teknologi og System. En sikkerhetshistorie*. Tapir Akademisk Forlag: Trondheim.

(2012) Dypt Vann i Horisonten. *Regulering av sikkerhet i Norge og USA i lys av Deep Water Horizon ulykken*. Nr 4, Senter for Teknologi, Innovasjon og Kultur. Universitetet i Oslo: Oslo.

14

Near Major Accidents

A Challenge for the Regulator and the Regulated

Ole Andreas Engen

14.1. INTRODUCTION

This chapter compares the underlying factors of the Snorre A and Gullfaks C incidents on the Norwegian continental shelf in 2004 and 2010. These two incidents challenged the basis of the tripartite institution and brought to the fore the presence of risk of a great accident on the Norwegian shelf (see Chapter 10). The latter incident also took place in the same period and in the same "ideological climate" as the Macondo incident. The accident in the Gulf of Mexico had a great impact on the regulator and the safety community in Norway and the Gullfaks C incident served as a strong reminder that a similar accident just as easily could happen on the Norwegian shelf.

Renn (Chapter 1) in this volume refers to how risk management faces complexity, uncertainty, and ambiguity. The merger of Statoil and Norsk Hydro creates a relevant organisational backdrop of alignment of these factors.[1] The Gullfaks C incident revealed challenges within the new organisational structure of Statoil and sounded strong warnings to the top management about the increasing uncertainty of new technologies of well drilling as well as enhanced complexity of the internal organisational safety regime. Because of the political climate surrounding the investigations of Gullfaks C incident, the ambiguity concerning safety regulation offshore increased. More stakeholders outside Statoil and the Norwegian Petroleum Agency became involved and the unions and NGOs such as the environmental organisation Bellona intensified their agitation.

In this chapter we will see how risk governance is reflected in an operating company such as Statoil. As a follow-up to the investigation of Gullfax C, the Norwegian Petroleum Safety Agency (PSA) requested that Statoil allow the International

[1] Statoil ASA, trading as Statoil and formerly known as StatoilHydro, is a Norwegian oil and gas company, formed by the 2007 merger of Statoil with the oil and gas division of Norsk Hydro.

Research Institute of Stavanger (IRIS) to conduct an independent investigation.[2] The main objective was to conduct an analysis of underlying causes of the incident at Gullfaks C related to governance, management and other organisational factors. An additional goal was to analyse why measures for improvement initiated after previous incidents, among them the blowout at Snorre A in 2004, have not had the desired effects concerning risk management in the new Statoil organisation. The content of this chapter is essentially based on this report, but other sources have also been used.[3] The chapter also explores the functionality of the semi-self-regulatory systems discussed by Kaasen (Chapter 5), Paterson (Chapter 6) and Kringen (Chapter 11), as well as the relationships between inspectorate and companies discussed by Ryggvik (Chapter 15), all in this volume.

The chapter is organised as follows. First we briefly introduce the concept of underlying factors. Thereafter we go through technical details about the two incidents. We then deal with each case individually, but the general discussion that follows is organised around the key concepts: organisational context, management and decision making, and compliance.

14.2. UNDERLYING ORGANISATIONAL FACTORS

"Underlying factors" may be defined as the combination of active failures and failure in barriers and defenses. According to Reason (1997), active failures occur when unsafe acts have direct impact on safety of technological systems. In many contexts barriers and defenses have the same meaning. The notion of barrier often refers to technical measures in safety analysis, such as well integrity, cement plugs and so forth, while defenses are employed in relation to organisation, technology and operational efforts in order to protect human, environment and economic values from accidents, losses and damages (Hopkins 2012). According to Reason (1997), defenses may be further divided in two subcategories of "hard" and "soft." Hard defenses refer to all types of technical devices that are installed in order to warn, protect and inform the organisational member of potential danger, for example, security alarms, fire alarms, entrance codes and so on. Soft defenses refer to all types of organisational arrangements that aim to reduce the risk of undesirable incidents. Typical examples are rules, regulations, routines and procedures and different

[2] The IRIS – report 2011. The IRIS report is based on interviews with employees at different organisational levels, both at Statoil and at central contractors. In addition, data from a survey carried out in Statoil D & W unit and with nine contractors are reported on.

[3] The main sources beside the IRIS report 2011 are: 'Granskingsrapport COA' (Internal Investigation report-Statoil), Petroleum Safety Agency, Norway-Audit Report. Statoil's planning of Well 34/10 – C – 06A. 'Organizational Accidents and Resilient Organizations: Six Perspectives. Revision 2. Sintef A17034 report. Rosness et al 2011.

ways to enhance the ability of the organisations to perform in accordance with such institutional characteristics.

The point is that people working in complex systems such as an oil company or a well-drilling company make errors that go beyond the scope of individual psychology. This is what Reason calls latent conditions:

> Latent conditions are to technological organisations what resident pathogens are to human body. Like pathogens, latent conditions – such as poor design, gaps in supervision, undetected manufacturing defects or maintenance failures, unworkable procedures, clumsy automation, shortfall in training less adequate tools and equipment – may be present for many years before they combine with local circumstances and active failures to penetrate the systems' many layers of defenses. (Reason 1997, p. 10)

Hence defenses are placed proximally to the accident while latent conditions precede the active failures. In an ideal organisational model, the defenses will capture the consequences of the active failures immediately and prevent the incident from escalating. In the real world, however, the defenses possess several weaknesses. At best, when active failures occur, the second defense will capture a beginning chain of consequences if the first one fails, but in a complex organisational structure with many defenses, the likelihood for many concurrent weaknesses or latent conditions is significant. When such weaknesses appear simultaneously and create a coherent chain of incidents, active failure and latent conditions can cause significant accidents. When all defenses fail, a single failure may spread in the entire technological system and a catastrophe becomes a matter of fact. In large technological systems, active failure and latent conditions as underlying factors encompass humans both at "the sharp end" and "blunt end" of the organisations. The sharp end of organisations denotes those people with direct contact with the production process. In this context practitioners at the sharp end interact with the hazardous process in their roles as roughnecks and drillers. Those at the blunt end of the system, affect safety through their effect on the constraints and resources acting on the practitioners at the sharp end. The blunt end thus includes the managers, systemdesigners, and suppliers of technology.

Modern organisations are multioriented. The relationship between hard and soft defenses constitutes the total robustness of the organisation or the technological system. Encouraging innovation capabilities, increasing robustness and focussing on cost efficiency simultaneously create dilemmas in all organisations. Innovations can enhance safety in one part of the organisation and induce new risks in other parts of it. Increased robustness can improve safety but also can increase the complexity. Cost efficiency and safety are normally considered contradiction in terms. Studies of underlying factors are thus not the only internal investigations of what takes place at the sharp end of the organisation; analysing the significance of management, regulation, legal factors and economical external conditions is equally essential. When this chapter compares the underlying factors on Snorre A and Gullfaks C, it calls for focussing on dilemmas in management, governing documentation, organisational development, risk regulation and economic factors.

14.3. THE INCIDENTS IN MERGING SURROUNDINGS

The Snorre A and Gullfaks C incidents both took place in organisational surroundings characterised by comprehensive changes. The root causes for these changes were to be found in organisational merge processes, first between Hydro and the smaller Norwegian oil company, Saga and second between the "new" Hydro Oil & Gas and Statoil. Both merger processes were considered painful experiences and demanded large efforts from management, working organisations and individual workers.

14.3.1. *Snorre A – Chain of Events*

Snorre A (henceforth SNA) is a combined living, drilling and production platform in the Tampen area, 150 kilometres west of the Norwegian city of Florø. The living quarters can accommodate approximately 220 persons and the platform is anchored to the sea bed by 350-meter tension legs. The template consists of forty-two wells with risers and export lines. Aside from the offshore installation, SNA also has a division on land.

SNA became operational in 1992, with Saga Petroleum as operating company. In January 2000, Norsk Hydro acquired the operator responsibilities, which they kept until December 31, 2002, when Statoil succeeded as operating company. Two changes of operating companies in a period of four years contributed to several organisational changes on SNA, particularly with regard to management systems and governing documentation concerning safety regulation. Until 2003, the SNA organisation was also subject to several efficiency processes and consequently under constant pressure to maintain high production levels.

It turns out that Hydro had not had much "cultural" impact on the SNA organisation during the two years it served as the operating company, but the change still implied a new management system and the insertion of new leaders. Negative attitudes to this organisational change revealed themselves in 2003, when Statoil took over as operating company, at which point the SNA organisation expressed a desire to "be left alone for a while" (Schiefloe et al. 2005). This request was partially taken into account, resulting in a lower pressure for immediate organisational integration of SNA into the new parent company. The 2004 report of the events on SNA underlined that although SNA formally was more integrated in the common organisational structure of Statoil than other parts of the Hydro organisation and although a new set of governing documents was implemented, SNA was less socially and culturally integrated both in the former Hydro organisation and in Statoil. Actually, the report concludes that some of the "Saga culture" survived on SNA until the incident on November 28, 2004.

In the summer of 2004, a project team for reservoir recovery planning of a well slot recovery was appointed (hereinafter SNA RESU). SNA RESU was a new way of organising such an operation. Employees from two separate units – Drilling and Well and Reservoirs – were merged into one unit. Until November 2004, several meetings of SNA RESU were held, but the drilling contractor was not present at these meetings. In 2004, it became known that the initial drilling operation on well P-31 had resulted in damage to several casings (both 13 3/8-inch and 9 5/8-inch casing were punctured). A long scab liner was therefore installed to plug the gaps and to strengthen the integrity of large portions of 9 5/8-inch casing and the plans for slot recovery of the well P-31A were changed immediately (see Figure 14.1).

In September 2004, because of deficiencies in the well's integrity, SNA RESU determined that the reservoir section should not be opened. This would ensure a more robust solution. In October, the plan changed again and reservoir sections were instead cemented to avoid possible communication with the subsequent sidetrack P-31B. In order to implement cementing of the reservoir section, it was planned to puncture the tail pipe before pulling the tubing. An early puncture operation would save time and implied establishing a link between the reservoir section and hydrocarbon-bearing formations during the rest of the operation, until the cementing of the reservoir was completed. There were different views concerning these decisions within the SNA RESU organisation.

The final planning meeting was held on November 11, 2004 and consisted of suppliers and core personnel from SNA RESU. The meeting went through the well's history and discussed each part of the operation. A risk review program for personnel with qualifications and skills in well drilling was planned but not implemented and four days later the entire program was signed off on by top management and management of SNA RESU. The operation consisted of puncturing the tail pipes, replacing brine with oil-based sludge, cutting and pulling of tubing, cutting and pulling of scab liner, cementing of the well's reservoir sections, as well as cutting and pulling casings.

On November 16, 2004, the rig was moved to the well P-31. The same day, exit meetings were held where the drilling program was presented for offshore drilling management and leading personnel from the drilling contractor.[4] The primary barrier consisted now of brine. Water-based drilling fluid (brine) was pumped from the well and replaced with oil-based drilling mud in order to form a new primary barrier. On November 23, 2004, the drilling supervisor sent a request by e-mail to the

4 Exit meetings refer to the final meetings before the operation begins.

FIGURE 14.1 Picture of Snorre A and image of the well.

software engineering and drilling operations, which expressed concern whether a formal permission to draw the scab liner was required, which implied no opportunities to use the safety valve's (BOP's) pipe ram and blind ram. He received an answer from the program engineer that this was not necessary, because the operation was not in the open hole. This was a misjudgement, as there was direct communication with the formation of the reservoir section through the perforated tail pipe.

It was decided to make holes in the scab liner to equalise the pressure behind the pipe before they pulled it out. As soon as the scab liner was pulled, gas that had

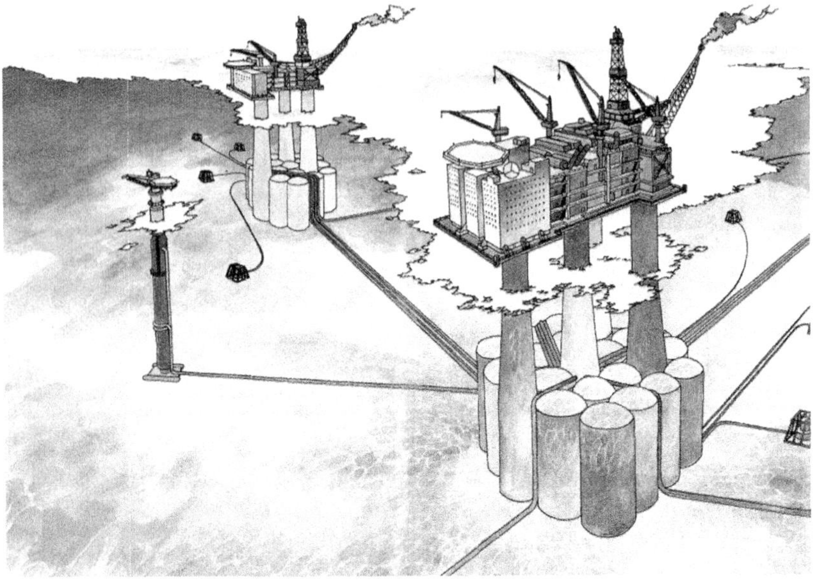

FIGURE 14.2 Picture of Gullfaks C and images of the well.

flowed into the bay was observed because of swabbing, or suction when the pipe is drawn up. A decision to circulate the gas out was not taken, however. The swabbing continued during winding and an attempt to circulate the gas was tried later on, but instead there was a leak at the top of the scab liner, which prevented normal circulation. Before the top of the scab liner could be drawn through the Blow Out Preventer (BOP), it was verified that well volume (flow check) was stable and therefore the scab liner was extracted and now blocked.

A few hours later a gas flow was observed in the well and the annular safety valve on the BOP was thus closed. The pressure under the BOP increased after it was closed, but soon after the well began to lose mud. Because of leaks in the 9 5/8-inch and 13 3/8-inch casings, the well pressure came in contact with formations behind the 13 3/8-inch casing, which then burst. There was gas under the platform outside the casing protection and the situation escalated with a gas blowout on the seabed. The well control operations became extremely difficult and dramatic, but ultimately the control of the well was restored.

14.3.2. *Gullfaks C – Chain of Events*

Gullfaks C is an oil platform on the Gullfaks field in the North Sea. The frame of the platform was built by Norwegian contractors, with a Condeep concrete substructure (see Chapter 13 for a detailed description of the Condeep design). Gullfaks C is a combined drilling, production and accommodation platform – an integrated oil platform. Production began on November 4, 1989.

In 2008, a sidetrack on well C06 at Gullfaks C was planned in order to plug an existing well and extend it to a new area. The original well had been an oil producer as far back as 1991. In January 2009, the conceptual design of the new well was selected and the sidetrack was planned in the high-pressure area of the Top Shetland formation. The Top Shetland formation consists of calcareous shale, which over time has experienced changed pressure conditions and is separated from the overlying formation of a dense lime scale that has acted as a natural pressure barrier.

In 2009, the measurements showed pressure in the annulus between two casings. Samples of the annulus showed that the well barriers against the reservoir were diminished. The conceptual design was changed and the lower part of the well was shut off to stop a beginning gas leak. In this operation, the old casing was removed in the upper part of the well and it was necessary to measure the pressure in the formation. A new sidetrack was planned higher in the well and above the limestone layer. The sidetrack was thus drilled to the top of the Top Shetland where there would be a new casing before it was drilled into the high-pressure area.

At a depth of 2,665 metres in the Lista-Formation, a well kick occurred on December 23, 2009.[5] This occurred because the formation pressure was higher than expected. The well was now at 37 metres vertical depth of Top Shetland. The well was plugged with cement and the casing was inserted to ensure the well integrity. In addition, the process owner – the appointed suborganisation in Statoil responsible for drilling and well – began to investigate the incident. Later seismic studies showed that lime comprising Top Shetland was broken up and discarded in the area where the well was drilled.

A new sidetrack was thus planned. This was the second time the plans for the well had to be changed and it was now planned to use managed pressure drilling (MPD) because of new information about pressure conditions.[6] Extensive tests of the formation strength showed that the pressure tolerance for further drilling was reduced. MPD involves extra pressure on light mud so that it is possible to regulate the pressure in the well at any time.

[5] The Lista Formation refers to the geological structures on the sea bed of the eastern part of the Norwegian North Sea, the area alongside the western part of Norway. The southern part of this area is associated with the Våle formation and the Top Shetland base (information from NORLEX).

[6] Managed pressure drilling (MPD) is an adaptive drilling process used to precisely control the annular pressure profile throughout the wellbore. The objectives are to ascertain the downhole pressure environment limits and to manage the annular hydraulic pressure profile accordingly. It is the intention of MPD to avoid the continuous influx of formation fluids to the surface. Any influx incidental to the operation will be safely contained using an appropriate process.

In April 2010, drilling in well C06 started. According to internal investigation reports, there already were several challenges only a few days after the drilling started. There were problems with leakage through a new gasket element (PCD) and there was also a new well kick that resulted from the pulling of the drill string. Finally there was a further increase in pressure in the C-annulus, which, however, has gone unnoticed.

After drilling was completed, a twofold critical situation emerged. There was leakage of gas into the environment in the well, accompanied by a loss of back pressure in the MPD system. Because of the loss of back pressure with deficit in the well, further influx from the high-pressure formation occurred in Lista/Shetland. The well was shut down and an attempt was made to implement well control procedures. Simultaneously large amounts of gas in the sludge-processing area were detected and as a result, an alarm was raised. Gas was observed several times on the rig while the crew tried to deal with the well control situation.

The crew worked hard to secure the well after the event began, but soon after it was discovered that the pressure in the C-annulus was abnormally high. Later in July, gaps in the 13 3/8-inch casing were discovered. The casing as the joint barrier element for the primary and secondary well barrier/defense had thus failed. During the two months the normalisation work was in progress, a set of three cement plugs was installed. The first one was a permanent barrier to the reservoir; the second plug was inserted as a barrier to high-pressure zone in the Lista Formation and the Shetland Group. Finally, there was a plug inside a 9 5/8-inch casing between 1,848 to 2,043 meters before being wrapped in a 13 3/8-inch casing at approximately 1,420 metres, which would be sealed with new casings.

14.4. UNDERLYING FACTORS AT SNORRE A

The PSA investigation report revealed a total of twenty-eight deviations or violations. According to PSA, all discrepancies would have been rectified if the barriers had not failed. Errors in individual barriers may occur, but it is unusual that so many barriers fail simultaneously in one operation. According to PSA, this was attributable to failure on several levels in the SNA organisation both offshore and onshore.

The serious failings and shortcomings in all phases of Statoil's planning can be summarised in three categories: compliance with governing documents, management's involvement and lack of risk assessment.

14.4.1. *Compliance with Governing Documents*

According to Statoil's own analysis, it was clear that failure in this category had to be attributed to failure of the system rather than failures of individuals. Replacement of the operating company meant that the employees had to deal with new sets of governing documents. In addition, training was inadequate, particularly onshore and contributed to making the procedures and new internal regulations difficult to understand for the employees. Another important factor was the hectic everyday life in the SNA organisation. Lack of time reduced the opportunities to become familiar with the new governing documentation, so the failures in defenses were not owing to lack of willingness or intent to comply, but rather to many urgent tasks that superceded such familiarisation (Schiefloe et al. 2005).

14.4.2. *Management's Involvement*

Multiple changes of operating companies, new drilling contractors, and the reorganisation process occupied management's attention and work capacity. The SNA organisation prior to the event was characterised by a high level of activity, with a focus on short-term production targets. In addition, the technical system was worn out and required maintenance and was therefore not very robust. This led to many unforeseen technical problems during a normal workday.

Another factor that affected management's involvement was an excessive number of administrative tasks. The management on SNA reported that there was not enough time to fulfill all the duties and that they had to prioritise between administrative and operational tasks. They reported that the administrative burden occupied most of their attention. When combined with a lack of clear deadlines, this has led to operational tasks being given lower priority. The main objective became to focus on efficiency, maintain production and ensure maximum profit.

14.4.3. *Lack of Risk Assessment*

Focus on production regularity and concerns over increased wear of equipment were among the constituting risk factors at the SNA installation. Production goals were high, especially for the drilling contractors, stemming from the fact that renewal of contracts was connected to the contractors' ability to demonstrate efficiency and profitability. Such one-sided effects produced ripple effects within the organisation. On the one hand, it contributed to hard work and creative solutions to ongoing problems. On the other hand, it caused negative consequences with respect to safety and compliance with regulations. To some extent it seems that Snorre A developed "a culture of risk tolerance," which implied an unambiguous focus on production, operating efficiency and progress at the expense of safety margins. Prioritising safety while upholding high production requirements is a dilemma for most organisations. When it is communicated that the work is to be done safely, but at the same time clear expectations are expressed that the highest possible level of production is to be maintained, safety issues get neglected most of the time. In case of Snorre A, this meant the postponement or even complete exclusion of obligatory risk assessments and consequential dismissal of ordinary precautionary principles.

Furthermore, the general organisational knowledge concerning risk assessment was low. The SNA organisation employed a large number of external consultants. This weakened the overall organisational competence because the knowledge and experience of the consultants were not staying in the organisation for a long time. Because of their unambiguous focus on their own contract renewals, the contractors also exhibited a low tolerance threshold towards constructive criticism and reflections concerning risk and safety. Meanwhile, the general organisational skills in Statoil were diminishing. Organisational expansion made it more difficult to know where in the company it was possible to find relevant experience and critical input. Internal relationships and networks are often of crucial importance for carrying out daily safety and risk assessments. In the new Statoil organisation there was at this time low experience of collaborating across organisational units and in SNA RESU in particular they were hardly aware of the expertise and resources that existed outside the unit.

14.5. UNDERLYING FACTORS AT GULLFAKS C

The merger of Statoil and Hydro Oil & Gas in October 2007 was an organisational event that created several of the underlying factors discussed in this chapter. The merger implied a full integration of all the activities, resources and management systems. All employees were embedded in new jobs where most of them were changing tasks and roles and most had to deal with new colleagues. Encountering new forms of work organisation and management was a general challenge for all employees of the new company. The underlying factors at Gullfaks C can be summarised in three categories: changes in organisational context, autonomous culture and external pressure and lack of compliance.

14.5.1. *Changes in Organisational Context*

In the drilling and well department (hereinafter D&W) there were two key challenges. The first concerned how to harmonise work processes and governing documentation. At the outset, Hydro and Statoil had two different systems and there were local work processes and installation-specific adaptations as well. After APOS was selected as the new system, the main challenge was to find the right level of detail of the work process descriptions, choose the best work process and make the employees aware of it in order to ensure compliance. Efforts were thus made to organise such processes with as little complication as possible.

The second challenge involved the replacement of staff with specific platform skills. In order to reorganise the workforce, the new company pension schemes for people older than fifty-eight years were introduced. As a result of this fifty-eight-plus provision, many experienced employees left Statoil, leading to a significant loss of valuable expertise. Many employees felt that platform-specific expertise was reduced and within the D&W organisation in Bergen, former Statoil-operated fields were now managed by former Hydro employees. Experienced employees felt that such new power constellations were unfortunate and that the company underestimated experience-based competence needed to handle the technical complexity of these fields. Finally, communication between offshore and onshore departments during the planning of operations became less satisfactory. Before the incident, parts of the D & W personnel working offshore were little involved in the planning and discussion of the well-drilling program. This also resulted in the exclusion of practical skills and led to a situation in which local knowledge was not being utilised in the general planning of the well or in risk assessments.

14.5.2. *Autonomous Culture and External Pressure*

The Gullfaks field has a long history and is characterised by a distinct culture concerning the operation, maintenance and drilling, although these work areas are separate organisations and deal with different reporting lines. Traditionally trade unions have had a strong position and from a management point of view, the Gullfaks culture is perceived as rigid and difficult to control. However, strong local cultures, elements of power struggles and conflicts are not a unique feature of Gullfaks. Many both inside and outside the Gullfaks organisation presented Gullfaks as quite unique – a view that underlines how the organisation is perceived as a "politicized" entity consisting of many powerful interest groups.

Accordingly, there were several reasons behind the decision-making processes and technological choices concerning drilling at well C06A. Two economic aspects can be highlighted. First, the desire for increased production contributed to the well being drilled in one section instead of two. Second, lack of capacity among the existing staff contributed to shortening the time of planning, with the necessary changes concerning the drilling program being postponed or cancelled.

Two factors related to competence are also highlighted. First, the relocation of personnel in connection with the reorganisation/integration process has led to the disappearance of important skill sets from Gullfaks C. Second, because of cost pressure, decision-making processes were poorly documented, which further led to increased dependency on personal knowledge and experience. Finally, the Gullfaks organisation felt pressure from PSA to make use of MPD technology, which also may have played a role in this particular technological choice. Whether this pressure was real or erroneously perceived, the circumstances related to the implementation of MPD as new technology led poor planning and use. External economic pressures played an important role, and as a consequence, adequate risk assessments were not conducted. Hence, confusion and disorderly communication about the technological options and their risk were added to the latent conditions behind the incident.

14.5.3. *Lack of Compliance*

The IRIS-report (2011) eveals latent conditions in defenses related to organisational processes such as management documentation, compliance and competence and communication and decision making. These factors are latent conditions that appeared synchronously when the incident at Gullfaks C happened. Furthermore, the merger between Statoil and Hydro has to be considered as an important backdrop for the event. Harmonisation of management systems, location of personnel and expertise, as well as the fifty-eight-plus provision created additional confusion and noise alongside organisational relocations and the creation of new units.

Both effective governing documentation and compliance were important for risk control. Altogether, organisational change and uncertainty stemming from complicated and difficult management procedures constituted clear latent conditions. This was further compounded by the fact that the actors in the organisation were unable to relate to the prevailing systems and thereby developed their own strategies and action patterns locally, most of which related to getting the job done on a hectic schedule and in a hectic environment. This is not a unique to the Gullfaks organisation; it is a phenomenon common to organisations with complex management systems. Such management systems tend to require so much space that they actually hinder the employees' ability to work effectively.

Altogether, the Gullfaks C incident showed significant weaknesses in the barriers and defenses and resident latent conditions. At the same time, the finding in the IRIS report (2011) underlined how the Gullfaks organisation was decoupled from the rest of the Statoil management structure. This is partly attributable to complex control systems and partly to the fact that Gullfaks at the outset had developed an autonomous culture. Such organisational phenomena are well documented by Snook (2000), who shows how subunits in large technological systems tend to be decoupled and develop their own logic of actions and their own ways of working. Such decoupling may be hazardous because it produces units that do not work according to same patterns and practices as the rest of the organisation where standardisation and harmonisation of some processes are of crucial importance. As a consequence, defenses deteriorate, latent conditions occur and major accidents become more likely to happen.

14.6. THE RELATIONSHIP BETWEEN SNORRE A AND GULLFAKS C

So far we have looked at conditions around gas blowouts at Snorre A in November 2004 and Gullfaks C in 2010 – that is, the sequence of events and analysis of causes and actions taken afterward. Now it is time to discuss whether the incident on Snorre A actually has relevance to what happened at Gullfaks C.

With regard to the technical conditions on Snorre A and Gullfaks C, there are obvious differences. Snorre A is a tension-leg installation (float), whereas Gullfaks C is a Condeep. On Snorre A they were plugging a well – extracting the scab liner, cementing the well and cutting and pulling the casing. Gullfaks C was in a completely different phase, as the well was already drilled and had a clear production profile. At Gullfaks it was decided to use MPD, which was regarded as a relatively new technology. This was indeed very different from what happened at Snorre A, where they went through a well-known operation. As underlined in the IRIS report (2011), on Snorre A the crew knew about the hole in the casing, whereas on Gullfaks C the crew was completely unaware of a similar hole. Conditions on the installations and drilling methods also contributed to the fact that there was seemingly a greater catastrophe potential on Snorre A than on Gullfaks C. On Snorre A, they could have risked losing the entire platform. Hence, a manager described the two events as follows: "There are no similarities between Snorre A and Gullfaks C beyond that both incidents happened on platforms which were surrounded by sea" (IRIS report 2011, p. 51).

This statement is quite biased. Technically there were not many similarities beyond the fact that they were both well control events, specifically related to well integrity, but the underlying organisational factors include strong similarities. We shall, therefore, thematically compare the following underlying factors: the merger process, the organisational context, leadership and decision making and compliance.

14.6.1. *Merger Process*

The crew of each installation had gone through a merger process. On Snorre A, Hydro had taken over as operator from Saga and the staff had to deal with Hydro's management systems instead of Saga's. Similar processes took place on Gullfaks C, which experienced a major reorganisation in connection with the merger between Statoil and Hydro Oil & Gas. At Gullfaks C, as elsewhere in the new Statoil organisation, the crew had to deal with APOS that to previous Statoil employees was an unknown system. Most of the Statoil employees were experienced in DocMap. Hence the confusion about governing documentation was similar in both cases.

14.6.2. *Organisational Context*

The IRIS report (2011) about Gullfaks C refers to different cultures that had evolved over time and distinguishes between two different cultural dimensions. The first cultural dimension is referred to as a D & W culture. In general, D & W communities have strong positions in the petroleum industry because they act as an economic and technological driving force for the entire industry: after all, no oil found means no money. Such communities consist of skilled professionals who for the most part remain in this community throughout their entire career. This could be considered as strength, because such communities possess special expertise and experience. The downside is that such communities will be difficult for newcomers to enter and harder for upper management to control and govern. Such communities easily nurture people who "know best." A relevant comparison could be the medical profession in the sense that it also sees itself as consisting of exalted and independent personnel. The similarities between the events on Snorre A and Gullfaks C are thus that they both occurred in operations that were planned and handled within the D&W community.

Cultural traits such as dominant professional cultures are fortified by field- or installation-specific cultures. On Snorre A such culture was referred to as the Saga culture – that is, a work culture that was strongly influenced by the previous operator, Saga, even after Hydro had taken over. Similarly, the reports also highlight a Gullfaks culture, which originated in the perception that Gullfaks is a special field. Gullfaks is often referred to as Statoil's golden calf or money machine, with many skilled employees, a hierarchical structure and strong unions (IRIS report 2011). The merger between Statoil and Hydro did not have immediate consequences for the crew on the Gullfaks installations and the Gullfaks culture also survived in the new organisation, in the same way as the Saga culture survived on Snorre A.

In the winter of 2010, however, almost the entire crew on Gullfaks C, including the management teams, was replaced. People who had been involved in planning the drilling of C06A and in the well kick in December 2009, were removed. On Snorre A, similar cultural reorganisation took place as a result of earlier reorganisations of the culture offshore (Saga to Hydro) and working organisation onshore (Hydro to Statoil). It had been revealed that Snorre A offshore had failed to follow orders that Hydro previously received from the PSA and that other fields and installations had taken into account. Snorre A also had different challenges associated with three different shifts. When it comes to cultural dimensions, both Snorre A and Gullfaks C had to handle different cultures on the installation and differing relationships between the fields and in the rest of the organisation.

14.6.3. *Leadership and Decision Making*

The IRIS report also highlights lack of management involvement related to the events on Snorre A and Gullfaks C. The Snorre A project was referred to as "not very attractive" from a management point of view. Clearing the stage for a new well was considered "not very prestigious." Those who did the "clearing job" were concerned about such neglect and tried to call into meetings. They did not achieve any breakthrough, however and thus were not assigned the resources they required. On Gullfaks C, lack of management involvement was also reported. It thus appears that at either location management was not sufficiently present in critical phases of the projects.

Both organisations were also simultaneously faced with tough demands. On Snorre A, the crew felt pressure to attain a high level of effectiveness associated with high production requirements. On Gullfaks C, time pressure was tightened during the winter/spring of 2010 whereby suppliers felt that things should happen very quickly. Some informants have related this to financial considerations and increasing fear of losses. Both conditions contributed to speed up the decision-making process that normally required more time and patience.

Analysis of both incidents reveals that risk assessment is a central issue. PSA accused Statoil of "lack of understanding and implementation of risk assessments." Further investigations of the Gullfaks C incident confirm this. PSA concluded that the risk assessments were not conducted in accordance with prevailing guidelines, that they did not reflect the difficult characteristics of well C06A, that there was a lack of system policy in regards to risk analysis and finally, that documentation on risk assessment was not stored. In other words, risk evaluations in both cases were insufficient. It undoubtedly created a higher risk for the subsequent operations, primarily with respect to well control and the plans for handling the potential consequences of the increased risk were not good enough.

PSA's arguments concerning "lack of understanding and implementation of risk" are therefore particularly relevant in the execution phase of the Gullfaks C operation. In this phase they could have made new assessments after the problems had risen. Statoil could have made the operation safe, even with MPD, in order to prevent dangerous situations from occurring; they also could have drilled the well with two sections and fixed the casing higher up. Such alternative procedures could thus have protected the deteriorated part of the casing.

14.6.4. *Compliance*

Compliance is a theme that recurs in all the investigation reports and it is clear that prior to both events, compliance was violated with regards to governing documents.

On Snorre A, there had been new requirements for risk assessments in connection with drilling and well operations. The incident on Gullfaks C shows that having the governing documentation is only half the job, because safety is also dependent on the requirements being taken seriously and followed by the personnel. Unfortunately, promises of compliance learned from past events (in this case Snorre A) were not kept. This resulted in a situation where many of the measures proposed after Gullfaks C are surprisingly similar to those measures that were proposed after Snorre A. Several informants pointed out the similarity in the aftermath of Snorre A and Gullfaks C and it is questioned whether the company has learned anything at all from what happened on Snorre A. One informant in the IRIS study who was involved both on Snorre A and Gullfaks C actually expressed embarrassment to have first been on Snorre A and thereafter on Gullfaks C (IRIS report 2011).

14.7. DISCUSSION

In the aftermath of the gas blowout on Snorre A in 2004, Statoil carried out a major process whereby large parts of the organisation were involved in the interventions. Both technical and organisational measures were implemented. Most recommendations, however, were implemented in general and the "Impact Report" (2007) gave predominantly good testimonials on how the measures were received and evaluated.

Previously we discussed whether the incident on Snorre A has relevance to what happened on Gullfaks C. Despite the obvious technical differences in installations, operations, techniques and the facts (known or unknown hole in the casing), there are so many similarities in the underlying factors that we cannot ignore the relationship between the two events. Moreover, several of the measures implemented after the Snorre A incident had significance for how to understand the Gullfaks C incident. Recommendations related to planning and risk assessment, management documentation, management training and reduction in administrative work for the operative management were all highlighted.

The same elements that were revealed after the Snorre A incident were still challenges when it came to the underlying factors connected to the Gullfaks C incident. Issues concerning governing documentation are highly relevant in this context. Analysis of the events on Snorre A reveals how the instructions from the PSA had countervailing effects. Increased focus on planning and risk assessment actually contributed to more detailed work processes and more bureaucracy instead of simplifying the governing documentation and reducing bureaucracy. Besides experiencing challenges related to the governing documents in general, Statoil still permitted development and preservation of closed and autonomous cultures, especially in D&W. Despite recognition of the closed Snorre culture in the Statoil organisation, equally different cultural traits continued and survived on Gullfaks C in the wake of the merger between Hydro and Statoil.

According to all investigation reports, lack of compliance is emphasised as a major explanation for incidents. Perhaps it is more important to explain the lack of compliance and the lack of learning ability in the company? In our analysis, there are two factors that may serve as key explanatory factors: autonomy and power. It is clear that the Gullfaks C organisation and the Snorre A unit constituted subunits that operated autonomously and therefore possessed their own logic of action. Such logic of action may have its benefits. Safety will often be better if the individual units in an organisation develop their own ways of accomplishing things with their own procedures, but this requires that management and other parts of the organisation accept this and that the remaining organisational context is stable and adjusted to such informal procedures and routines.

For Statoil this has not been the case in recent years. Instead, the merger with Hydro, changes in organisational structures, as well as rotation and early retirement challenged the local logic of actions both on Gullfaks C and Snorre A and resulted in conflicts of interests and power struggles. This influenced the degree of compliance and the ability to learn. The degree of autonomy of Gullfaks C and related constellations of power are important underlying factors in our analysis in terms of lack of compliance. The Gullfaks culture tended to act autonomously in relation to other parts of the organisation. Such power constellations and professional hegemonies are important explanations for this, but overall may not be a negative factor. On Snorre A, the independent working culture was also an important factor when the subunit actually managed to stop the escalating gas leak and prevent an explosion. The Gullfaks organisation had a long history in Statoil and had been a significant factor in the economic performance and competence in the company. In both cases, local competence was unique and important in order to maintain safety. The problem is that a complex system like Statoil cannot only trust local expertise and the willingness of local crews to take responsibility when unforeseen incidents happen. There has to be a certain consistency between general governing rules and regulation and actual behavior at the sharp end of the organisation.

The management of Statoil proclaimed that much effort has been made to achieve good communication at all levels of the company (IRIS 156/2011). Communication is a prerequisite for effective safety management and effective communication is also closely linked to efficient knowledge transfer and knowledge management. The problem associated with communication and information occurs when the complexity of information systems increases. According to the IRIS report, it was difficult to find information and use information systems in an effective manner. This represented a significant bottleneck within the company, particularly in situations of time pressure and contributed to increased distance between the various units; subsequently, a local logic of action was nurtured and further developed in the single units. This has led to a trend towards a further imbalance between those who design governing documents and the different units that must operate on their basis. Accordingly there were considerable uncertainties about the functionality of governing documents. Complexity connected to governing documentation led to parts of the organisation more or less beginning to "live their own lives" and consequent reduction in confidence in the system.

Competence is a crucial variable concerning the introduction and implementation of new technology. On Gullfaks C, PSA's pressure to use MPD, combined with lack of risk assessment and monitoring organisational performance related to technology, resulted in special challenges to competence. Lack of competence is described as the underlying causes both at Snorre A and Gullfaks C, but lack of competence has a more general role in this. Earlier we have mentioned large rotations in the organisation because of the merger between Saga and Hydro and Hydro and Statoil, as well as the consequences of the fifty-eight-plus retirement program. Expertise, both related to how technology works and especially to how competence profiles match the organisational levels in general, is important. There is little doubt that the combination of skills, poor coordination and uncertainty in regard to new teams induced instability in different Statoil organisations. From Snorre A to Gullfaks C, a long reorganisation process took place. The underlying factors connected to these organisational changes were especially related to whether new people were systematically supervised or not and how existing and tacit knowledge was communicated.

Management is particularly responsible for the development of governing documents and procedures. In addition, management must develop communication procedures and competency profiles. From the incident reports it is obvious that management in Statoil had not generated sufficient consistency in the organisation. On the one hand, this is expressed through lack of trust and understanding of how the governing documentations were to be used, but it can also be explained by the notion that the distance between the workers on the platforms (the "sharp end") and managers and experts onshore (the "blunt end") has increased. In some contexts, it seems that management explained failures solely as lack of compliance – which means that someone has not followed the governing documentation. As a top manager at Statoil expressed it: "Incidents happen when someone does not follow governing documentation."[7]

Isolated, this quotation can be interpreted as a statement that the only way to safeguard against accidents is by enforcing a greater degree of compliance. In context, however, it was more a recognition that many departments within Statoil operated

7 Executive Vice President of Development and Production, Norway, Øystein Michelsen, at the internal release of the IRIS report at Statoil, September 15, 2011.

in accordance with other logics of action than prevailing governing documents and procedures suggested.

During the past five years, safety programs with different names have been applied at Statoil in order to increase compliance. One confusing discussion that has evolved is whether these programs mix compliance with the principles of behaviour-based safety (BBS). The ideological demarcation line concerning such compliance programs has traditionally gone between the employers' and employees' organisations (Lindøe and Engen 2007). Some variants of compliance programs may imply that management categorises what kind of behaviour leads to accidents and determines how such behaviour can be observed and measured. Considering that positive behaviour should be encouraged, effects and changes have to be measurable as well. All factors mentioned imply an unambiguous focus on undeviating causes and the single employee. Moreover, it may lead to a pursuit for scapegoats. In this case, an unambiguous focus on compliance will by its nature be ignorant when it comes to underlying causes and the impact of the organisational environment. Because such compliance system assumes that the individual workers behave in accordance with certain given standards, the responsibility for accidents is placed individually. Governing documents may therefore serve as instruments to relieve managerial responsibility of organisational accidents and put the onus on the workforce.

According to the unions, such use of governing documentation is incompatible with the Nordic model, mainly because of the lack of democratic influence on program design and practice (see Chapter 10 in this volume). According to the unions, BBS-inspired programs lead to a deterioration of the working environment and the scapegoating of workers in case of accidents. The unions fear that existing collective and legitimate rights that have been developed in the Norwegian petroleum industry will be disposed and that the Nordic model is going to be replaced by management-dominated, top-down governance principles. The investigation reports produced in the wake of the Gullfaks C incident support that the unions still perceive the management's unambiguous focus on compliance as a hidden agenda, with the ultimate goal of implementing BBS at Statoil. Whether or not there is any truth to this, such suspicions create ambiguity, undermine trust and confidence in the governing documents and nurture local adaptations concerning safety behaviour.

14.8. CONCLUSIONS

This chapter has identified several underlying causes of the incidents at Snorre A and Gullfaks C. In 2007, Statoil and Hydro Oil & Gas merged and full integration of activities, resources and governing documents was initiated. All employees were given new job positions and new functions, tasks and responsibilities were assigned. As part of the integration process, large parts of Statoil's system for governing documents (DocMap) were included in the Hydro's system (APOS). This resulted in an increase in complexity of governing documentation. In this respect, the chapter shows how the entire system of procedures consisted of too many documents and that it was challenging to distinguish between processes, requirements and methods. More specifically, the procedures were perceived as cumbersome, hard to relate to and difficult to comply with as there were contradicting requirements related to a single operation. Factors related to *governing documents and compliance* therefore represented relevant underlying causes of the incident at Snorre A and Gullfaks C.

A fundamental premise in this respect is that the incident at Snorre A (and the subsequent improvement measures) was relevant for Gullfaks C. In this chapter we emphasise several important similarities between the underlying causes of these two incidents. Factors related to organisational context, management and decision making and compliance are central in both cases. The merger between Statoil and Hydro Oil & Gas contributed to the occurrence of the same type of challenges revealed in the Snorre A incident investigation. In particular, aspects related to rotation, knowledge transfer and governing documents are emphasised in this respect.

There are also several factors indicating that the change of manning of the platforms, effected in 2009 and 2010, had consequences for *management and decision making* at Gullfaks C. A large part of the management team at Gullfaks C was replaced and data from the IRIS report 2011 indicate that this process was characterised by a low degree of transfer of experience. This resulted in a lack of *field- specific competence*. In addition, this chapter shows that there was a low involvement of Statoil's central professional resources (MPD experts) during operations. Several aspects related to *planning* of the drilling operation are also emphasised. For example, it is a common perception that critical comments and remarks were muted. There were also important weaknesses of the *risk assessment* related to use of MPD. Haste in carrying out the operations is highlighted in both operations and in this respect, economic aspects and internal pressure also are pointed out as important explanatory factors.

From related studies of accidents and incidents in complex industries we know that such processes are not unique. Challenges concerning complexity of regulatory systems and uncertainties related to the degree of compliance are more or less general. Ambiguity concerning legitimacy of governing rules and documentation is another recurring factor. Similar cases also show how risk evaluations were carried out without being documented and indicate lack of (or insufficient) routines when it comes to documentation of relevant discussions concerning risks in the projects (e.g. minutes), which also means that decisions were based on selective and limited data. We also have numerous examples of lack of systems that ensure appropriate handling of whistle blowing. In other words, aspects related to organisational context, leadership and decision making and compliance as underlying causes of incidents (and accidents) are not unique to Snorre A and Gullfaks C, but rather are general challenges for the entire petroleum industry.

REFERENCES

Hopkins, A. (2012) *Disastrous Decisions. The Human and Organisational Causes of the Gulf of Mexico Blowout*. CCH: Sydney.

IRIS-report (2011) *Læring av hendelser i Statoil* (Learning from incidents in Statoil). Report no. 156, International Research Institute of Stavanger, Stavanger.

Lindøe, P. and Engen, O.A. (2007) "Behavior based safety and the Nordic model", in T. Aven and J.E. Vinnem (eds.): *Risk, Reliability and Social Safety*. Taylor & Francis: London, pp. 1705–1712.

Reason, J. (1997) *Managing Organisational the Risks of Organisational Accidents*. Ashgate: Aldershot.

Schiefloe, P.M. et al (2005) *Årsaksanalyse etter Snorre A hendelsen 28.11.04* Statoil rapport til Petroleumstilsynet: Stavanger (*Analysis of the Snorre A incident 11/28/04* Statoil report submitted to the Norwegian Petroleum Safety Agency: Stavanger)

Snook, S.A (2000) *Friendly Fire. The Accidental Shootdown of U.S. Black Hawks over Northern Iraq*. Princeton University Press: Princeton, New Jersey

"The Impact report" (2007) Oppsummeringsrapport av effektmålingene på tiltakene etter Snorre A Hendelsen 28.11.2007 (Summary Report of the effect measurements of interventions by Snorre A incident 28/11/2007). Statoil: Stavanger.

15

Inspections, Independence and Intelligence

Helge Ryggvik

15.1. INTRODUCTION

The aim of this chapter is to describe the essence of the Norwegian regulatory regime offshore by showing how inspections, audits and several enforcement measures are conducted. The main empirical base is qualitative discussion of several audits of companies and installations in the autumn of 2009, the year before the Deepwater Horizon accident.

Both immediately after the accident and when different commissions made their reports, the Norwegian regulatory regime together with the British Safety Case (Chapter 6.2.3) has often been held up as a positive alternative. However, it can be difficult for outsiders to understand how this seemingly contradictory combination of self-regulation and a strong public regulatory agency that characterise the offshore sector is working. Even though the main focus is the Norwegian experience, it is partly a comparative analysis. The broad approach to audits, inspections and enforcement measures in the Norwegian oil sector is mirrored by the very different way inspections were conducted in the Gulf of Mexico up until April 2010. Bang and Thuestad (Chapter 10) and Rosness and Forseth (Chapter 12) in this volume have shown how important tripartite cooperation between the state, unions and companies is in the Norwegian offshore safety regime. This chapter shows how conversations and conflicts between involved groups also to a large extent influence where, when and how audits are conducted. The chapter supports the claim that the broad, qualitative audits one can find in the Norwegian system represent a better way to uncover safety breaches than the narrower, technically oriented approach to inspections one could find in the Gulf of Mexico. However, both systems must be understood in their respective historical contexts. In the Norwegian system, years can go by before regulators visit an installation, which is why it is essential to uncover potential dangers in a timely manner. There is no guarantee that regulators always will have the necessary experience, independence and diligence essential for the system to work.

As part of the research project titled "Robust Regulation" in late March 2010, I visited the Minerals and Management Service's (MMS) headquarters for offshore safety in New Orleans. It was three weeks before the Deepwater Horizon accident. There I met with the then-leader of all offshore safety inspectors in the Gulf of Mexico. From the comments he made, it appeared that he was frustrated because he felt pressure to reduce the number of inspections of offshore installations. He told me that in internal MMS discussions, the Norwegian system of regulations – with much fewer but more system-oriented audits – was often mentioned. My immediate response was that if others should copy the Norwegian system, it is essential to understand the conditions that make it work. In the wake of the Deepwater Horizon accident, the MMS was criticised for having reduced the number of inspections on installations in the Gulf of Mexico. The National Commission's main report, which considered the reduction of inspections to be a sign of the gradual weakening of an already weak regulatory system, put the main responsibility on politicians.[1] Not only had authorities in Washington reduced the necessary funding; the political signals had also been that civil servants should avoid regulations that could in any way slow down the pace of development of the strategically important deep-sea drilling. At the same time both the National Commission and the report from BOEMRE (the former MMS) made recommendations for improvements that pointed in the direction of the more system-oriented British and Norwegian offshore regulatory regime.[2]

A regulatory regime can be defined by two, equally important elements: the *regulations* themselves and the system of *enforcement*. Enforcement is partly a question of uncovering violations of these regulations and partly a question of what sort of sanctions, punishment or even positive stimuli like explaining and advising are in place to ensure that operators adapt in the preferred direction.

The differences between an orientation towards prescriptive written regulations in the Gulf of Mexico and a more performance-based orientation on the Norwegian continental shelf are described by Baram (Chapter 7) in this volume. However, despite its prescriptive orientation, the U.S. regulations also have overreaching paragraphs with formulations placing an operator as responsible for maintaining all places of employment: "…free for recognized hazards to employees…"[3] Why is it that in the U.S. system this seems to have been dead letters, whereas similar general formulations in the Norwegian system have more practical implications? This can only partly be explained by a juridical approach. To some degree it is of course

[1] National Commission on the BP Deepwater Horizon Oil Spill and Offshore Drilling, Deep Water, The Gulf Oil Disaster and the Future of Offshore Drilling. Washington, DC (2011), p. 63.

[2] The Bureau of Ocean Energy Management, Regulation and Enforcement, Report regarding the Causes of the April 20, 2010 Macondo Well Blowout (14 September 2011).

[3] Hopkins, A. (2012) *Disastrous Decisions: The Human and Organisational Causes of the Gulf of Mexico Blowout*, p. 148.

possible to read the underlying regulatory philosophy into certain paragraphs. Here, however, we explain the fundamentally different regulatory approach by describing how inspections or audits are conducted and regulations are enforced.

15.2. U.S. INSPECTIONS

The historical background of the offshore-related regulatory system in the United States is well described by the commission established by President Barack Obama in the wake of the Deepwater Horizon accident.[4] As Baram shows in Chapter 5, before April 2010, the main U.S. institution with regulatory responsibilities offshore was the MMS.

Seen from a Norwegian perspective, it would be wrong to envision the regulatory system maintained by the MMS as a typical U.S. system. There are, of course, "American" elements in the approach to safety that are fundamentally different from safety in general in Norway. Driving along the wetlands, canals and swamps around the oil industry bases at Port Fourchon in Sothern Louisiana, one can regularly see large billboards advertising legal services. The lawyers offering their services seek those who have suffered an injury or a loss, ready to sue the transgressors. This litigious spirit translates into the U.S. system of safety inspections as well. With an established welfare state, free health care and collective pensions and insurance systems, it is less likely that safety-related issues would end up as court cases in Norway.

The system regulating risk in the offshore oil and gas industry in the United States, as well as in Norway, is influenced both by specific historical conditions that prevailed when it was established and developed and by peculiarities within the local industrial context. Scarlet et al. (2001) have compared the MMS with the U.S. Federal Aviation Administration (FAA) and Environmental Protection Agency (EPA).[5] Compared to these, the MMS uses voluntary industrial standards to a greater extent. Being an industry totally dominated by private firms, the offshore oil industry also is unique because companies do not have to adhere to regulations made by the Occupational Safety and Health Administration (OSHA). This is different both from Norway and all comparable industries onshore in the United States. In Norway, both the concrete paragraphs and the underlying philosophy in Work Environment Act of 1977 make up an important part of the regulatory system offshore, as shown by Bang and Thuestad (Chapter 10) in this volume. This could have been compensated for in the United States if the MMS had incorporated work environment and health issues in its regulations. Up until April 2010, this was never done.

4 National Commission (2011), chapter 3.

5 Scarlett, L. Fraas, A. Morgenstern, R. and Murphy T. (2001) "Managing Environmental, Health, and Safety Risks, A Comparative Assessment of the Minerals Management Service and Other Agencies", Resources for the Future, DP 10–64.

Based on its use of prescriptive regulations, the MMS's inspection model was more intensive than was the case for most other U.S. regulatory institutions, at least when it came to resources. Even though OSHA was a much larger institution than the MMS, its responsibility was so great, with far more companies to inspect per inspector employed, that its inspections had to be sporadic. The MMS's aim was to have all offshore installations inspected regularly. In such a perspective, the presidential commission's general description of the MMS as a rather weak regulatory institution must be qualified. The commission goes a long way in rejecting the image of the MMS as a corrupt institution, the one that flourished in the U.S. media after the accident.[6] It points out that the much-talked-about scandal could not be related directly to the MMS department concerned with safety. The commission describes the root problem as follows:

> [P]olitical leaders within both the Executive Branch and Congress have failed to ensure that agency regulators have had the resources necessary to exercise that authority, including personnel and expertise, and, no less important, the political autonomy needed to overcome the powerful commercial interest that have opposed more stringent safety regulation.[7]

The commission is mainly blaming the political system. Nevertheless, given that so many MMS safety activities were based on concrete inspection, the preceding quote should imply that there are shortcomings in how these have been conducted.

Even before the accident in April 2010, the MMS implemented several measures to ensure that inspections could be carried out with the greatest possible degree of independence from the companies that were to be inspected. In the Gulf of Mexico the concrete inspections were carried out from regional offices in small towns like Lafayette, Houma, Lake Jackson and Lake Charles. Each districts had their own helicopters. That meant that the inspector did not have to rely on transport from the operator they were inspecting. The aim was that each of the about ninety drilling rigs operating in the area should be inspected once a month. More than 3,000 production installations should be inspected once a year. The inspectors wear uniforms. To avoid too close personal relations developing between inspectors and personnel on the installations, the inspectors were rotated, so that no single one would inspect the same installation consecutively for more than two years.

Although activities in the Gulf of Mexico were increased from the 1990s to 2010, the number of inspections was reduced slightly.[8] They reached a peak in 1994, with 7,000 inspections in total that year. In 2009, the number of inspections was reduced to approximately 5,000. Through the 2000s, unannounced inspections were still in

[6] National Commission (2011), p. 77.
[7] Ibid., s. 67.
[8] National Commission (2011), p. 76.

use. However, compared to the 1990s, they represented only a small proportion of the total numbers. On the Deepwater Horizon rig, eighty-eight inspections in total were carried out between the onset of the drilling in 2001 to the accident in 2010.[9] One criticism of the MMS in the wake of the accident has been that the number of inspections in the last years was not in line with the aim of one inspection every month.

Apart from checking that all required equipment was in place and in order and used in the right way, the companies were also required to show the inspectors some documentation. They were, for example, required to submit documentation of when and how a blowout preventer (BOP) was tested, if the crew had conducted emergency drills and what sort of drilling mud was in use.[10] The visual part of an inspection could take care of everything from relevant drilling equipment and electrical equipment to general order on board rigs and on platforms. Both visual tests and review of documentation were conducted by filling in a written checklist. The MMS called the list "potential incidents and non compliance" (PINC).[11]

This detailed approach to inspections, using a written checklist, is of course well known in different types of safety institutions. Apart from a broader focus, including health and work environment issues, a similar kind of inspections have for years been conducted by the Norwegian Labour Inspection Authority and its U.S. counterpart, OSHA. However, by checking every installation at least in theory so regularly, the MMS was more reminiscent of the kind of certificatory activities that were conducted by private classification institutions like Det Norske Veritas(DNV), Lloyds and Aamerican Bureau of shipping (ABS) in the shipping industry.[12] Similar to the Norwegian regulation, it was clear that operators always had the main responsibility for all activities. One problem with this detailed approach was that it could create and appearance that the inspectors, by approving the activities through controlling the PINCs in the checklist, could then be held responsible if incidents or accidents took place.

Neither the presidential commission nor the report from BOMRE conducts a detail study of how inspections were conducted in general in the MMS. In a chapter about regulatory findings there is a detailed description of the inspections on the Deepwater Horizon prior to the accident, which gives valuable insights.[13] Given that BOMRE to a certain degree has been investigating its own organisation, it is not surprising that the panel making the report found "no evidence that the MMS

9 "BP rig inspections were fewer than advertised", *Associated Press*, 16 May 2010.
10 BOEMRE Report (2011), p. 162.
11 National Commission on the BP Deepwater Horizon Oil Spill and Offshore Drilling, Cheif Counsel's Report (2011), p. 260.
12 Det norske Veritas, Lloyds Register of Shipping, American Bureau of Shipping.
13 BOMRE Report, chapter XIV, p. 157.

inspection on the Deepwater Horizon conducted in 2010 were incomplete" and that it found "no evidence that the MMS inspection of the Deepwater Horizon while on location at Macondo were a cause of the Blowout".[14] However, the report does conclude that the likelihood of a blowout may have been decreased by some regulatory improvements.[15]

In his book *Disastrous Decisions*, Andrew Hopkins uses BOMRE findings for a more fundamental critique of the MMS approach to inspections. "The fact that the Deepwater Horizon could pass these inspections with flying colours, yet fail 'Well Control 101', as some commentators have put it, highlights the failure of MMS inspection strategy".[16] Hopkins's main criticism is that the MMS, by concentrating on hardware-related issues, completely ignored human and organisational factors. He also implies that the MMS, by confining inspection to doing checks against a list, may have undermined the competence in the institution.[17] Carrying out inspections using a list can be mechanical work that does not need very much experience. For an agency starved for resources, this was a benefit.

When the MMS inspection was limited to painstaking review of technical details, it was of course not only a question of an established practice or culture in the institution. Many leaders in the MMS were familiar with the approach to regulation on the British and Norwegian continental shelf. The elements of workers' participation by unions and health and safety representatives that were most strongly found in the Norwegian system may not have seemed very relevant. In the Gulf of Mexico, trade unions had no presence at all. However, as shown by Baram (Chapter 7) in this volume, the risk-based performance approach that followed from both the Norwegian "internal control system" and the British "Safety Case"[18] was followed closely. So, when in the early 1990s the MMS discussed a more system-oriented approach to safety, a key question was whether a management system based on systematic risk analyses should be introduced in regulations. A few years later, the American Petroleum Institute (API) introduced a risk-based management system as a voluntary standard (API RP 75 SMEP).

To make evaluations of management systems and the quality of related risk analyses against this standard part of MMS's inspections activities, it was necessary to make management and risk analyses a central part of regulations. Based on Norwegian experiences, such a fundamental change in regulatory approach must mean some retraining of MMS's employees.

14 BOMRE Report 2011, p. 164.
15 Ibid, p. 172.
16 Hopkins (2012), p. 141.
17 Ibid., p. 140.
18 Paterson (2000), p. 188. "Health and safety at work offshore", in Paterson, J. and Gordon, G. (eds.): *Oil and Gas Law, Current Practice and Emerging Trends*, p. 188.

15.3 NORWEGIAN AUDITS AND INSPECTION

Even with the reduction of inspection activities, for which MMS was criticised after the Deepwater Horizon accident, it still conducted a far larger number of inspections than the Norwegian Petroleum Directorate (NPD) or, after 2004, the Petroleum Safety Authority (PSA) had ever conducted. Independently of the efficiency of MMS inspections, the U.S. regulators have been much more visible on offshore installations than regulators in Norway. The Norwegian oil and gas sector has never had the kind of detailed governmental "hands on" offshore activities such as the United States had after the establishment of the MMS, not even in the years before the introduction of the internal control system in the 1980s (see Chapter 10 in this volume).

In fact, in the summer of 1974, just eighteen months after the establishment of the NPD, Arne Flikke, the then-leader of the new institution's safety department, left the organisation in protest because he felt he did not have the necessary resources to visit every offshore platform regularly enough.[19] Flikke wanted a system of inspections similar to what the MMS established later. The next year, the NPD's resources were increased but were still not enough to cover every installation with regular technical inspections. The challenge for the NPD was not only to hire the right number of inspectors. One major obstacle was to organise the trip out to the Ekofisk and Frigg installations, approximately 300 and 200 kilometres from the coast.

With the introduction of the internal control regime in the 1980s, it was often stated that, theoretically, the companies' safety performance could be checked from their headquarters onshore. According to the regulation, the operator was required to document all relevant safety activities. The essence of the performance-based approach was that companies should have an "internal" safety system that reduced risk effectively. The underlying idea was that the regulator should focus on whether or not a company had such systems in place, not on the details of how a company achieved its results.

In the first years after the new regime was introduced, the balance between simple audits of documentation in the companies' headquarters onshore and more hands-on inspections on installations shifted somewhat, based on experience. Even if one with documentary evidence could check a lot, in practice it was sometimes necessary to inspect actual activities out in the ocean. However, there were some dilemmas that were built in as contradictions in the system. One essential part of the internal control system was to make it unequivocally clear that it was the operator who was responsible for ensuring that accidents do not happen. When something went wrong it should not be possible for a company to hide behind the fact that it

[19] Ryggvik and Smith-Solbakken (1997), p. 163.

has been working in accordance with regulations. In the same way it was important to avoid that inspection or audits could take the form of an endorsement of an operator's safety work. The more detailed such an inspection was, the easier it would be for a company to share blame with regulators if something went seriously wrong at the actual site. Nevertheless, the aim of any regulatory regime must be to uncover serious flaws in time, before an accident happens. To uncover inconsistency between an operator's "system" on paper and the practice out on the platforms, on-site inspections on particular installations can be necessary. But sometimes the underlying weaknesses are neither company- nor installation-specific, but more a reflection of a general pattern in the industry. In that case the intervention from the regulator must be more general as well. Certainly, coping with these and other challenges, making this regulative approach to succeed, requires both experience and intelligence from the regulators involved.

In accordance with this approach in the regulations, neither NPD nor later PSA kept statistics on the number and the scope of visits to offshore installations. However, it is easy to count the number of recant inspections on the basis of easily available information from the PSA's Web site.[20] Because of both historical and technological differences, counting the number of inspection in the Norwegian offshore sector and the Gulf of Mexico does not tell much in itself.

From the 1990s, the production on the Norwegian shelf was larger than in the Gulf of Mexico (2.3 million barrels per day against 1.7 million barrels per day in 2009). These numbers are changing fast, with the production in the United States increasing and the production in Norway falling. In 2009, the Gulf of Mexico had approximately 4,000 installations, compared to Norway's approximately 100 installations. This is because the activities in the United States originally were an extension of oil activities onshore, with much smaller, often unmanned installations, many in very shallow water. Recently built deepwater installations on the outer U.S continental shelf are much more similar to Norwegian installations than the older ones were.

The best indication of the level of activities is the number of hours worked. In the United States, total hours worked in 2009 amounted to 140 million, compared to Norway's 42 million.[21] The numbers of wells drilled gives a similar impression. While 150 to 200 wells were drilled per year on the Norwegian continental shelf in the years after 2000, the numbers in the same period in the Gulf of Mexico was between 600 and 1,000 wells.[22]

[20] Petroleum Safety Authority, *Status og Signaler* (2009), p. 41.

[21] International Regulators' Forum, Offshore Safety (IFR), Country Publication Data. www.irfoffshore-safety.com/conferences.

[22] RNNP (2009), p. 18. MMS, Incident Associated with Oil and Gas Operations.

In any case, the difference between the number of U.S. inspections and the number of Norwegian audits is so great that it only can be explained by at completely different approach. While the MMS, as noted, conducted approximately 5,000 inspections in 2009, the PSA in the same year conducted 99 audits. Several of these audits did not involve any hands-on inspection on offshore sites at all. To get an impression of how often the PSA agents visit offshore installations, one can use the number of the so-called "offshore supplement" (extra pay for working offshore) they receive during a year. The numbers show that in 2009, PSA staff stayed offshore a total of 291 days.[23] Of these, 259 stays included accommodation. Most inspections were conducted by delegations of two or more inspectors. Measured in time, this presence does not correspond to more than the annual working days of one to two ordinary oil workers. Hence, more than 30,000 oil industry workers on the Norwegian continental shelf will seldom or never meet PSA inspectors offshore.

However, the qualitative differences between U.S. inspections and Norwegian audits are so great that a comparison of the number of inspections does not in itself tell us much. Therefore, the best way to describe these differences is to show concretely how PSA decides where and when to audit and inspect and how these audits and inspections are carried out.

15.4. INSPECTIONS IN OCTOBER 2009

When internal control was introduced, the goal of the NPD's safety department was to limit itself to inspecting the companies' overarching safety systems. In the beginning, the term "system reviews" was often used to describe the kind of audits or inspections one expected to conduct. In recent years, the PSA began using the English-language version of the term "audit" to describe a process where a team of inspectors review certain safety or work environment issues at a specific company or installation, ending in a written report. A review (made by the author) of all audits made by the PSA and completed in 2009 shows that the focus is often at this overarching level. However, most inspections are targeted at fairly specific issues. One specific audit can consist of meetings, documentary research and hands-on inspections stretching over a long period of time. In many cases it will take more than six months from when a company is notified of an upcoming audit until a report from the PSA is published. A list of every audit reported in October 2009 gives a good impression of the scope of the kind of regulations the PSA inspects. Here scope refers both to what kind of installations and companies are inspected and to the focus of these inspections. The point here is to describe how the PSA's qualitative

[23] The numbers are based on offshore supplements paid for PSA employees. These numbers were sourced with the assistance Ole-Johan Faret in the PSA's information department.

approach is being implemented in practice. The month of October was chosen simply because it is generally a good working month in Norway, without any public holidays. The following fourteen audits were conducted:[24]

1. The Snøhvit LNG plant near Hammerfest.
2. The drilling rig Ocean Vanguard. B
3. The production platform Veslefrikk.
4. The state-owned Gassco's gas pipeline between Heimdal and St. Fergus in the United Kingdom.
5. Production equipment on the Alvheim field.
6. Petro-Canada Norge AS's drilling activities on the Norwegian shelf.
7. Emergency drills on Ekofisk.
8. Risk perception in diving in ConocoPhillips and Subsea 7.
9. Service Jack 1 (delivery of an accommodation module and other equipment on the Ekofisk field).
10. The drilling rig Rowan Gorilla VI.
11. ExxonMobil's use of risk analyses.
12. Verification of an earlier review of logistics (lifting from cranes etc.) on Rowan Gorilla VI.
13. Maritime systems on Norne and Njord.
14. StatoilHydro's emergency drills on Njord A.

Of these fourteen reports, only the audit of ExxonMobil's risk analyses was limited to meetings in the company's headquarters and documentary research, and was thus like the kind of overarching "system revision" which was aspired to in the 1980s. The use of risk analyses is, as shown, a very central part of the Norwegian regulations as discussed by Bang and Thuestad (Chapter 10) and Kringen (Chapter 11) in this volume. The actual inspection in 2009, however, was narrower in its scope in the sense that it was not a full review of ExxonMobil's safety system. It consisted of a pre-planned meeting in ExxonMobil's headquarters in Stavanger, where the company described how it conducted risk analyses, while five PSA inspectors asked questions.

Most other audits also started with pre-arranged meetings in the relevant company's headquarters. These were, however, followed up by pre-announced visits to installations. The audit of state-owned Gassco started with a review of its management systems at the company's headquarters in Karmøy, north of Stavanger. Gassco was split off from Statoil when Statoil was partly privatised in 2001. With a staff of only 300, Gassco was the official operator of most pipeline networks on and from the Norwegian continental shelf. However, Gassco had contracted out a substantial

[24] All audits are available on the Petroleum Safety Authority's Web site at www.ptil.no.

part of its area of responsibility to private companies. The focus for PSA's audit in 2009 was how Gassco was supervising the French company Total E&P, which operated the receiving terminal in St. Fergus near Aberdeen. After a meeting in Gassco's Norwegian headquarters, the PSA regulators went for a hands-on inspection at the facilities in Scotland. Both during the meeting in Norway and the visit to the St. Fergus plant, representatives from the British Health and Safety Executive (HSE) also participated. This kind of practical international cooperation was natural because it involved inspecting a company structure and technical equipment which interacted across national borders.

The audits and inspections on the fields of Ekofisk and Njord focussed respectively on ConocoPhillips' and StatoilHydro's emergency preparedness. In both cases the PSA was observing an emergency preparedness exercise. The purpose was to verify whether the preparedness met with both PSA's own standards and the companies' rulebooks. These exercises focussed both on evacuations of crew and technical aspects. For StatoilHydro, the maritime system on the floating production platform at Njord and the production ship on the Norne field were also inspected. The maritime part of the inspection was conducted in cooperation with the Norwegian Maritime Authority. Even though the PSA was responsible for all safety issues on both floating and fixed installations, it used expertise from other agencies when it was needed.

The audit of the Snøhvit LNG plant near Hammerfest, close to the North Cape, was part of the regulatory responsibility which the PSA acquired when it became responsible for onshore oil installations from 2004 on. In the United States, OSHA, not the MMS would have conducted similar inspection on onshore facilities. For the Norwegian PSA, the inspection of the onshore LNG terminal was different from others in the sense that it followed regulations that demanded yearly inspections of activities where dangerous chemicals were involved. The PSA had inherited regulations and inspection routines that had been developed by the Norwegian Labour Inspection Authority. The regulations and the inspection routines would most likely have been different if they had been developed by the PSA itself. If the PSA did not carry out similar regular inspections offshore, it was partly because it was feared that such reviews could become a kind of approval procedure, making the PSA responsible if something went wrong. However, when PSA regulators visited the LNG plant, they did not walk around inspecting technical installations. As in other audits offshore, the three-day visit focussed on whether Statoil had conducted acceptable risk analyses.

The audit of the Service Jack 1 project is the only one of the fourteen reports to focus on the design and construction phase. Some new cranes, an office module and a larger accommodation module at Ekofisk field were to be constructed. A large part of the actual work was to be done in Indonesia. The inspection was conducted by interviews and documentary research at the supplier Master Marine's project offices in Oslo and Singapore, and through a visit to the production plant in Batam, Indonesia. PSA's approach was wide in the sense that it covered a broad range of safety aspect – from correspondence between documentation and reality, HSE questions, design and safety procedures in relation to cranes and lifting operations. The audit was unique in the sense that the PSA was inspecting installations in a country where it had no jurisdiction. The PSA had neither any legal authority to inspect nor any possibility to sanction actors in Indonesia and Singapore. The main purpose of the inspection was to gain a more comprehensive idea of how the installations were constructed. However, the inspections resulted in a long list of demands for improvements. Some of these had to be conducted at the plant in Indonesia. PSA's power was, of course, in the right to prevent the operator PhillipsConoco from using the installations on the Norwegian shelf if it did not meet the necessary standards.

The broadest overall review of a single production field was PSA's audit of the Alvheim oilfield where oil is produced from a production ship and several underwater installations in the nearby British sector. The operator is the medium-sized U.S. oil company, Marathon. Typical of many other small and medium-size oil companies, Marathon had handed the operation of the field to a contractor, the Danish Maersk Drilling. Here also, the PSA's approach was different from what the MMS would have done in the United States. Before the Deepwater Horizon accident, the MMS limited its inspection to the company formally defined as an operator. When inspecting technical details on a drilling rig, U.S. regulators would meet all kinds of personnel. However, the formal communication would be between the MMS and operator's representatives on board the rig. In this sense, the U.S. system goes further than the Norwegian one does to underline that the operator is responsible for everything its contractor firms do. Even if the same formal responsibility is made clear in Norwegian regulations, the inspections of Master Marine and Maersk are examples where the PSA explored contractors' activities in some depth.

The Alvheim audit started with a meeting onshore, followed by a four-day preannounced visit on Marathon's vessel in the North Sea – in other words, a vessel operated by workers from Maersk Drilling. In PSA's report, Maersk was criticised for poor maintenance of critical technical equipment. The report also discussed more organisational issues like the lack of oversight of safety barriers. At the same time, the report focussed on lack of expertise and training of the crew. The PSA wanted to check whether Maersk, which was traditionally a construction and transportation firm, had the necessary expertise to operate a drilling installation on an oilfield. Where HSE questions were concerned, the PSA showed that there seemed to be a culture of using too much overtime on Alvheim. The inspector pointed to the Working Environment Act, where working overtime was only acceptable when unforeseen circumstances left no other alternative. The PSA did not go as far as to issue fines or initiate a criminal investigation. However, the report had to be read as a warning.

As with the MMS in the United States, and as a consequence of the potential risk, inspection of drilling installations is overrepresented among PSA's inspections. Four of the fourteen reports from October 2009 discussed drilling-related aspects. The inspection of the Veslefrikk field was partly related to an application from the operator StatoilHydro for extended use of the existing drilling installations. This time the inspections were conducted in cooperation with the Climate and Pollution Agency (Klif). This is another example of the PSA using expertise from other government agencies when needed. The focus was on the health and environmental consequences of increased use of chemicals which were used to increase production when fields were in the so-called tail-end production phase. The PSA concluded that StatoilHydro had not conducted the necessary assessment of the potential risk for oil workers and the environment. In the report, several deviations both from more general paragraphs of the Work Environment Act and from PSA's own safety regulations were identified. The audit was partly a response to objections raised by environmental movements concerned about the consequences for the environment of various measures to increase production from old fields. However, this was an example where health, environmental and safety-related questions were intertwined.

The audit of the drilling rig Ocean Vanguard started with a three-day meeting in the operator Diamond's office near Stavanger in December 2008. The inspections ended with a four-day visit to the platform in August 2009. Three PSA employees participated. Both during the meeting at the headquarters and on the platform, the main focus was on the working environment. Diamond was given a directive to make sure that a work environment committee was established in accordance with regulations. Workers' exposure to noise was also discussed.

The audit of the oil company Petro-Canada can be seen as a consequence of its role as a recent drilling operator on the Norwegian shelf. The inspection focussed on several fundamental safety aspects related to drilling. Questions were asked as to whether the company had conducted the necessary risk assessment before starting drilling. In the light of the subsequent Deepwater Horizon accident, one can note that PSA's inspectors doubted whether Petro-Canada had the necessary expertise "in connection with challenges resulting from drilling under high pressure and high temperature wells". In the report the PSA did not draw any conclusions about whether Petro-Canada had appropriate expertise and experience concerning well design and blowout simulations, because the company intended to use external contractors.

In a general audit of the jackup rig Rowan Gorilla VI, the PSA discussed several safety issues relevant to a drilling operation. Again, however, the main focus was on whether the company was operating in accordance with the Work Environment Act. The rig was operated by the British company Rowan Offshore. The inspection is of general interest in the sense that it was typical for several other foreign rigs which were just starting operating on the Norwegian shelf. The company did not have any experience in working under Norwegian regulations. As with similar other instances, the rig was inspected in a foreign shipyard or harbour, in this case a shipyard in Dundee, Scotland.

As one can see from the preceding examples, PSA's inspections were conducted using a broad range of approaches. In several cases the main focus was purely on questions of the working environment. Other inspections were more focussed on questions of safety alone. Instead of focussing on technical details, however, these inspections were also targeted at the company's safety management systems. A common thread was the identification of an absence of risk analyses from the company's side.

15.5. STATEMENTS OF COMPLIANCE, CONSENTS AND ACTORS' EVALUATIONS

Apart from regular audits that take up a significant part of PSA's regulators' time, the institution has a range of others, partly related instruments to influence the industry. Although the Norwegian system puts a great emphasis on safety being the operator's responsibility, both the actual installations and equipment and the operators themselves have to go through a formal approval process before being allowed to operate on the Norwegian shelf. These procedures are slightly different depending on whether floating rigs and other maritime equipment or fixed installations must be assessed. The differences can partly be traced to the period when the regulatory responsibilities were divided between the maritime regime and the Petroleum Directorate.

A floating vessel has to get approval from PSA in the form of an Acknowledgment of Compliance (AoC)[25] before it can operate on the Norwegian shelf (Moen and Lindøe 2008). The AoC system was introduced in 2004 and later reframed. The October 2009 audit and report on the rig Rowan Gorilla was part of such a AoC process. In fact, Rowan Gorilla was the first rig to be rejected after the new system was introduced. The rejection was partly because of technical failures and partly because the rig's staff was not working in compliance with Norwegian labour standards The aforementioned inspection in 2009 was part of a new review where the inspectorate wanted to check whether earlier directives had been carried out. In accordance with PSA's philosophy of internal control and purpose-based, functional regulations, an AoC also focusses on general matters like the company's safety management systems and whether or not the company has accepted safety representatives and work environment committees. However, since the circumstances when errors take place are by definition specific, the subsequent directives can also focus on technical details, like access to safety equipment and the like.

[25] In Norwegian: *Samsvarsuttalelse* (SUT).

PSA's handling of the AoC for Rowan Gorilla also had a politicised side, which could not be read explicitly in PSA's report. When Norway joined the European Enlarged Economic Area (EEA) in 1993, oil industry workers feared that a free market for services in the oil sector would lead to "social dumping".[26] After a political strike in 1992, the oil industry workers were assured that Norwegian wages and general working conditions like safety and union recognition would apply when foreign companies were operating on the Norwegian sector. The problem of social dumping was considered to be particularly significant for drilling rigs and other kinds of services which could easily cross borders with the same crews. Seen from a union perspective, these were challenges in which wages interacted with safety-related questions. Norwegian workers feared that if foreign rigs with foreign crews did not adhere to Norwegian safety demands, they could undermine or outcompete Norwegian rigs. This explains the attention in the AoC to safety representatives and other similar measures. Workers' wages were, of course, not a main concern for the PSA. But making sure that the election of health and safety representatives was not only a right on paper was considered to be important. Workers' representatives should be able to put forward a safety concern without fearing for their career.

To strengthen its position in its second attempt to get the necessary approval to operate in Norway, Rowan Drilling assured the PSA that the company would join the Norwegian Shipowners Association. The company also contacted the oil workers' union LO.[27] A union secretary in IE responded in the autumn of 2009 by stating that his impression was that Rowan Drilling "would do what they could to follow Norwegian regulations and tariffs and working hours". This was a signal that could be interpreted as if IE was endorsing the company and therefore accepting that it should be allowed to operate in Norway. The union secretary, however, modified his statement subsequently, saying that he was sceptical of the company and that the union would follow up on it closely. Because there were several unions competing for the same members on the Norwegian continental shelf, there was some room for political games here. By accepting one union, the company not only accepted certain working condition, but also made it easier for that particular union to get members. When the PSA finally agreed to give Rowan Drilling an AoC, it was PSA's decision alone. However, an important part of the informal side of the Norwegian system was to listen to the union's statement.

To perform a given operation, being approved as a competent company by an AoC is not enough. As in the United States, all activities, from planning, drilling and productions to the removal of old installations, have to be approved by the PSA in a form of "consent" (*samtykke*). The most comprehensive of these is the operator's plan for the

[26] Ryggvik and Smith-Solbakken (1997), p. 314.

[27] Industri Energi (IE). Petro.no, 26 November 2009. "Will follow Rowan Gorilla tight".

development and operation of a petroleum deposit (PDO) and the plan for the installation and operation of facilities for the transport and exploitation of petroleum (PIO).

In October 2009, the PSA approved twelve such PDOs and PIOs. These included approval for a reconstruction of the Oseberg C-platform. Another was an approval for Statoil's plan to accommodate workers on Sleipner B, previously an unmanned installation. Six of the approvals were drilling permits. One was an approval for Marathon to drill in the Alvheim and Volund fields from the Transocean Winner rig. The rig was operated by Transocean Offshore (North Sea) Ltd. The rig had a AoC from 2006.

In the United States, the MMS administered approval procedures that were very similar to corresponding procedures in Norway. Some of them are described by Baram (Chapter 7) in this volume. Again the main difference can be found in how procedures and regulations are implemented and enforced. This can be illustrated by the permit to drill and the much-discussed oil spill plan for Deepwater Horizon's Macondo well. As part of the MMS's detailed prescriptive regulatory approach, all such plans should be followed up in detail. The practice of using copy/paste buttons in producing documents for approval by regulators was demonstrated with an oil spill plan describing dangers to sea mammals like sea lions and walruses – mammals that do not exist in the Gulf of Mexico.[28]

Norwegian regulators would certainly hope that, in the aftermath of a possible accident, they are not accused of similar mistakes. However, the main difference with the U.S. system will be that some AoCs, PDOs and PIOs can be issued without inspections and formal paperwork, based on trust in the company involved. Again the apparent contradiction to the internal control principle is relevant. Detailed procedures for approvals might imply that the PSA endorsed the company's activities and hence must also take some of the responsibility if something goes wrong. However, just as regular inspections are only carried out on some installations, the level of detail leading to approvals varies from case to case. The companies are legally obligated to deliver documentation of their activities, including risk assessment. However, in cases where PSA's inspectors have no reason to believe that there is anything wrong, they can approve without checking the details in the documentation or conducting any hands-on inspections. The idea is to set detailed audits and inspection only in circumstances where there are causes for concern.

15.6 THE ENFORCEMENT REGIME

The MMS and the PSA have rather similar systems for how accidents and serious near-accidents are to be followed up by investigations. In both countries and for some instances, these investigations could be conducted in parallel with or in

[28] National Commission (2011), p. 84.

cooperation with criminal investigations. Both countries have a set of increasingly more serious enforcement measures where they, as a last resort, could stop production and demand that a company lose its right to a concession. The last is more of a threat than a real practice. The MMS has levied fines for individual breaches of specific regulations on a more regular basis than the PSA has done in Norway. From 1990 to 2008, the MMS has levied 553 fines connected to offshore activities in the Gulf of Mexico.[29] The amount of fines varied between $800 and $810,000. The total amount of all fines in the same period was approximately $20 million.

Again, even when it comes to these kinds of strict enforcement strategies, the PSA has tried to incorporate the same performance-based regulatory philosophy.[30] Even if PSA's philosophy is that smaller deviations from regulations should be corrected through dialogue, the institution has several coercive measures at its disposal.[31] After investigative reports or serious violations of regulations, PSA can issue an enforcement notice (*pålegg*). An enforcement notice will, in most instances, follow a warning or a notification of order (*varsel om pålegg*). If the response to an enforcement notice is not satisfactory, it can be followed by stopping operations, compulsory fines, penalties and special compulsory measures. In 2009, a total of eighteen notifications of order were made. Some orders may make the use of certain technical solutions legally binding on companies. Here too, however, the PSA tries to operate within the principle of functional regulations. Hence most orders concentrate on criticising what is wrong; they do not often specify what the companies must do to improve, but leave it to them to decide.

One example is a notification of order which BP received from the PSA in September 2010.[32] The notification focussed on injuries because of workers' exposure to noise. When the PSA did not get any response to the notification, the following enforcement notice was issued. BP was instructed to

> review its system for managing noise damage risk for employee groups and taking necessary steps to achieve compliance with regulatory requirements. This shall include a review of procedures and systems to identify necessary improvement measures, and also the development of schedules for risk-based corrective measures related to the risk of noise injuries.[33]

In accordance with PSA's idea of cooperation and dialogue, very few enforcement notices were followed by stricter enforcement measures. In Norway in 2009, a total

29 www.bsee.gov/Inspection-and-Enforcement
30 Granskingsrapport etter kollisjon mellom Big Orange XVIII og Ekofisk 2/4-W. Petroleum Safety Authority, 13 October 2009.
31 Petroleumstilsynet, Sikkerhet, Status og signaler (2008) "Oftest nok med dialog".
32 Petroleum Safety Authority, Notification of order to BP, 9 July 2010.
33 Petroleum Safety Authority, Order to BP after follow-up of noise damage risk, 3 September 2010.

of eight new audits of serious incidents were initiated.[34] The only report completed in October 2009 discussed an incident where the vessel *Big Orange XVIII* collided with one of the installation on the Ekofisk field. Following a serious gas leak incident at the Snorre A platform in 2004, (presented by Engen in Chapter 14 of this volume), Statoil received a fine of US$13 million. This fine was issued as late as 2008, after an extensive investigation process. This was the highest fine ever for an operator on the Norwegian shelf. For Statoil it was certainly a question of prestige: the same year it had a turnover of about US$100 billion and its profits were about US$7 billion. As we saw, the fines issued by the MMS in the United States were small compared to the companies' income. In Norway, fines were even more peripheral to the safety system.

15.7. CRITERIA FOR ATTENTION

The preceding examples demonstrate a distinct difference between how regulatory authorities audit and inspect offshore activities on the Norwegian and U.S. continental shelves. Whereas the Norwegian regulators take on a much broader set of safety and working environment issues, the Americans are focussed on narrower technical issues. Whereas the Norwegian regulators often focus on an overarching management level and are concerned with outcomes, the U.S. system is oriented towards whether or not companies work in compliance with detailed procedures.

But precisely because of the Norwegian system's broad, overarching approach – in contrast to the detailed approach that dominated in the MMS before the Deepwater Horizon accident – a substantial proportion of activities on the Norwegian shelf are carried out without any direct government oversight at all. As we have seen with the AoCs, PDOs and PIOs, in theory every installation and activity has to be approved. On paper, therefore, the authorities appear to have a hand on the wheel in every critical activity. In practice, however, as we have shown, many of these approval procedures are based on trust between the regulator and the regulated. This determines whether the detailed documentation is analysed and whether or not the PSA follows up with inspection, checking whether the company is actually following their documented plans. As an example we saw that the Transocean Winner rig was recognised as fit to operate in the Norwegian shelf by an AoC in 2006. When the PSA issued a drilling permit in the autumn of 2009, involving the use of the same platform, it concentrated on Marathon's capabilities as its operator, not the condition of the platform. Without maintenance and constant education of the crew, a rig can deteriorate in less than three years. In this instance, however, the PSA trusted Transocean to keeping the rig fit to operate safely.

[34] Petroleum Safety Authority, *Status og Signaler* (2009), p. 45.

A decisive precondition for the Norwegian regulatory system to be workable is that at any given time it captures the significant safety challenges. As long as the PSA does not conduct systematic regular comprehensive inspection routines, it is essential that the approval procedures and the relatively few inspections that are conducted have as much impact as possible, both on the inspected companies and on their uninspected fellows. Apart from identifying weak points in the companies' performance at any given time, it is essential to generalise from all new experiences, making sure that companies that have not been subject to inspections learn before they make the same mistakes themselves. These are challenges that demand a high level of competence and experience from PSA inspectors. At the same time, it is a system that relies on the tripartite arrangement – embedded in the Work Environment Act – being effective at every level and area within the industry. The health and safety representative system can be seen as an extended arm of the official regulatory system. While checks by the PSA will be sporadic, the health and safety issues, in theory, will be represented permanently on every work site.

Even if an ordinary oil industry worker seldom or never meets PSA representatives during a work year offshore, the contact between PSA's employees and union and health and safety representatives is tight. The Labour Inspection Authority, union representatives, platform chiefs and others active in safety work regularly meet PSA representatives in safety conferences and different arenas of tripartite cooperation onshore. This is very different from the U.S. continental shelf, where regulatory representatives are very visible on installations but seldom or never meet elected representatives of oil industry workers.

Like other institutions that do not conduct inspections on a regular basis, covering all installations, the PSA will use safety indicators both directly and indirectly when it decides when and where to make audits. As described by Blakstad (Chapter 9) and Bang and Thuestad (Chapter 10) in this volume, the RNNP project has been developed as a rather advanced tool for monitoring the risk level on the NCS, combining statistical indicators with qualitative measures. Nevertheless, the focus of most of the audits mentioned in October 2009 can be related to issues that in the previous period had formed part of discussions in these arenas of cooperation described by Rosness and Forseth (Chapter 12) in this volume. In fact, even the development of the RNNP program is a more or less direct consequence of the initiative taken in these arenas.

The inspection of Petro-Canada and Maersk's activities on Alvheim can be understood as a response to concerns expressed by unions as to whether small oil companies and companies with a contracting background without experience in production had the necessary expertise. The follow-up of construction work in Singapore and Indonesia can similarly be seen as a response to both trade unions and some competing Norwegian companies which questioned the quality of the work, something that may, of course, have consequences for safety. To the extent that PSA's inspection is a response of discussions between organised workers, company associations and other involved actors, the selection criteria for where to focus the audits acquire a certain political character. If union representatives and company associations bring up certain HSE-related questions, these will very probably be reflected in PSA's inspections. In terms of the underlying rationale for the Norwegian tripartite cooperation, it is an advantage that discussions between the parties and about current political challenges are reflected in the criteria for how one should conduct audits, inspections and investigations.

15.8. CONCLUSIONS

However, there are no guarantees that active safety representatives and open discussions between unions, companies and regulators will uncover all weak spots in the industry's approach to safety challenges. Even in a historical period with strong unions concerned about safety issues and well-educated health and safety representatives, there may be workplaces where local workers' representatives are more afraid of becoming unemployed, weakening career opportunities and similar negative outcomes if they take up issues that might be expensive for a company. A generation of strong, well-educated workers can be followed by a less confident generation for several reasons. In the same way, a generation of employers that accept that workers can constitute themselves as a collective can be replaced by employers with a neglectful or even hostile approach to the legitimacy and proactive role of unions.

The Deepwater Horizon accident shows that important operational decisions concerning safety may be taken in areas where workers who naturally organise themselves in unions have little or no presence at all. Some of the main mistakes in the run-up to the accident were made by BP leaders onshore, BP's drilling chief offshore and Transocean's toolpusher, with the crew not knowing what kind of risk they were exposed to.[35] The toolpusher – a leading position on the rig – who wrongly concluded that the well was under control did not himself have full insight into the serious concerns and discussions there had been between BP leaders onshore and cementing experts from Halliburton during a difficult drilling process. Because major operational functions can be conducted by using electronic devices, personnel considered as part of management can have important operational functions. Here, of course, workers can win back some control, demanding more open oversight of processes essential to their safety. Nevertheless, in a system largely based on trust, there will always be potential for that part of the system to erode, without regulators recognising this in time. At the same time, as documented by Engen in Chapter 13 in this volume, new technological and historical conditions will create

[35] National Commission (2011), Chief Counsel's report (2011), BOEMRE report (2011).

challenges that require new solutions. This will require intelligent regulations and audits, something that will require a lot of the PSA as an institution.

A necessary precondition for such a continuously intelligent approach will always be a high degree of expertise and experience among PSA's employees. To have the necessary will to address the weaknesses identified, both integrity and independence are required. At the same time, to achieve the necessary respect in the industry, both professional and formal authority is required from the regulators. These are questions which can be approached with some objective measures like high wages, good educational programs, experience requirements, institutional formations that underpin independence and a focus on ethical rules. In the end it also comes down to more subjective factors like the background and attitude of every single regulator. As shown by Bang and Thuestad (Chapter 10) and Rosness and Forseth (Chapter 12) in this volume, there are several historical examples where earlier the NPD and the PSA have acted as a very confident, independent institution in its relations with the industry. But it would be naive to think that future regulators in the PSA might not develop the kind of links between themselves and some company interest for which the MMS was criticised in the aftermath of the Deepwater Horizon accident.

REFERENCES

Hopkins, A. (2012) *Disastrous Decisions: The Human* and *Organisational Causes of the Gulf of Mexico Blowout.* CCH: Sydney.

Moen, A and Lindøe, P.H. (2008) "Acknowledgment of compliance (AoC) – A mix of regulatory principles in the petroleum industry". Presentation at conference *Working On Safety*, September 30-October 4, 2008. Crete, Greece.

Paterson, J. (2007) "Health and safety at work offshore", in J. Paterson and G. Gordon (eds.): *Oil and Gas Law. Current Practice and Emerging Trends.* Dundee University Press: Dundee, pp.187–230.

Ryggvik, H. and Smith-Solbakken, M. (1997) *Norsk Oljehistorie.* Bind 3, Blod, svette og olje (Norwegian oil history. Vol. 3, Blood, sweat and oil). Leseselskapet: Oslo.

Scarlett, L., Fraas, A., Morgenstern, R. and Murphy, T "Managing Environmental, Health, and Safety Risks, A Comparative Assessment of the Minerals Management Service and Other Agencies. *Resource for the Future, Discussion Paper, DP 10-64, 2011 January 2011.*

16

Advancing Robust Regulation

Reflections and Lessons to Be Learned

Andrew Hale

16.1. INTRODUCTION: OBJECTIVES AND STRUCTURE

Attempting to summarise what has been said in such a complex, densely argued and extensive book as this one is impossible. The summary would need to be as long as the book to do it justice. This chapter, therefore, is a personal reflection on what has been written in the preceding chapters. It draws out and discusses some common themes and issues, the choice being necessarily idiosyncratic and reflecting the concerns and biases of the author. It also tries to define some lessons which might be learned both within and across the regimes which form the backbone of the book. These are the Norwegian, U.S. and British regimes, the latter supplemented by Hayes's contribution (Chapter 8) on the Australian system which shares many of the characteristics of its British sister or even stepmother. The proposed EU regulation also makes a brief entry in Chapter 6 by Patterson.

16.1.1. *What Is 'Robust'?*

One aim of the book is to consider what constitutes robust regulation and how that can be supported. This consideration requires an answer to the question: What are the signs that regulation is robust? This is the same sort of discussion as has emerged around the word 'resilience' in recent writings on major hazard prevention (e.g. Hollnagel et al. 2006). Indeed, the two adjectives (robust and resilient) might be seen as cognate if not as alternatives for each other. 'Robust' has a connotation of standing up to and resisting attack and weathering storms unchanged, whereas 'resilient' has a more flexible feel to it, bending and adapting and bouncing back. It is clear from the chapters on all of the regimes studied that the second metaphor is more appropriate for them than the former. The histories of all three regimes show early periods of major change and adaptation, followed by reappraisals and modifications at intervals. The chapters on the Norwegian regime in particular show how that regime has responded to major challenges from accidents, economic downturn leading to cost-cutting and loss of trust between major parties. The main question that reading the book seems to pose is whether the adaptations which followed those challenges are a proof of robustness or, in contrast, indications that the earlier regime was flawed and needed reform. The Norwegian authors present them as indications of robustness because the basic principles underlying the regime, notably its tripartite nature and reliance on trust, were preserved or restored. If we accept this argument, it can lead to a definition of robustness as a regulatory regime whose basic design principles stay the same over time, or are restored after a challenge, but whose detailed operationalisation adapts to changing demands and situations.

16.1.2. *Indicators and Comparisons*

Amongst the themes of the book, as announced in the Introduction, are comparisons and lessons. They might even seem to suggest the unspoken objective of assessing which of the regimes is 'better' than the others, with the implication that the less good should then learn from the best. One thing which the book as a whole, and particularly Blakstad's chapter (Chapter 9), shows is that this unspoken objective is on the one hand optimistic and on the other needs to be treated with extreme care. It is optimistic because the indicators of safety performance are too idiosyncratic and disparate to make between-regime comparisons very meaningful, although there is some indication in Baram's description of the U.S. system (Chapter 7) to suggest that it performs worse than the British and the Norwegian systems. One lesson to learn from the book might be that it would be valuable to work on defining and gaining acceptance for a minimum set of indicators across regimes, which would allow for such a comparison to be made with more confidence. However, this bypasses the care which needs to be taken in such a comparison, if it is fuelled by a wish to take over elements of the 'more effective' regime to improve the less good. Many of the chapters, in tracing the history of the regimes, show that each country's regime is such a complex product of its technology, history, political institutions, legal system, industry structure, culture and management that unquestioning adoption of one regime's elements in another country could be an expensive disaster. Great care always needs to be taken to see the elements of the regime in their context and to understand what other elements and characteristics of the regime are needed to feed and support the success of the one of interest, which is being considered for adoption.

What some chapters (Chapters 8, 10 and 12 in particular) do show, however, is that there is some evidence, at least in Britain and Norway, that there has been an improvement in major hazard and injury indicators which coincides with the introduction or modification of the regulatory regime. It would be too optimistic to draw strong causal links on such evidence, given that the causal chain runs through the technology, actions and processes of the companies and their workforces before hitting the proximal causes of accidents situated at the workplace or in the design office, but it does mean that there is encouragement to make some comparisons, at least to understand how the regimes work and what their strengths and weaknesses are seen to be. The Norwegian regime goes further with its RNNP project (Chapters 9 and 10). It recognises that there needs to be a respected and legitimated process which measures the trends in risk and safety performance in the industry under a given regime, both as a whole and in segments of the regime, in such a way that all parties can see if things are becoming safer or more dangerous over time. This was seen to feed and restore the necessary consensus to reverse a negative trend from the 1990s. Other regimes might usefully see how they could create a broad spread of indicators and collect sufficient data to fulfil this function in their own situations.

Despite the caveats about the care needed in taking lessons across from one regime to another, it is worthwhile to compare them in a much more descriptive way to see how the different regimes work and how they fill in the various steps in the risk governance process. To do this we can make use of the risk governance model that Renn sets out in Chapter 1. This helps us to understand better what the functions are which have to be fulfilled at each step and what the alternatives are in fulfilling them, even though we must realise that choosing one alternative at one step carries with it a restriction on what may be compatible alternatives at another step. For example, the choice of goal-based regulations at the management step is incompatible with frequent and detailed external inspections of the technical details of risk control measures at the monitoring and control step.

Baram and Lindøe (Chapter 2) emphasise that the regime needs to be seen as consisting not just of the formal regulator, but more broadly of also insurers, liability law and compensation, legal challenges to rules, industry standards, pressure groups, issues of corporate governance and information about financial risks, market forces and the factors affecting company image and 'license to operate'. They suggest that where the role of government is less prescriptive and more at arm's length, these other 'social controls' play a greater role and raise issues of their legitimacy and the need to coordinate them (even to regulate their collaboration) and to decide how to manage their conflicts (such as the effect of tort law cases, or threats of cases, suppressing information necessary for learning)

In the remainder of this chapter I will use Renn's framework from Chapter 1 as a structure to locate a number of issues which stand out to me in the chapters of the book. I start with a summary characterisation of the offshore oil and gas industry and what this means for the focus of the book within that model. I end with a more detailed discussion of the strengths and weaknesses of the different regimes featured. Throughout the text I suggest lessons to be learned and work to be pursued, both by researchers and practitioners.

16.2. THE CHARACTERISTIC OF THE OFFSHORE OIL AND GAS INDUSTRY AND ITS REGULATION IN RELATION TO RENN'S FRAMEWORK

The offshore oil and gas industry is a mature one, having left its 'Wild West' phase behind it. Like all major hazard activities, it is multi-layered in its structure and multi-played in its actors. A number of the chapters trace briefly that history of maturing and the effect it has had on the industry and its regulatory frameworks. The attained maturity means that the chapters of the book cluster around the steps in Renn's model which deal with management (step 4) and monitoring/control (step 5). The earlier steps, particularly that of pre-estimation (step 1), lie largely in its history, although the results of that step still impinge strongly on present-day practice. Notably the issues of the frames and focus which different actors had and have reflect still in the concerns they prioritise in the interdisciplinary estimation step (step 2) and when faced with choices in the evaluation, management and monitoring steps (steps 3–5). The risks of offshore oil and gas winning are both simple and complex in Renn's classification, but relatively well understood, so there is not much uncertainty or ambiguity. There are still some situations which require the industry to go back to that first step of pre-estimation, notably when it moves into new and ecologically sensitive areas such as the polar region, or starts to push the boundaries of its technologies as in deepwater drilling. However, these more recent moves are not prominently discussed in the chapters of this book (see Hasle et al. 2009 for an industry example of how that assessment was done by one oil company).

The interdisciplinary estimation step (step 2) is represented, amongst other things, by the issues of the incorporation of human, organisational and cultural factors in risk assessment and the different weighting of their 'concerns' by different actors in the system in understanding and tackling the different risks of offshore oil and gas operations (major hazard, environmental and occupational safety and health and even stress and well-being) (see e.g. Chapter 10). The evaluation step (step 3) is reflected amongst other things in the issue of the use of cost-benefit analysis and the problems of trading off the different risks just mentioned against each other (reflected in Chapters 10 and 13)

The management and monitoring steps (steps 4 and 5) are the main focus of the majority of the chapters. The central theme in management is the different choices made by the three types of regimes as to the form of the rules, their targets and the way in which these are developed and implemented. The enforced self-regulation regime of Norway, based on internal control, sits at one end of a continuum from goal-based regulation to detailed prescriptive regulation, with the U.S. regime situated at the other end. The UK (and Australian) safety case regime sits in between with its emphasis on the companies, rather than the government, specifying how they will control their risks, but with less concern than the Norwegian regime for the issues of safety management and culture (Chapters 6 and 8). The proposed EU regulation as articulated so far is to be found between the UK and U.S. positions on these dimensions.

There are striking parallels in the history of regulation in the three regimes of United States, the United Kingdom and Norway, but also striking differences. Both Norway and the United Kingdom rejected the detailed prescriptive regimes they had used earlier, which themselves were a reaction to the free-for-all, 'Wild West' era of the 1960s, in favour of goal- and performance-based regulation filled in with industry standards. Norway did this in 1987 in the wake of the Alexander L. Kielland disaster, and the United Kingdom in 1992 the Piper Alpha incident. Only in the wake of the Deepwater Horizon blowout has the U.S. offshore begun a flirtation with that same step, but, as Baram records in Chapter 7, it is not yet clear whether the flirtation will lead to marriage and how successful and long-lasting that marriage might be, given the enormous weight of the tradition of prescriptive regulation in the United States in all mature areas of safety and health (Baram 1997; Hale et al. 2011). Baram and Lindøe (Chapter 2 in this volume) also warn that we should not be too blinkered by the formal structure of a given regime. They point out that the U.S. regime, despite its prescriptive surface, hides much collaboration through its use of industry-based standards.

All three regimes have been challenged by major accidents leading to reassessment of their structure and operations, and all three have had to face the effects of economic recession and low oil prices, together with calls to reduce the burden of regulation. The question of whether and how to incorporate safety culture into the regulatory framework has challenged them all. But the outcomes of these challenges have been different, sometimes resistance and reaffirmation of the status quo, sometimes detailed or even radical change.

They have all, albeit at very different moments in their history, made the decision to split the agency responsible for safety from the ministry responsible for developing the industry and licensing exploration and production, in order to safeguard the safety oversight from the undue influence of the imperative to earn revenue. Norway made the decision to separate the reporting lines of the NPD between two ministries in 1979 post Bravo accident, and to split the agency into two independent entities (NPD and PSA) in 2004. The United Kingdom made both decisions at once post Piper Alpha (1988). The United States has split the functions post Deepwater Horizon (2010), but left the two reporting to the same ministry. Thus the United Kingdom and the United States ignored the lesson that Norway drew from Bravo, whilst Norway and the United States ignored the one the United Kingdom drew from Piper Alpha, as did the Netherlands. The latter commissioned a report to study the question in the wake of Piper Alpha (Hale et al. 1992), but decided not to move the oil and gas safety inspectorate out of the economic ministry, based on its finding that no major party favoured the move, partly because they felt that there was no viable existing health, safety and environment regulator with an understanding of major hazards for it to move to. That decision has not (yet) been modified despite the triggers which persuaded Norway and the United States to make the move. So issues repeat themselves under the different regimes, but the response to them is different per regime and seems much stronger to their own accidents than to ones in other jurisdictions. I return to a more detailed discussion and comparison of these and other management issues later in this chapter.

The monitoring and control step (step 5) links closely with the management step, as the management regime determines how and by whom the monitoring should and will take place and whether and how the whole system will be a learning regime which adapts its own workings, or a regime that resists lessons which others see and respond to. In the sections which follow I pick up on a number of the issues identified earlier to look for lessons.

16.3. COMPARING THE REGIMES

16.3.1. *Different Frames, Different Foci*

The offshore oil and gas industry is characterised by a complex actor-network involving the oil companies, their supply chain on the upstream design and purchase side and on the downstream operations and maintenance subcontracting side, their workforce and its unions, the customers for the oil and gas, but also the public, the regulator and various interest and pressure groups. It is important in steps 1 and 2 of Renn's framework that all relevant actors in the system feel they are suitably engaged in the risk assessment process and feel that their frames and focuses and their concerns are treated seriously and suitably.

There are striking differences in the involvement of these actors in the three countries. Norway is the most explicit and articulate in placing its tripartite structure as one of the key foundations of its regime. The role of the unions in particular is powerful compared to its more limited nature in the tripartism of the United Kingdom (Chapter 6) and its almost complete absence in the offshore industry in the Gulf of Mexico (Chapter 15), which is effectively bipartisan. This means that there is a frame of reference and a set of concerns relatively or completely missing in the latter two. The role (or potential role) of unions and of workplace involvement in safety and health is perceived differently. It is seen in essence as constructive and creative in the social democracy of Norway, and any disturbance to that positive role is seen as a threat to the regime (Chapters 12, 13 and 15). One such threat had to be repaired by the creation of a new Safety Forum to debate issues that had not been resolved in the existing Regulatory Forum. The strength of the unions is seen to be crucial to keeping a balance of the three parties, and that tripartism allows shifting alliances to form and steer the regime; safety representatives are seen as extensions of the eyes and reach of the regulator at one moment, but the unions ally themselves with the employers in some cases where they feel that poorly formulated regulations or too intensive enforcement are threatening the future of the industry. The stability of a three-legged stool with equal-length legs compares very favourably to one with only two legs or three uneven ones.

In fact the Norwegian 'stool' has more than three legs, as Chapter 12 relates when it talks about the recruitment of other actors by the three primary bodies. It has given a role to the rhetoric of the public and injured parties, and above all to the research community through the RNNP project to track the trends in risk in the industry. These might be seen as extra supports for a robust regulation, given that all of these frames of reference and concern are aired and can have their influence relatively quietly in the normal process of regulation. They do not need to build up enormous pressure from outside the regime in order to exert influence, with the risk of breaking the regime when it does give in to that pressure. The Norwegian oil and gas companies also realised that there was a need for a forum with their suppliers and contractors set up in the Working Together for Safety project. There is much less transparency about the actors involved and the way in which their different frames and focuses are articulated and taken note of.

The balance of frames is also different between the regimes. The Norwegian is dominated (Chapters 5 and 10) by the engineering frame of technical and organisational problem solving and systems thinking, driven by the dominant engineering background of PSA's regulators and the additional practical input of the trade unions. The lawyers' frame of prime cause seeker, of clarity of requirements and judicial review is relatively lacking. The U.S. regime has the opposite bias; the United Kingdom, as often is the case, sits in between, with the safety case emphasising the rational engineering frame but failing to ensure enough workplace involvement in its construction to ensure the practical input. In particular, the U.S. regime has the possibility of a legal challenge in the courts to any regulations on the grounds of fairness and lack of bias. Baram and Lindøe (Chapter 2) present this as a positive aspect of the U.S. regime, which the other two do not possess, providing as it does a way for parties to check the legitimacy of rules as well as that the government is taking its appropriate moral responsibility to safeguard safety and to provide a level playing field for companies. Parallel to this difference in frames is another dimension, which runs from the dominance of conflict in the U.S. regime to the dominance of collaboration and trust in the Norwegian one, with the United Kingdom, as ever, in between.

With different frames go often different focuses. The inevitable conflict between safety and cost is seen in all regimes. The question, therefore, is how it can be given its proper place and resolved. The tripartite structures of the United Kingdom and Norway were set up partly to provide that place and a constructive sphere to let it operate. In the U.S. regime, that debate is played out between the industry and the regulator, with the former eliciting the authority of Congress and of the president's office to try to have the regulator put on a leash, through executive orders requiring restraint in new regulation and the imposition of a cost-benefit test on all rules. The Office of Information and Regulatory Affairs (OIRA) was set up to carry out that resolution of concerns between cost and safety (Chapter 7).

Another conflict of focus is on the type of risk. All regimes are now aware of the danger of thinking that occupational injuries in day-to-day workplace activities are a good indicator of major hazard process safety. Deepwater Horizon illustrated this with suitable irony with the presence of a senior management group praising the reduction in lost-time injuries on the rig at the same time the blowout and gas explosion occurred (Hopkins 2010). A focus on driving down minor injuries was also a factor in the Texas City explosion, also at a BP installation, indicating that the lesson was particularly hard to learn in the U.S. regime (although not easy either in the UK regime; see Chapter 6 in this volume). Hopefully this lesson is now learned and will be reflected in the safety indicators used by all in future (Chapter 9). Deepwater Horizon has also shifted the focus more from safety to environmental issues. This has brought to the forefront again the question of how to trade off, or balance, the attention to those two only partially overlapping sets of risks. This is a reinforcement of the conflict of focus involved in the opening of the polar and other ecologically sensitive regions to exploration and production and the clash between the concern for fuel stock and costs and the damage to those ecologies. This conflict is played out in particular with pressure groups which also need to find their place in the forum for resolving and balancing these different concerns. Such trade-offs are seen to be difficult to discover even for the Norwegian regime (Chapter 13).

16.3.2. *Quantification and Cost-Benefit*

The previous section identified, amongst other things, the dominance of an engineering frame amongst Norwegian and British regulators. This is reflected in the similar frame of the employers with their 'economic-operational' discourse, emphasising measurable indicators, especially ones measuring the effort put into risk control, alongside, or even in preference to, those measuring the output of accidents and near-misses (Chapters 9 and 12). This is contrasted with the much more personal frame of the unions, based on lived experience and focussing on precursors and qualitative indicators.

The UK safety case regime is the clearest manifestation of the engineering frame. It requires quantitative risk analysis (QRA) in order to prioritise risks. An extension of the risk quantification into the quantification of the costs and benefits of taking different risk control options takes this even further, but is seen to challenge the feasibility of the whole enterprise (Chapter 6). Patterson's analysis shows that such a QRA was and still is beyond the competence of the oil companies themselves and is therefore contracted out to consultants. As with so many similarly outsourced works of documentation, the QRA goes from the keyboard of the consultant to the shelf of the commissioning company and then that of the regulator without passing through the brains or in the practices of the operating company. This makes it stillborn and it never has a chance to become a living document. Moreover, it is hard to incorporate into it in a convincing way the realities of the lived experience of the workforce, the good human factors involved in good risk control and the lessons from the bad involved in disasters. Above all, there are still no wholly convincing incorporations into QRAs of the organisational factors (leadership, competence, judgement, learning) which make up good safety management systems, let alone the subtleties of safety culture. The result is that those issues tend to be ignored in the review of the safety cases and played down in debates between the regulator and the companies.

The Norwegian regime, with its far more even-handed concern with both the quantitative and the qualitative, gives these issues a clearer place, even mandating companies to tell how they intend to create a suitable safety culture. The in-depth audits conducted by the PSA are largely focused on how the actors in the system manage their risk control and safety management. This is in stark contrast to the much more frequent inspections in the U.S. regime, which are focused on compliance at the detailed technical level. The British safety case system does at least recognise that management and culture are relevant, whereas they have difficulty finding their place in the prescriptive and technical regulations of the U.S. system.

16.3.3. *Management and Monitoring*

We now turn to the management and monitoring steps of Renn's model, which are dealt with in the bulk of the chapters of this volume.[1] They are characterised by their different styles of rules, which I examine first from a theoretical and a cultural perspective. I then move on to consider what the prerequisites are for a regime with that type of rules to work successfully and examine the strengths and weaknesses of the three regimes. Finally I look at the way the regimes appear to respond to challenges, to adapt and to learn.

[1] Because so many chapters refer to the same issues, we have only referenced the previous chapters in this section very lightly, to avoid disturbing the flow of the text. We assume the reader will recognise or can search out from where the points are taken.

16.3.3.1. Rule Types and Culture

The most striking difference between the regimes of Norway, the United Kingdom and the United States is in the way in which their rules are formulated. Hale and Swuste (1998)[2] defined a dimension stretching from goals at one end to specific actions to be carried out or states to be achieved at the other. Goals simply state the outcome to be achieved, sometimes in concrete, measurable terms (such as a tonnage limit on polluting discharges), but more often in generic terms (such as using the state of the art of risk control to achieve as low a risk as reasonably practicable) and leave it up to the company to decide how to operationalise that, subject to having to justify that choice to the regulator. Action/state rules specify exactly what must be done in terms of behaviour (e.g. wearing of specified personal protective equipment [PPE]) or put in place in terms of hardware (the provision of the specified PPE, scaffolding, pressure release systems, machinery guards etc.). In between is a category of rules which specifies the processes of risk management to be carried out to arrive at the specific, locally appropriate risk control specifications; in other words, they tell how to translate goals into actions and states. At which point on that dimension the regulations of a regime are phrased defines the freedom of choice left to the regulated person or organisation in choosing how to meet the regulation. At the action/state end there is no room for choice, at the goal end there is the maximum.

To be clear, the framework does not argue that those actions or states should not be defined by someone, given that all risk control actions and states must ultimately be made specific. It simply indicates that the translation from goals down through processes to action/states must take place and that translation can be placed at different points – by the regulator, in the company at professional, management or supervisory level, or by the person carrying out the actions of risk control. The decision about where to lay that responsibility of translation is affected by such issues as the perceived competence and trustworthiness of the parties, the degree of harmonisation needed in the detailed specifications and the relative costs of decentralising or centralising the translations.

Using this framework, it can be seen that the Norwegian regime is predominantly and explicitly goal oriented. There are conscious efforts not to specify in the regulations how to achieve the goals. Some generic processes are specified – risk assessment, for example – but not the methods to be used to carry them out. There is a concern about the regulator giving detailed approval for proposed risk controls for fear that this would entail a transfer of responsibility (and liability) to the regulator should they go wrong. Guidance notes on what safety management processes to put

[2] Papers on the use of this dimension to examine safety rules at the workplace and the regulatory level can be found in Hale and Borys (2012a, 2012b) and Hale et al. (2011).

in place to drive the risk control are based on industry standards and emerge from the tripartite processes.

At the other end of the spectrum lies the United States, where the vast majority of the regulations are formulated at the action/state level, giving no freedom of choice on how to comply. Again industry standards may be cited or mandated, but if mandated they are usually also expressed at the technical, action/state level. If not mandated, but left voluntary, they are seen in a number of chapters (Chapters 7 and 15) in this volume to be toothless and easy to ignore.

The safety case regime of the United Kingdom and Australia is typical of the midpoint of the dimension whereby a process (risk assessment) is specified by which the goal of risk control is to be met, and acceptable methods of risk analysis are prescribed, but it is then left to the regulated company to convince the regulator of the way it is led to control risk by that risk assessment process. Further guidance is in the form of codes which have the status of acceptable translations of the goals, but leave the choice open to the company to comply in other ways.

Action/state rules are only suitable for aspects of risk control that have fully crystallised out. That means they are impossible to apply to issues such as safety culture where consensus over its elements is still in an early stage. Only a goal-based rule is tenable there. This also makes prescriptive rules problematic for new technologies or new situations where risk controls must be deployed (Chapters 3 and 11).

There is a cultural dimension to this difference in rules (Hofstede 1980, 1991). One of Hofstede's four, or latterly five, main dimensions of culture is uncertainty avoidance (UA). One of the defining aspects of high UA is a great perceived need for rules made by experts to remove ambiguity, especially detailed prescriptive ones, governing all situations. At the low end of the scale, situations are left open-ended, allowing pragmatic conventions to emerge over time, with a high tolerance of uncertainty. The contrast between the British common law, based on precedent and the changing perceptions of judges, and the Roman law systems in continental Europe with their more prescriptive nature grounded in written statute conforms to this dimension. Braut and Lindøe (2010) argue that the traditionally statute-based Norwegian law on safety regulation has been showing increased influence of common law over recent decades.

Scandinavia scores amongst the lowest on the UA scale, with Britain a little higher and the United States somewhat higher still. There are, however, many countries, for example in Southern Europe, Latin America and Asia, which score much higher. This may explain the greater use of prescriptive rules in the EU regulatory proposals (Chapter 6), based on the relative dominance of the high-UA Mediterranean European countries over the low-UA north-west European countries.

Anyone wishing to apply the Norwegian, or even the British, regime in one of those high-UA countries should bear this national cultural dimension in mind.

Even transplanting Norwegian elements to the United States is likely to meet cultural resistance and attempts to bend that element to fit the existing, highly prescriptive regime. Baram (Chapter 7) describes some of that adaptation (or distortion, depending on your viewpoint) post Deepwater Horizon.

With the management regime goes a monitoring and control regime. As would be expected, the U.S. regime goes with close supervision and punishment (fines) for non-compliance with little use of discretion (Chapter 5). The Norwegian regime uses great discretion, eschews detailed inspection in exchange for auditing on themes of concern to one or more of the tripartite parties and defines a hierarchy of ascending enforcement, ranging from dialogue (requests) through orders to coercive fines, stopping of activity or prosecution and finally exclusion from operation in Norway. Each step is dependent on the response to each earlier step. Little use is made of even the third and fourth steps, let alone the last two.

16.3.3.2. Prerequisites, Strengths and Weaknesses

The three regimes described in the book are very different, not only in the types of rules they specify, but also in many other aspects. It is illuminating to try to summarise the main characteristics or prerequisites for operation of each and the strengths and weaknesses indicated in the range of chapters of the book. Baram and Lindøe (Chapter 2) discuss these in the context of the shift from prescription to goal-based self-regulation. Table 16.1 draws on that chapter and fills that out with the issues raised in the remaining chapters of the book, to give a summary which the reader can criticise and modify, and also compare to the table in Chapter 9, which overlaps in content to an extent. The columns are only indicative and not equivalent to each other, in the sense that the same issues have not been dealt with systematically across all three regimes in the book. This means that the lack of mention of a particular topic by one or more of the three rows does not mean that it would not appear there if such a comprehensive review were to be conducted. The Norwegian regime is the topic of more chapters than the other two regimes and hence has more characteristics, strengths and weaknesses mentioned.

The Norwegian regime comes over as the most rounded and well thought-through. This may partly be because there are more chapters examining how it has worked and responded to challenges. However, it still has issues which it is struggling to deal with and incorporate into its workings. I return to these in the next section. The weaknesses identified in Table 16.1 are often identified in the chapters as weaknesses that opponents identify and level at the regime, and the chapter contributors go on to refute them or at least indicate that the strengths outweigh the weaknesses. In particular it is the acceptance by the tripartite parties that the regime is legitimate, despite those potential shortcomings, which is its principal claim to strength.

TABLE 16.1. *Characteristics, strength and weakness of the Norwegain, UK and U.S. regimes*

Regime	Norway	United Kingdom	United States
Type and Pre-requisites	Goal-based regulations, based on Internal Control under general duty of 'prudent operations'. PSA requires to be informed of, but does not approve plans, etc. Tripartite, strength of parties must be equally balanced. Oil companies must operate from Norway. Forum(s) for debate needed to 'regulate' social controls and participation. Much left to discretion of the regulated. High trust between parties needed. High competence of inspectors and union reps needed. Skill in negotiation needed. Use of industry competence to train inspectors. Use of industry standards as basis for guidance. Regulator needs high legitimacy. Enforcement by audit of concerns arising partly in tripartite forums. RNNP sets baselines and establishes trends.	Centred on safety case under general duty of care. Focuses on risk control and safety management system (SMS). Uses QRA to assess priorities in safety case. Review of safety case has suggestions for improvement and statement of acceptance, not approval. Tripartite, strength of parties should be equal. Trust needed between parties. Use of industry standards as guidance. Key Programmes (HSE) monitor trends and new issues.	Prescriptive rules. Directed at primary barriers and controls. Narrowly technical. Enactment of many industry technical standards. Some inspectors uniformed and rotated regularly through jobs to minimise regulatory capture. Needs high competence in technical aspects of risk control by regulator. Response to Macondo to enact an SMS standard based on existing voluntary code (still prescriptive). No systematic performance data collected. Indicators reactive.

(continued)

TABLE 16.1. *(continued)*

Regime	Norway	United Kingdom	United States
Strengths	Clear definition of involved parties. Regulatory and Safety forums as places to air and adjust perceptions. Self-regulation engages the risk-maker. Strong place for risk assessment. Emphasis on learning and improvement High trust and legitimacy present and restored when threatened. Emphasis on competence of all parties. Goal oriented. Repairs itself when challenged, by enrolling other parties/discourses. Flexible to meet local differences of conditions. Incorporates safety management well and culture issues to an extent. Safety reps augment inspection role. Some proof of effectiveness.	Self-regulation engages the risk-maker. Strong place for risk assessment. Risk-based with worker involvement (by design). By design safety case is a proactive, living document. Indicators used are partly predictive. Emphasis on learning and improvement	Clarity of requirements. Uniformity of requirements leads to level playing field. Legal certainty backed by due process and access to courts for rule challenge. Close supervision by regulator needed. High legitimacy as government visibly takes moral responsibility for safety Incorporation of industry standards

Weaknesses	Lack of sharpness of requirements, too open-ended, leading to high uncertainty. Low predictability of what to do. Few inspections and objectives unclear. Soft enforcement. Law made by negotiation. No level playing field created to ensure fairness Poor transparency No judicial review. Issues of poor legitimacy for self-regulation Challenged by cost-efficiency drive, change in technology, BBS, mergers and take-overs. SMS rules float free of industrial reality. Safety culture not clearly integrated. Has difficulty trading off different risk types. Safety reps cannot assess design issues so the focus on these may be less than on operations.	Too much emphasis on QRA, too little on qualitative factors (adjusted in 2005). Hard to quantify where money involved. Organisational issues hard to incorporate in QRA. No QRA competence in companies, so contracted out to consultants and result does not live in company. Use of safety case is reactive and static not proactive and dynamic. Criteria for evaluating safety cases not explicit Worker involvement weak. Has not adapted to modern management and culture issues (leadership, competence, learning, etc.). Monitoring of maintenance not done.	Regulations complex and uncoordinated from several regulators and subject to appeal. Inter-regulator conflict. 'One-size-fits-all' nature of prescriptive regulations not appropriate and discourages innovation. Regulator does not possess competence to do risk assessment and make detailed rules. Regulator easily pushed into taking responsibility for rules. Industry standards made without union or public participation. Low trust between parties. No worker involvement (should change with new enactments). Attempt to move to SMS regulation still at early stage and meeting problems. No mandating of learning. Difficult to recruit technically qualified inspectors. No performance data

The U.S. regime comes in for the most criticism, with its prescriptive rules being seen as outdated and inefficient and forcing the regulator into an untenable role of rulemaker, for which it does not have the manpower or competence. The regime itself, even before Deepwater Horizon, seems to have regarded itself as out of step in this respect (Chapter 15) and to have started moves to incorporate attention to the SMS and more use of elements of the Norwegian regime. The question is how far such a move can go, even with the stimulus of the severe criticism after Deepwater Horizon.

The greatest criticism of the British regime is about getting the safety case regime to fulfil its potential, which would move it nearer to the Norwegian regime. The shortcomings are in its poor engagement with labour and its overreliance on QRA. The implication is that it is less the regime itself which is to be criticised and more the current implementation of it.

16.4. CHALLENGES, LEARNING AND CHANGE

The book chapters provide clear histories of how the three regimes have changed over time. It is clear that major accidents and, more recently, near-misses in their region have been one of the biggest stimuli for challenge and change. Norway has reacted to Ekofisk Bravo, Alexander L. Kielland and, more recently, Gullfaks C and Snorre A; the United Kingdom had Sea Gem and Piper Alpha; the United States had Deepwater Horizon. The large disasters such as Piper Alpha and Deepwater Horizon have clearly had effects beyond their immediate hinterland, but there would appear to be an attenuating effect as the accidents are assessed and digested in other jurisdictions so that major overhauls of a regulatory regime only occur when the accident falls under that regime. This may be because there are always arguments to be advanced that the regime where the accident happened was different in significant ways, and those differences were crucial to its aetiology. If Renn's framework is used to analyse the lessons from such accidents, it should be clearer in what ways the regime where it happened is functionally similar to one's own and in what ways different, and hence where there are real lessons to be learned.

But accidents are not the only challenge to a regime. For the Norwegian regime there have been a number of threats to the equilibrium of the tripartite process, the balance of power within those interactions. A serious challenge was the NORSOK process, with its background of low oil prices and increased need for cost-efficiency. That strengthened the employer's discourse and legitimised cost-cutting in the name of the survival of the industry. It led to more reliance on cost-benefit analysis (CBA) and QRA to argue priorities, which derive from the rationalist, engineering discourse common to employers and the regulator but more alien to the trades unions. The opposition to that approach is also driven by the fact that it is always easier to quantify the costs of risk control than its benefits. The disturbance of the balance led to union militancy and conciliation by the regulator who restored the balance partly by threatening the employers with a return to prescriptive regulation (here seen as a bogeyman figure, pace the U.S. regime) unless they tempered their perceived attack on safety rules and procedures. This breach of the consensus was fed by differing views amongst employers and unions of whether risk was increasing or decreasing at the time. This was resolved by the regulator by initiating the RNNP project to bring in independent researchers to study the risks and trends and re-establish the consensus. The regulator in particular seemed to see this threatened breakdown of tripartism as a challenge to the robustness of the Norwegian regime, which justified going outside its formal role to rescue the regime; indeed, all three parties are seen as exceeding their normal roles in the rescue attempt.

The introduction of behaviour-based safety (BBS) in some of its manifestations threatens to disturb that balance in the same direction by laying the responsibility for accidents too one-sidedly on the workforce. That is an issue of implementation, because BBS is in principle just as applicable to the behaviour of managers as to that of the shop floor workers and could be extended from occupational safety to cover process safety, which would focus on design as well as operations and maintenance. Hopkins (2012) recommends just such an extension in his book on the Deepwater Horizon accident.

The UK regime appears to be facing the same challenge to find the right place for QRA and for worker involvement in its regime. Some attempts were made to correct these perceived imbalances in the revamp of the regulations in 2005, but the result is not seen as completely successful (Chapter 6).

The UK regime does, however, seem to have drawn the same conclusion as Norway about the need to have a measure of the performance of the industry – operationalised in its Key Programme monitoring activities. It is worth consideration to see if these two monitoring actions and their constituent indicators could be harmonised so that, in addition to assessing the trends in one jurisdiction, they could allow comparison across jurisdictions and regimes, however carefully such comparisons would need to be made.

The U.S. regime faces more challenges arising partly from Deepwater Horizon, but much more deeply embedded in its modus operandi. It is one of the few regimes which has not made the transition from detailed prescriptive regulations to goal- and process-based rules. The cultural inertia which holds back that shift is the long and almost universal reliance on such prescriptive rules in all other mature areas of safety regulation.[3] This reliance is partly driven by the culture of compensation relying on legal cases, which can only be settled relatively easily if

[3] Although in areas of emerging technology such as genetic manipulation in agriculture (Baram 2002, 2011), the U.S. approach to regulation is notably relaxed and nonprescriptive, thus favouring innovation.

there are detailed rules specifying who is responsible for doing what in all conceivable circumstances. To break out of that culture would require enormous effort, as is shown by the fact that the first step to enact an SMS has resulted in the imposition of a rule that prescribes what that SMS should consist of (Chapter 7), rather than leaving those decisions to the company. Baram and Lindøe (Chapter 2) discuss that move and its problems for the U.S. regime. As they point out, such a move would also require a higher level of competence from the inspectors, which is not easily realised, especially when recruitment of suitable candidates for the current lower-level positions is already difficult (see e.g. Hale et al. 2002). The involvement of labour in the regulatory regime is also only just being considered for mandating in the Safety & Environmental Management System (SEMS 2) regulations.

If the more process-oriented rule requiring an SEMS were to be enacted alongside the prescriptive technical rules currently governing risk control, the United States would find itself in the difficult position described by Paterson (Chapter 6), in which the United Kingdom found itself when it had to reconcile the highly prescriptive (at that time) offshore regulation with the goal-based Health and Safety at Work etc. Act of 1974 governing occupational safety. That difficult period for the regulator only ended when the safety case regime was introduced to replace the prescriptive rules. It is probably too much to hope that a similar transition could be driven in a similar way in the United States, when prescriptive rules are not being strongly challenged in other activities with mature technology.

There are some challenges all the regimes face, notably the availability of enough suitably qualified inspectors, whatever the regime, to carry out the audit and enforcement activities. Üşenmez (Chapter 4) discusses this issue and evaluates the potential of privatisation to tackle it. However, his analysis largely rejects that approach on the grounds of cost and issues of how to avoid a private monopoly replacing the current state monopoly.

16.5. CONCLUSIONS

16.5.1. *Robustness*

I have defined robustness at the start of this chapter as a regime that has survived for a considerable period with its principles intact, but with adaptation in its detail to changing situations and priorities. We need to expand that definition to refer to Renn's framework. A robust regulatory regime must have dealt with all of the steps in that framework explicitly and achieved a stable balance in each before we can call it robust. This means in particular coping with the different frames and focusses of the different actors, so that none of them mobilises to derail the regime.

Based on the chapters in this book, we can tentatively conclude that the Norwegian regime meets the most criteria. It has lasted in roughly its present form, within its principal functional characteristics, for almost twenty years. Its balancing act between the different actor groups has been largely successful. It has survived several challenges and has mobilised the forces and discourses to retain those principles. At the same time it has shown itself flexible enough to adapt in a more detailed aspect to those challenges. It is clear, however, that this has only happened because the regulator in particular, but also the other parties to the tripartite approach, have consciously managed that robustness in response to the challenges and made it a learning system. That is no mean feat when we consider how few individual organisations can really claim to be learning organisations (Senge 1990; Koornneef 2000).

The UK regime cannot be assessed very confidently on its robustness, represented as it is by only one chapter in the book (plus the one on the similar Australian regime). Its safety case regime has survived, with detailed modifications, for ten years, despite the fact that the balance of the interests of the different actors has been less effective than the Norwegian regime. So the signs are positive, but the regime is being buffeted by the winds of deregulation (Young 2010; Löfstedt 2011), and it is still not clear what challenges the proposed EU regulations will pose for it, especially if they turn out to be much more prescriptive than the current UK regime.

The U.S. regime is portrayed as the one with the least balanced input from the different parties and the most conflictual way of working. It appears to be going through a phase in which it is questioning its basic design principles and flirting with the adoption of some aspects of the other two regimes. If it rejects the transplant of those elements and preserves its prescriptive nature, it will in one sense be robust in defending those principles. If the comparative performance figures quoted by Baram (Chapter 7) are correct, and the performance of the current U.S. regime is significantly worse than those European regimes, this robustness will be counterproductive to safety. We might therefore choose to add to the definition of a robust regulatory regime that it should achieve the best practicable safety performance, otherwise we should portray it as rigid and not robust.

16.5.2. *Learning from Self and Others*

As for transplanting elements of one regime to another, I repeat the warning that this must be done with much prior analysis of how each of Renn's steps works in the 'donor' regime and what characteristics of the adjacent steps are instrumental in making that step successful, and comparing that with the characteristics of the 'recipient' regime. Are the tripartisms of the UK and Norway regimes similar enough to copy the Safety and Regulatory Forums as places for debate and collaboration? Could the RNNP from Norway be exported to the United States? Could the way in which industry standards are used in the United States be modified to copy more closely how they are dealt with in Norway or the United Kingdom?

To be a learning system, the regulatory system in any country must be open to criticism and be critical itself of its own performance, structure and functioning. Above all the learning and change must be managed. For example, instead of waiting for the next accident to re-examine itself, it needs to trigger that re-examination at regular intervals. It needs to mandate and encourage the reporting of performance indicators, both of the system as a whole (as RNNP does) and of the regulator as a constituent party. Blakstad (Chapter 9) shows that there is potential for agreeing indicators of the whole regime, but that the current indicators of the regulator are confined to ones dealing with its activity rather than its influence. As Mearns (Chapter 3) shows, reporting of performance is very sensitive to the regulations governing it and particularly the protection of confidentiality and even the guarantee of immunity from prosecution for breaches of the law. She cites the Danish Air Traffic Management (ATM) law giving immunity to those air traffic controllers who report breaches of rules governing air space, which resulted in a 65-fold increase in those events being reported and available for learning. Guaranteeing immunity may be perceived as breaching the requirements for a just culture (Reason 1997), but this example does highlight how much learning material is hidden in most systems.

The comparison with ATM leads to a final area for learning, namely what the chapters in this book have to say for regulators and regulated in other industries and activities. There is clearly great potential for learning by onshore oil and gas, which differs only in its accessibility and is often regulated by the same inspectorate and may have the same legal framework. A step further away is the chemical industry with its major hazard regulation, and parts of the aviation industry, given the use of helicopters under the offshore regulations. Another step away is the energy industry, nuclear, conventional and emerging solar and wind power, and yet further away are the areas of conventional onshore occupational safety and health and specialised industries such as health care. It would take another book equally as large as this one to make the detailed comparisons and indicate the specific lessons, but the basic descriptions and data presented here could be used by regulators and regulated in those industries to make that comparison themselves.

An important issue in deciding which of the three regimes described in this book should be the template for another industry in a specific country is the one of trust. The offshore oil and gas industry at least in Norway and the United Kingdom has relatively few organisations involved (oil and gas companies, contractors, unions etc.). These are counted in the hundreds, and close relations between them or their representatives and the regulator are feasible. Trust can therefore be built up and managed. This is in stark contrast to the number of organisations falling under the labour or factory inspectorates in all industrialised countries, which commonly runs into the hundreds of thousands (Johnson 2012). For such numbers there can never be enough inspectors to make inspections in a significant sample of companies, whether those inspections are of the in-depth nature of the Norwegian oil and gas regime or the compliance-based inspections characteristic of the U.S. regime (Chapter 4). Building up trust in such a huge jurisdiction, especially with the SMEs, is impossible, and even coercion struggles to cope. The industries, such as railways, aviation, nuclear and major hazards, with much smaller numbers of actors, have a much greater affinity with offshore oil and gas in this respect and could be candidates for learning from the high-trust Norwegian regime.

There are other specific issues where learning across industries could be valuable. An example is the split between ministerial responsibility for promoting the industry and for regulating its safety. In Section 16.2 of this chapter the history of that split in oil and gas was briefly summarised from the relevant chapters. There are other industries which are still struggling with that issue. An example is the maritime regulation in Norway where the Maritime Directorate responsible for ship safety sits in the Ministry of Trade & Industry responsible for promoting the shipping industry (Størkersen 2012). Inspectorates responsible for pollution control and for railways in the United Kingdom have also been transferred into and out of the Health and Safety Executive, from ministries and then into independent agencies. Evaluation of the results of such different structures for inspectorates could be valuable. Doubtless the readers of this book will be able to think of more issues in offshore oil and gas which offer lessons to their industry or activity.

REFERENCES

Baram, M. (1997) "Shame, blame & liability: Why safety management suffers organisational learning abilities", in A.R. Hale. B. Wilpert and M. Freitag (eds.): *After the Event: From Accident to Organisational Learning*. Pergamon: London and Amsterdam, pp. 161–178.

(2002) "Biotechnology and social control", in B. Kirwan, A.R. Hale and A. Hopkins (eds.): *Changing Regulations: Controlling Risks in Society*. Pergamon: Oxford, pp. 217–232.

(2011) "Governance of GM crop and food safety in the United States" in: M. Baram and M. Bourrier (eds.): *Governing Risk in GM Agriculture*. Cambridge: Cambridge University Press, pp. 5–56.

Braut, G.S. and Lindøe, P.H. (2010) "Risk regulation in the North Sea: A common law perspective on Norwegian legislation", *Safety Science Monitor*, **14** (1): Article 2.

Hale, A.R. and Borys, D. (2013) "Working to rule, or working safely? Part 1: A state of the art review", *Safety Science*, 55: 207–221.

(2013) "Working to rule, or working safely? Part 2: The management of safety rules and procedures.", *Safety Science*, 55: 222–231.

Hale, A.R., Borys, D. and Adams, M. (2011) *Regulatory Overload: A Behavioral Analysis of Regulatory Compliance*. Working Paper No. 11–47. Mercatus Center, George Mason University, Arlington, VA.

Hale, A.R., Goossens, L.H.J. and Timmerhuis, V.C.M. (1992) Staatstoezicht op de mijnen – een verkenning van rol en toekomst (State Supervision of Mines – an assessment of their role and future). Report to the Dutch Ministry of Economic Affairs. Safety Science Group. Delft University of Technology, Delft, The Netherlands.

Hale, A.R., Goossens, L.H.J. and v.d. Poel, I. (2002) "Oil & gas industry regulation: From detailed technical inspection to assessment of management systems", in B. Kirwan, A.R. Hale and A. Hopkins (eds.): *Changing Regulations: Controlling Risks in Society*. Pergamon: Oxford, pp. 79–108.

Hale, A.R. and Swuste, P.H.J.J. (1998) "Safety rules: Procedural freedom or action constraint?" *Safety Science*, **29** (3): 163–178.

Hasle, J.R., Kjellén, U. and Haugerud, O. (2009) "Decision on oil and gas exploration in an Arctic area: Case study from the Norwegian Barents Sea", *Safety Science*, **47**(6): 832–842.

Hofstede, G.R. (1980) *Culture's Consequences*. Sage : London.

(1991) *Cultures and Organisations: Software of the Mind*. McGraw-Hill: London.

Hollnagel, E., Woods, D.D. and Leveson, N. (eds) (2006) *Resilience Engineering: Concepts & Precepts*. Ashgate: Aldershot.

Hopkins, A. (2010) *Failure to Learn: The BP Texas City Refinery Disaster*. CCH: Sydney.

(2012) *Disastrous Decisions: The Human and Organisational Causes of the Gulf of Mexico Blowout*. CCH: Sydney.

Johnson, C. (2012) "The competency gap: The failure of regulation in workplace safety", Paper in the Proceedings of the 6th International Working on Safety Conference: *Towards Safety through Advanced Solutions*. CIOP: Warsaw.

Koornneef, F. (2000) *Learning from Small-Scale Incidents*. PhD thesis. Safety Science Group. Delft University of Technology: Delft.

Löfstedt, R.E. (2011) *Reclaiming Health and Safety for All: An Independent Review of Health and Safety Legislation*. Report to the Secretary of State for Work and Pensions. Command 8219. HM Stationery Office: London.

Lord Young (2010) *Common Sense, Common Safety: Report to the Prime Minister*. Her Majesty's Stationery Office: London.

Reason, J.T. (1997) *Managing the Risks of Organisational Accidents*. Ashgate: Aldershot.

Senge, P.M. (1990) *The Fifth Discipline: The Art and Practice of the Learning Organisation*. Doubleday: New York.

Størkersen, K.V. (2012) "Dealing with double standards. Maritime regulator's handling of political paralysis", Paper in the Proceedings of the 6th International Working on Safety Conference: *Towards Safety through Advanced Solutions*. CIOP: Warsaw.

Index

Lightning Source UK Ltd.
Milton Keynes UK
UKOW05n0610230317
297334UK00014B/383/P

9 781107 025547